Chemistry
and
Pharmacology
of
Anticancer
Drugs

Chemistry
and
Pharmacology
of
Anticancer
Drugs

David E. Thurston

CRC Press
Taylor & Francis Group
Boca Raton   London   New York

CRC Press is an imprint of the
Taylor & Francis Group, an informa business

CRC Press
Taylor & Francis Group
6000 Broken Sound Parkway NW, Suite 300
Boca Raton, FL 33487-2742

© 2007 by Taylor & Francis Group, LLC
CRC Press is an imprint of Taylor & Francis Group, an Informa business

No claim to original U.S. Government works
Printed in the United States of America on acid-free paper
10 9 8 7 6 5 4 3 2 1

International Standard Book Number-10: 0-8493-9219-5 (Hardcover)
International Standard Book Number-13: 978-0-8493-9219-1 (Hardcover)

Library of Congress Cataloging-in-Publication Data

Thurston, David E.
    Chemistry and pharmacology of anticancer drugs / David E. Thurston.
        p. ; cm.
    Includes bibliographical references and index.
    ISBN 0-8493-9219-5 (alk. paper) -- ISBN 1-4200-0890-0
    1. Antineoplastic agents. I. Title.
        [DNLM: 1. Antineoplastic Agents--pharmacology. 2. Antineoplastic
Agents--chemistry. QV 269 T544p 2006]

    RS431.A64T48 2006
    616.99'4061--dc22   2006008937

Visit the Taylor & Francis Web site at
http://www.taylorandfrancis.com

and the CRC Press Web site at
http://www.crcpress.com

This book is dedicated to cancer patients worldwide who participate in Phase I clinical trials. Their valuable contribution furthers our knowledge of the treatment of cancer. Also, to members of my research group, past and present, who are a constant source of inspiration.

Foreword

Cancer is one of the most serious health problems in the Western Hemisphere. Significant progress in the development of novel drugs and therapies has occurred within the last 60 years and, thanks to the discovery of drugs such as cisplatin, the taxanes, and the nitrogen mustards in the last century, treatment of some forms of the disease has a high success rate, although patients must often tolerate unpleasant side effects. Only in the last decade has there been some success in developing "targeted" drugs and therapies with fewer side effects. For example, the recent discovery and rapid licensing of imatinib (Gleevec™) for the treatment of chronic myelogenous leukemia (CML) is often heralded as the start of a new era in the development of noncytotoxic agents targeted toward distinct biological pathways. However, despite these advances, the treatment of most types of solid tumors remains problematic and survival rates are, in general, disappointingly low.

This book attempts to bring together a broad spectrum of information relating to the chemistry and pharmacology of anticancer drugs and therapies. The first chapter provides an introduction to cancer and includes a discussion of its causes, the problem of metastasis, its general modes of treatment, and the philosophy of drug discovery. The rest of the book provides a comprehensive survey of the various families of anticancer agents (including biological agents) in present use, some that are still at the research stage, and others that might be important in the future. Where possible, the different classes of anticancer agents are described in terms of their discovery, chemistry, mechanism of action, and some elements of structure activity relationships (SARs) through to pharmacology, clinical use, mechanism-based toxicity, and relevant aspects of formulation and dose scheduling. The book is unique in providing the chemical structure of every drug discussed, including both those in clinical use and those at the research stage, and also in providing information on the side effects of most agents. A number of recent developments in research tools and methodologies that may help cancer researchers are also discussed, as are the relatively new areas of *personalized treatments* and *chemoprevention*. One of the difficulties in compiling a book of this nature is in deciding how to differentiate between agents in clinical use and those experimental agents at an earlier stage of development. Inevitably this has led to some difficult decisions as to whether certain agents should appear in Chapter 9 (The Future) or in other chapters relating to their class or mechanism of action.

It is hoped that this book will have wide appeal for undergraduates, postgraduates, and practitioners in the health sciences (e.g., medicine, pharmacy, biomedical sciences, and nursing), and also for those studying pharmacology or specific areas of the natural sciences (e.g., the medicinal chemistry component of a chemistry course). Because the book contains sections on new research areas and tools, it should also be of interest to cancer researchers in both academia and industry. The text stops short of providing detailed clinical aspects of treatments but contains sufficient information regarding dosing and side effects to be useful to medical, pharmacy, and nursing practitioners.

After reading this book, it might be concluded that cancer chemotherapy has made spectacular progress since the discovery of the nitrogen mustards in the 1940s. However, in reality progress has been very poor given that more than 60 years have passed and still one in three of the population (at least in the West) develop cancer at some time in their lives and one in four die from the disease. Hopefully, this book will help to educate future generations of cancer researchers who will go on to discover more effective drugs and therapies.

Acknowledgments

There are two origins to this book. First, it is based on lecture notes started at the College of Pharmacy at the University of Kentucky (Lexington, KY, U.S.A.) in 1981 (inherited from Professor Laurence Hurley), improved upon at the University of Texas at Austin College of Pharmacy (U.S.A.) from 1982 to 1986, and then consolidated at the Pharmacy Schools of the Universities of Portsmouth, Nottingham, and London (U.K.) from 1987 to the present time, where they are still in use. These notes were initially summarized in a chapter in the book *Smith and William's' Introduction to the Principles of Drug Design & Action* (now in its fourth edition [CRC Press, Boca Raton, FL, 2005]), and I am grateful to the editor, Dr. John Smith, and to Taylor & Francis for allowing me to use some of the material for this book. The other origin of this book is based on my period of service on the Cancer Research Campaign Grants Committee (now Cancer Research U.K.) from 1994 to 2005 and the Cancer Research U.K. New Agents Committee from 1998 to the present time. Throughout the years, membership on these committees has provided me with a wealth of background material and new concepts regarding both existing and novel therapeutic agents and strategies. Therefore, I would like to thank Professor Gordon McVie (then Director General of Cancer Research Campaign) for involving me with the charity's scientific committees.

I would also like to thank those mentors and colleagues who have influenced my career and thus contributed either directly or indirectly to this book. Although it is not possible to mention all by name, I would particularly like to thank Gerald Blunden, Colin Richards, Gary Thomas, David McIntyre, and the late Edwin Crundwell, all of whom first sparked my interest in research as an undergraduate. I am especially grateful to Laurence Hurley, who mentored me in my formative years and fostered my interest in anticancer drugs and DNA-interactive agents. Others who supported my career so generously in subsequent years and laid the foundation for this book to be written include Trevor Crabb, Ken Douglas, Sandy Florence, Angela Galpine, Ken Harrap, Chris Martin, Malcolm McVicar, Neil Merritt, John Midgley, Colin Monk, Stephen Neidle, Malcolm Stevens, and the late Tom Connors. I would also like to thank my close colleagues and collaborators John Hartley, Phil Howard, and Terry Jenkins for their friendship and support over the years. Phil Howard deserves a special mention as, without his tireless efforts throughout the years to supervise researchers and look after our laboratory so effectively, this book would never have been completed. John Smith and Clare Simons are also thanked for providing a significant input into Chapter 6 (Hormonal Therapies). I extend my gratitude to the British National Formulary (BNF) for providing information on dosing and adverse reactions. Sam Kneller is gratefully acknowledged for providing chemical structures and other material in support of parts of the text, as is Ruth Pickering for her outstanding assistance with preparing the manuscript, and particularly for dealing with copyright issues. Colin James also deserves a mention for his input to the jacket design, and Sylvia Alban for help with other aspects of the manuscript.

Finally, I would like to thank my wife, Kim, and daughters, Sarah and Caroline, for their patience and understanding in allowing me the many hours of solitude required to complete this book.

David Thurston
London

The Author

Professor David E. Thurston is a pharmacist (B.Sc., M.R.Pharm.S.) and medicinal chemist (Ph.D.) who has lectured and carried out research at the University of Kentucky (U.S.) and the University of Texas at Austin (U.S.) and at the universities of Portsmouth, Nottingham, and London (U.K.), where he is presently Professor of Anticancer Drug Discovery, Head of the Department of Pharmaceutical & Biological Chemistry, and Director of the Cancer Research U.K. Gene Targeted Drug Design Research Group in the School of Pharmacy. His research interests involve the design, synthesis, and development of novel anticancer drugs, and one agent from his laboratory (SJG-136) is currently being evaluated in clinical trials. He has supervised more than 50 postdoctoral research fellows and Ph.D. students, is author of more than 100 research publications in medicinal chemistry and chemistry journals and books, and is frequently invited to speak at national and international conferences and in universities and research institutes worldwide. He has been a member of a number of national committees in the U.K., including the Committee on Safety of Medicines (CSM), the grants committees of Cancer Research UK (CR-UK) and the Association for International Cancer Research, the New Agents Committee of CR-UK (which evaluates new drugs for Phase I clinical trials), and the U.K. Government's University Research Assessment Exercise Panel for Pharmacy (1992 and 1996). Professor Thurston is also one of the scientific founders of the biotechnology company Spirogen Ltd. He is married with two daughters and lives in Hampshire, U.K.

From the Author and Publisher

Knowledge regarding cancer chemotherapy and best practice in the field (including correct information relating to drug licensing for specific diseases throughout the world) is constantly changing. Inevitably, a book of this type that strives to be as broad and as current as possible will be out of date in some subject areas before it is published. Therefore, readers are advised to check the most current information available on all drugs and therapies covered in this book (preferably with the manufacturer of each product) to verify the suitability for use in a particular disease type, the recommended dose or formulation to use, the method and duration of administration, the contraindications and the adverse effects expected. It is the responsibility of the health care practitioner, relying on his/her own experience and knowledge of the patient, to make diagnoses, decide on the best treatment for each individual patient, to determine dosages and to take all appropriate safety precautions. To the fullest extent of the law, neither the Author nor the Publisher assumes any liability for any injury and/or damage to persons or property arising from the use of any information contained in this book.

Note that various cancer statistics and the brand names of drugs quoted in this book reflect the U.K. and European regions where the author is based. However, the generic names given in the book can be used to search drug databases to obtain brand names for other parts of the world, including the U.S.

Contents

1 Introduction to Cancer

1.1 INTRODUCTION

Cancer is a disease in which the control of growth is lost in one or more cells, leading to a solid mass of cells known as a tumor. The initial tumor, known as the *primary tumor*, often becomes life-threatening by obstructing vessels or organs. However, death is most commonly caused by spread of the primary tumor to one or more other sites in the body (by a process called *metastasis*), which makes surgical intervention impossible. Other types of cancers known as *leukemias* involve a build-up of large numbers of white cells in the blood.

Using the United Kingdom as an example, in the first three decades of the 1900s, cancer accounted for less than 10% of all deaths, with infectious diseases being the main cause of mortality. However, although dramatic progress has been made in controlling infections, similar progress has not been made in the treatment of cancer. Improved diet, living conditions, and health care have now increased the average life span to the point where cancer, which is a disease of advanced years (e.g., 70% of all new cases of cancer in the U.K. occur in people older than age 60), has become more prevalent. Consequently, about 300,000 new cases of cancer occur each year in the U.K. The annual number of deaths from cancers of all types is approximately 160,000, which represents approximately 25% of all U.K. deaths. Statistics show that approximately one in three of the population will suffer from some form of cancer during their lifetimes, and one in four will die from the disease. Furthermore, 1 in 10,000 children will be diagnosed annually in the U.K. as suffering from cancer, totalling 1,300 new cases each year.

It is thought that exposure to an ever increasing number of chemicals (carcinogens) in the environment and the diet may be a significant contributing factor. Occupational factors are thought to account for 6% of cancers, while lifestyle and diet may account for up to 30%. Recognition of these contributing factors is currently leading to major changes in social behavior and legislation (e.g., reduction in cigarette consumption, curbing of smoking in public places, healthier diets, more exercise, more restrictive health and safety rules in working environments). Genetic predisposition is also a factor in some types of cancer.

1.2 TERMINOLOGY

A tumor (or neoplasm) is an abnormal tissue mass or growth that results from neoplasia, a state in which the control mechanisms governing cell growth become deficient, leading to cell proliferation. In the early stages of tumor growth, cancer cells often resemble the original cells from which they derive; however, they can lose the appearance and function of their origin at a later stage. Normal healthy adult tissues in most parts of the body tend not to grow but to maintain a steady number of cells. In some tissues (e.g., liver), this is achieved without proliferation because so little cell loss occurs. In the bone marrow, however, a steady number of cells are maintained by a fast rate of cell division balancing the rate of cell loss. It is worth noting that only a slow increase in the rate of proliferation of cancerous cells is required to gradually outgrow normally controlled cell populations and form a tumor mass.

Cancers are generally named in relation to the type of tissue from which they arise. For example, the term *sarcoma* describes a tumor arising from mesodermal tissue, which includes connective or supportive tissue, bone, cartilage, fat, muscle, and blood vessels. *Osteosarcoma* refers to bone cancer, and *carcinoma* refers to tumors of epithelial tissues such as the mucus membranes and glands (including cancers of the breast, ovary, and lung). Bone marrow cell cancers are referred to as *myelomas* and in *multiple myeloma* (the most common bone marrow cancer), a clone of plasma cells is involved.

Cancers of the blood or hemopoietic tissue are generally known as *blastomas,* and these tumors can involve erythroid, lymphoid, or myeloid cells. In particular, cancer of the erythroid stem cells is known as *primary polycythemia. Lymphosarcoma* is cancer of the lymphoid cells, whereas *Hodgkin's disease* is an example of a lymph adenoma that, although mainly affecting reticulum cells, can extend to eosinophils, fibroblasts, and lymphocytes. *Leukemias* describe those cancers originating in leukocytes, which can be classified as *myeloid, lymphatic,* or *monocytic.* The reticuloendothelial system is also susceptible to cancer. In addition, all of these different cancer types may be described as *chronic* or *acute.*

1.3 METASTASES

Metastasis is the term used to describe the ability of solid tumors to spread to new sites in the body and establish *secondary* tumors. Many patients who die of cancer do so as a consequence of *metastatic* spread to vital organs rather than from their primary tumors. Tumor cells commonly penetrate the walls of lymphatic vessels, distribute to draining lymph nodes, and move to distant sites. They can also directly invade blood vessels because capillaries have thin walls that offer little resistance. Furthermore, both primary and secondary tumors can expand in size and infiltrate surrounding tissue. When nerve endings are affected, pain and discomfort are experienced. A tumor can also spread across body cavities from one organ to another (e.g., stomach to ovary).

At the time of diagnosis of a cancer, curative surgical or radiological treatment is usually only possible if metastasis of the primary tumor has not occurred. Since about 50% of malignant tumors have already metastasized prior to diagnosis, the condition is commonly beyond the reach of curative surgery or radiotherapy alone. Therefore, early diagnosis is essential. For small, nonmetastasized tumors, systemic chemotherapy or radiotherapy can often reduce the risk of formation of secondary tumors. Recognition of the significant benefits of early diagnosis has led to an emphasis on mass screening programs, and some examples of these are described next.

1.4 DIAGNOSIS AND SCREENING

All too often, cancer is only diagnosed when a solid tumor has grown to sufficient size to become noticeable to the patient either directly by its size and location, where it may cause obstruction; a feeling of pressure or pain (e.g., in breast, esophagus, head and neck tumors); or through indirect symptoms, such as traces of blood in the urine or feces (e.g., in bladder or bowel cancer) or severe coughing (e.g., in lung cancer). Even worse, a secondary tumor may be detected, by which time it may be too late for effective treatment. Similarly, cancers of the blood (e.g., leukemias) are usually detected only when patients report indirect symptoms (e.g., irregular fever; hemorrhaging from the gums and mucous membranes or under the skin; anemia; enlargement of the spleen or lymph nodes), which vary depending on the type of cancer and whether it is in the chronic or acute phases.

Screening programs can be highly effective in preventing late diagnosis. If tumors are diagnosed early (i.e., before they have metastasized), effective treatments can be initiated, often leading to cures. For example, cervical and breast cancer screening in women has become routine in several developed countries during the last decade, and prostate cancer screening for men is presently being introduced in many countries. In the U.K., a two-yearly check for bowel cancer in all men and women from age 60 onwards was introduced in the National Health Service (NHS) in 2006. Screening consists of a fecal occult blood test, in which patient-collected stool samples are examined. The samples are tested for minute traces of blood, a possible symptom of bowel cancer that affects around 34,000 Britons each year. (Prostate cancer is the U.K.'s second main cause of cancer deaths after lung cancer.) Research suggests that this screening may prevent at least 5,000 cases and 3,000 deaths per year. However, although this type of screening has many benefits, a rise in the number of bowel cancer cases diagnosed and the number of false alarms expected will significantly increase the financial pressure on the NHS (e.g., more endoscopy procedures will be required). False alarms also cause significant unnecessary distress for patients; therefore, all screenings must be as accurate as possible for both financial and ethical reasons.

During the next decade, the development of cheaper and more-widely available high-resolution whole-body scanners based on techniques such as magnetic resonance imaging should allow more effective mass screening for many different types of solid tumors. However, the most exciting prospect is the development of highly sensitive analytical techniques to detect tumor markers (e.g., nucleic acids, proteins, or glycoproteins) in easily obtained body fluids, such as blood, urine, and saliva. Such techniques have the potential advantage of being able to detect tumors at the earliest possible stage of their development, whereas the best imaging techniques may only allow tumors to be detected when they are significantly advanced. Major advances have already been made in this area through the relatively new field of *proteomics,* in which the ultimate goal is to use highly sensitive techniques such as mass spectrometry to identify (or profile) the entire spectrum of proteins produced by single cells.

1.5   FORMATION OF CANCER CELLS (TUMORIGENESIS)

At the genetic level, cancer involves constantly changing modifications to the genome of cells brought about by both internal and external (e.g., environmental) factors. This process has been established by the discovery of mutated genes with dominant gain of function (known as *oncogenes*) and some with recessive loss of function (known as *tumor suppressor genes*). At the cellular and behavioral levels, Weinberg and others (see *Further Reading* for reference) have proposed a set of simple rules that explain the transformation of healthy cells into tumor cells. They postulate that there are a defined number of cellular or biochemical characteristics (so-called *acquired* or *hallmark* traits) that are common to most and perhaps all types of human cancers.

Tumorigenesis is a multistep process, with each step reflecting genetic changes that promote the progressive transformation of healthy cells into tumor cells. Studies show that the genes of tumor cells are frequently modified at many different sites, ranging from disruptions as subtle as point mutations (i.e., one deoxyribonucleic acid [DNA] base-pair change) to more obvious problems such as chromosomal translocations. It is worth noting that the transformation of cells in culture is multistep, with cells derived from mice or rats requiring at least two genetic modifications. Interestingly, human cells are even more complicated to transform in culture.

Based on these observations, scientists have proposed that tumor development occurs through a process similar to Darwinian evolution, in which a sequence of genetic modifications, each providing a different type of growth advantage, leads to the progressive change of healthy cells into tumor cells. Weinberg has suggested that the large catalog of tumor cell genotypes may result from just six essential modifications to cell physiology that collectively induce malignancy. These are:

- Self-sufficiency in growth signals
- Insensitivity to growth-inhibitory signals
- Evasion of programmed cell death (apoptosis)
- Limitless replicative potential
- Sustained angiogenesis
- Tissue invasion and metastasis

Weinberg has also proposed that the need for a cell to acquire all of these traits prior to complete transformation may explain why tumor cell formation is relatively rare during the average human life span.

The ability to study human tumors at the biochemical and genetic levels has undergone dramatic changes during the last 10 years (e.g., DNA arrays, genome sequencing, proteomics) and is likely to benefit from further developments in the future at an increasingly rapid pace. At present, the ability to understand a newly diagnosed tumor in terms of its genetic defect remains in its infancy. However, in 10 or 20 years' time, the evaluation of all somatically acquired DNA modifications in the genome of a tumor cell is likely to become common practice, along with gene expression and proteomics profiling of tumor cells. Rather than the relatively primitive approaches to chemotherapy now available, in the future combinations of highly specific antitumor agents may be used to target each of these cancer hallmarks. In concert with sophisticated diagnostic technologies, identifying and treating all phases of disease initiation and progression should then be possible.

1.6 MECHANISMS OF GENOMIC DAMAGE

As previously discussed, it is now accepted that, in general, cancer is a "genetic" disease resulting from changes to DNA sequence information in one or more genes, or from more profound structural changes (e.g., translocations). These alterations can occur through internal, external, or hereditary processes.

1.6.1 INTERNAL FACTORS

Tumor formation may result from changes to DNA sequence or structure (i.e., mutations, addition or loss of DNA, or epigenetic changes) brought about by malfunction of the normal DNA processing systems within a cell.

1.6.1.1 Mutations

Genetic mutations can take several forms. In a *point mutation,* only one base is altered, and the resulting new codon can lead to insertion of an incorrect amino acid at the corresponding position of the protein. Should the protein be critical for proliferation (e.g., a tumor suppressor protein), then tumorigenesis may result. In a *translocation mutation,* an entire segment of DNA may be moved from one part of a gene or chromosome to another. In this case, loss of the proteins corresponding to the two original DNA sequences or production of a new protein corresponding to the novel fusion sequence may lead to tumorigenesis. The original genes involved in this process are known as *proto-oncogenes* — that is, genes that do not cause cancer themselves unless suitably activated (i.e., by translocation to form an *oncogene*). The concept of proto-oncogenes and oncogenes has been validated in such cancers as Burkitt's lymphoma and chronic myelogenous leukemia (CML), in which the precise sequences involved in the translocations have been identified.

1.6.1.2 Addition or Loss of Genetic Material

During normal DNA handling processes, such as replication and repair, DNA bases may be accidentally added or deleted. This can have a similar effect to a point mutation as it can alter the codon reading frame and lead to the production of faulty proteins critical for control of cell growth.

1.6.1.3 Epigenetic Changes

Epigenetic control represents a mechanism to regulate gene expression independent of any changes to the DNA sequence. It is an important new area of cancer research and, with many genome-sequencing projects coming to fruition, there is now significant interest in gene function and regulation and in understanding how epigenetic factors can control gene expression.

It is now understood that the genome contains two types of information, *genetic* and *epigenetic*. The genetic information (i.e., the DNA sequence) provides the plan for the production of all the proteins required to form a living organism; the epigenetic information (i.e., chemical modifications to individual DNA bases such as methylation or acetylation) provides additional instructions on how, when, and where the genetic information should be used. Such epigenetic modifications to the genome can influence processes such as chromatin regulation, transcriptional repression, X-chromosome inactivation, DNA repair, genomic stability, and imprinting. Epigenetic changes may also suppress the potentially harmful effects on genomic integrity of repetitive and parasitic DNA sequences.

The most important way in which epigenetic information is transmitted in mammalian cells is through methylation of the C5 position of the pyrimidine cytosine of CpG dinucleotides ("islands") mainly found in the promoter region of genes (although methylation can also occur in other areas of DNA). It is important to note that, because methyl transfer occurs on only one strand of DNA at a time, the formation of fully methylated double-stranded DNA results from two separate single-stranded reactions. Changes in gene expression involve the interaction of control proteins with methylated regions of genes to suppress expression, although few specific proteins have yet been identified (methyl-CpG-binding protein 2 is an example).

DNA methylation patterns are often modified in cancer cells, and global hypo-methylation with accompanying region-specific hypermethylation are proving to be the most consistent molecular modifications across a number of tumor types. Hyper-methylation within the promoter of a tumor suppressor gene can silence expression, thereby providing the cell with a similar growth advantage to that resulting from deletions or mutations. Conversely, hypomethylation in the promoter of an oncogene may lead to overexpression and then tumorigenesis. This area of research is growing, and a large number of gene types, including oncogenes, tumor suppressor genes, and tumor-associated viral genes, are thought to be controlled by epigenetics information. Examples include APC, p15, p16, p73, MGMT, TIMP3, ER, RAR, DAPK1, VHL, E-cathedrin, GSTP1, and LKB1.

Histone acetylation and deacetylation are also thought to play a role in epigenetic control by modulating chromatin condensation and transcription. These dynamic processes are regulated by enzymes known as *histone acetyltransferases* and their counterparts the *histone deacetylases*, and the equilibrium between them can be modified by exogenous influences known as *epigenetic* agents. These can modify methylation or acetylation and therefore change the phenotypes of cells epigenetically, without their DNA sequences being altered. Some examples include chemicals such as bromobenzene and butyryl cyclic adenosine monophosphate, the hormone estradiol, and metals such as cadmium, arsenic, and nickel. In addition, radiation and reactive-oxygen species can act as epigenetic agents. It is noteworthy that zebularine and 5-aza-2'-deoxycytidine are capable of inhibiting DNA methylation and these agents are being investigated in the area of cancer chemoprevention. In addition, 5-aza-2'-deoxycytidine can reverse the loss of imprinting in cancer cells.

1.6.1.4 Modified Gene Expression

A problem with a cell's transcription or translation machinery may lead to uncontrolled expression or amplification of a gene or sets of genes. Should, for example, genes associated with growth factors or the proteins responsible for growth factor receptors be involved, then tumorigenesis can result.

1.6.2 EXTERNAL FACTORS

1.6.2.1 Viruses

A link between viruses and cancer was first recognized in 1911, when Peyton Rous demonstrated that avian spindle cell cancer could be transmitted from one bird to another by a cell-free filtrate containing the virus (which now carries his name — the *Rous sarcoma virus*). Since then, other viruses have been linked to human cancers. Well-known examples include the involvement of the Epstein-Barr virus (EBV) in Burkitt's lymphoma and the human papillomavirus (HPV) in cervical cancer.

Viruses may be either ribonucleic acid (RNA) retroviruses, such as human T-cell leukemia virus (HTLV-1), or DNA viruses. RNA viruses contain DNA polymerases, which facilitate the production of double-stranded viral DNA. On being incorporated into the host genome, the viral DNA may cause tumorigenesis via a number of different mechanisms, including production of an oncogene from an existing proto-oncogene, damage to a tumor suppressor gene, or insertion of a completely new gene. For example, HTLV-1 introduces a gene known as *tax* that results in the overexpression of interleukin-2. This can lead to adult T-cell lymphomas and leukemias with an increase in the number of activated lymphocytes, although these may take years to develop in susceptible individuals. HTLV-1 is endemic in Southeast Asia and the Caribbean; in the Far East, it is also associated with nasopharyngeal cancers.

It is now known that approximately 90% of the global population is infected with EBV, which is considered to be an "initiator" of cancer as opposed to a specific cause. For example, 90% of Burkitt's lymphoma cells test positive for EBV, and the infection allows lymphocytes to become immortal, leading to a potentially cancerous state. Burkitt's lymphoma is endemic in those parts of Africa with chronic malaria, suggesting that the latter may be a cofactor in lymphoma development.

Hepatocellular carcinoma has been linked to the hepatitis B virus (HBV) and is endemic in Southeast Asia and tropical Africa. The risk of tumor formation is greatest in those who are infected from an early age, and males are four times more likely to develop the cancer than females. It is believed that the X-gene in HBV codes for proteins that promote transcription.

Finally, more than 50 different types of HPV exist, and HPVs 16 and 18 have been linked to cervical cancer. The virus produces several proteins, some of which enhance mitosis while others interfere with p53 (a tumor suppressor gene) or modify the interaction between cellular proteins and transcription factors. An experimental vaccine has recently been developed for HPV that, when administered to young girls, can protect from cervical cancer. This vaccine is described further in Chapter 8.

1.6.2.2 Bacterial Infections

Certain bacterial infections can also lead to tumorigenesis. The best-known example is *Helicobacter pylori,* which colonizes in the stomach of some individuals and is associated with peptic ulcers. Infection with *H. pylori* has also been linked to a greater risk of developing stomach cancer and mucosa-associated lymphoid tissue, or MALT, lymphoma (a rare form of stomach cancer). The precise mechanism of tumorigenesis is unclear at present. Eradication of the infection is now routine using antibiotic therapy, which reduces the risk of developing cancer.

1.6.2.3 Chemicals

Certain chemicals in the environment and some encountered through diet, lifestyle, and occupation can lead to tumorigenesis. For example, the link between cigarette smoke and lung cancer is now well-established. It is also known that carcinogenic polycyclic aromatic hydrocarbons, which form when red meat is overcooked (especially through frying or barbecuing), can lead to colon tumors. Carcinogenic amines form in the stomach as a result of the bacterial degradation of nitrites used as preservatives in meat and fish. Potent carcinogens called *aflatoxins* (Structure 1.1) are also found in low concentrations in peanut butter; these carcinogens are secondary metabolites produced by a fungus that infects the peanuts during growth.

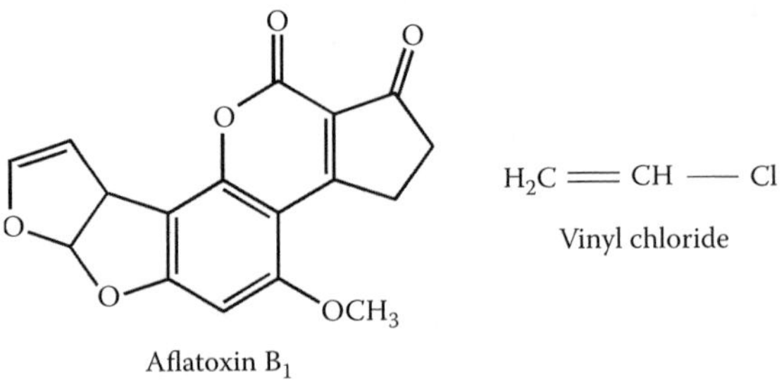

$$H_2C = CH - Cl$$

Vinyl chloride

Aflatoxin B$_1$

STRUCTURE 1.1 Structures of some known carcinogens.

Occupation-associated cancer is not just a feature of recent times. In 1775, Sir Percival Pott noted the high frequency of scrotal skin cancer in young chimney sweeps. Poor hygiene in this population meant that tarry deposits from coal fires were in contact with the sensitive skin of the scrotum for long periods of time. More recently, vinyl chloride (Structure 1.1) used by workers in the plastics industry has been associated with angiosarcoma of the liver, and furniture industry workers have been prone to nasopharyngeal malignancies induced by the inhalation of small particles carrying organic compounds that arise from leather and wood polishing processes. Many of these organic carcinogens exert their effects by covalently modifying DNA (either before or after metabolism).

(a)

(b)

FIGURE 1.1 Asbestos in its native state (a) and in its carcinogenic blue fibrous form (b).

In addition to these organic carcinogens, certain dusts and minerals are known to be tumorigenic. For example, the link between asbestos and pleural and peritoneal tumors (particularly mesothelioma) is well established and thought to be due to the physical damage of chromosomes by microscopic fibers of asbestos, a type of insulating material (Figure 1.1). Asbestos, which was commonly used after 1945, was popular in the building, ship, and car construction industries, and mesothelioma was eventually found to be closely linked to these work environments. This association was not realized for some time, as the first symptoms of mesothelioma do not become obvious until up to 10 to 40 years after exposure. There are three types of asbestos which have different colors, i.e., blue, brown, and white. The blue and brown types are most often linked with mesothelioma, and so were banned by most countries in the 1990s. Their carcinogenicity is thought to be due to the fact that they are comprised of very small fibers which, after inhalation, can travel to the outside lining of the lung (the *pleura*) where they can penetrate cellular and nuclear membranes and cause physical damage to the cells. This problem is compounded because the fibers are small enough to be carried unnoticed on clothing, leading to exposure of other family members as well.

1.6.2.4   Radioactivity

Malignancies have also been linked to exposure to α or β particles (and γ- or X-rays [see Section 1.6.2.5]), which are known to damage DNA by fragmentation through the formation of free radicals. A link between nuclear fallout and cancer was firmly established after atomic bombs were dropped on the cities of Hiroshima and Nagasaki, Japan, in 1945 at the end of World War II. It has also been postulated that children living close to nuclear power stations are at increased risk for developing certain forms of leukemia as well as brain tumors, although statistical analyses of these data remain controversial. However, the risk of escape of radioactive materials from nuclear reactors was highlighted by the Chernobyl incident in the Ukrainian republic of the Union of Soviet Socialist Republics in 1986. The accident caused widespread contamination of the food chain and led to a variety of different cancers in populations exposed to the radiation. It is also known that a buildup of radon gas produced by certain types of granite can endanger the occupants of houses built from this material. Radon is a naturally occurring radioactive gas that, once inhaled, enters the bloodstream and delivers radiation to all tissues. Bone marrow is particularly sensitive and so leukemias predominate. In the U.K., some local councils have been obliged to offer grants to affected householders so that buildings can be structurally modified to improve ventilation. The ability of radioactive particles to damage DNA and kill cells can also be put to good use by targeting them to tumors, where they can selectively kill cancer cells. In this context, neutrons are also being used in an experimental therapy called *boron neutron capture therapy*.

1.6.2.5   Electromagnetic Radiation

It is now well established that electromagnetic radiation from the higher-energy part of the electromagnetic spectrum (Scheme 1.1), such as ultraviolet (UV) radiation and γ- or X-rays, can damage cellular DNA and lead to tumorigenesis. More recently, there has been a growing concern about the potential for radio waves and microwaves to cause cancer.

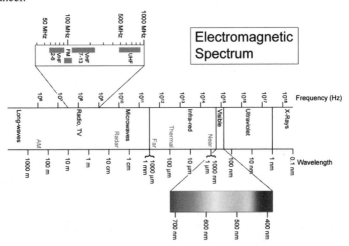

SCHEME 1.1 The electromagnetic spectrum.

The UV band, which is invisible to the human eye, constitutes one particular part of the spectrum of sunlight and makes up approximately 3% of all the solar radiation reaching the Earth's surface. Three types of UV light have been identified. One of these, UVC (200 to 290 nm), is generally thought to be the most carcinogenic. UVB (290 to 320 nm) causes the most sunburn, and UVA (320 to 400 nm), which can be up to 1,000 times stronger than UVB, is able to penetrate underlying tissues of the skin leading to skin damage, including "photoaging." Fortunately, the ozone layer absorbs most of the more carcinogenic UVC radiation, although there is presently concern that depletion of the ozone layer through the release of chloro-fluorocarbons used in the air conditioner and refrigeration industries and in the production of Styrofoam insulation may increase the intensity of UVC at the Earth's surface in the future. UV radiation occurs at a similar wavelength to the region of maximum absorbance by DNA (i.e., 260 nm), and the major damage is intrastrand linkage of adjacent pyrimidines (usually thymines) to form so-called *thymine dimers*. These thymine dimers create distortions in the DNA helix and can block replication and transcription, thus leading to tumorigenesis.

γ- and X-rays are more damaging to DNA because they have more energy and can penetrate further into tissues. Although both are used as diagnostic tools (i.e., scintigraphy, X-rays, and computed tomography scans), exposure time of a patient is normally limited and closely monitored. γ- and X-rays can cause a variety of different types of damage to the DNA helix. Single-strand breaks can occur, although these are usually rapidly joined by enzymes known as *DNA ligases* and so are not necessarily lethal for the cell. However, double-strand breaks can also occur. These are more serious for the cell; they usually lead to cell death because they cannot be joined by ligases and are instead degraded by nucleases. Modifications to DNA bases that do not lead to strand cleavage can also occur. For example, oxidized bases can occur that block transcription and replication and are usually lethal for the cell.

The DNA-damaging effects of X-rays (predominantly double-strand DNA breaks) are also put to good use in radiotherapy, in which an X-ray beam is focused on a tumor in order to kill cancer cells by fragmenting their DNA.

1.6.2.6 Cancer Treatments

The tumorigenic effects of chemicals and radiation have been previously discussed. Unfortunately, this association means that the chemotherapeutic agents and radiation therapy given to patients (particularly children) with cancer can themselves raise the risk of developing cancer later in life. This appears to manifest as soft tissue sarcoma, a disease affecting mostly fat and muscle tissues. According to one study reported in the *International Journal of Cancer* in 2004, soft tissue sarcoma is one of the most common new types of malignant disease appearing in young adults and teen-agers who were treated with anticancer drugs during childhood. Researchers from the Gustave Roussy Institute (France) followed more than 4,000 patients who had survived a first cancer during their childhood and observed the occurrence of 16 soft tissue sarcomas at least 3 years after diagnosis of the first disease. Although this rate of occurrence appears low, it is more than 50 times greater than that observed in the general population. More significantly, 14 of the 16 sarcomas were found in close proximity to the area of treatment of the first cancer. Furthermore, the probability of soft tissue sarcoma occurring as a secondary cancer was found to increase in proportion to the dose of radiation originally administered. The researchers also found that treatment with the DNA-methylating agent procarbazine appeared to raise the risk of sarcoma developing later in life.

This poses an ethical dilemma in that the original treatments giving rise to these later soft tissue sarcomas also provide benefits in terms of increased survival rates for the childhood cancers. On balance, most families are willing to accept this risk in order to treat the childhood disease. In the case of radiation therapy, the risk of sarcomas developing later can be minimized by restricting exposure of healthy areas of the body to high doses of radiation, and the technology to do this through the use of more-focused beams of radiation is continuously improving.

1.6.3 HEREDITARY FACTORS

A number of genes have now been identified that, if inherited, can predispose individuals to certain types of cancer. For example, two genes (BRCA1 and BRCA2) have been identified and sequenced that are inherited and closely associated with breast cancer. Other genes associated with colon and bowel tumors are known to be inherited. This knowledge has lead to the introduction of diagnostic screening, with subsequent genetic counseling for affected individuals. In some cases, women who discover that they are carrying BRCA1 or BRCA2, and thus have a high risk of developing cancer, elect to have their breasts removed at an early age as a prophylactic measure.

This research area is growing phenomenally as information from the Human Genome Project and advances in molecular biology and genetics converge. Although presently controversial, the ultimate application of advances in this area would be the screening of human embryos to ensure that no hereditary risk factors are being passed on.

1.7 TREATMENTS

Cancer treatment often encompasses more than one approach, and the strategy adopted depends largely on the nature of the cancer and how far it has progressed. The main treatments are still surgery, radiotherapy, and chemotherapy. However, other approaches such as photodynamic therapy (PDT), antibody- and vaccine-related approaches, and gene therapy are in development.

1.7.1 SURGERY

If a tumor is small or reasonably well defined, it can sometimes be surgically removed. However, additional treatment with chemotherapy or radiotherapy is often required to eliminate any cancer cells that may have remained behind or metastasized. Alternatively, radiotherapy or chemotherapy may be administered prior to surgery in order to shrink (or "debulk") the tumor, thus facilitating its removal. Where possible, a large area of surrounding healthy tissue (including neighboring lymph glands through which cancer cells can pass) is removed as well to ensure complete eradication of cancer cells from the site.

1.7.2 RADIOTHERAPY

Radiotherapy involves the use of X-rays or radiopharmaceuticals (radionuclides) that act as sources of γ-rays. Neutron beams are also utilized in some experimental procedures, although this is less common due to the expense and availability of the equipment required. In X-ray therapy, radiation is delivered locally in a highly focused beam to avoid damaging healthy tissue. Although X-ray therapy is a well-established technique, ongoing research is looking into the most effective treatment regimes in terms of the duration and frequency of exposure. The latest technological developments include the use of multiple beams under computer guidance that converge and focus on a tumor with a high degree of accuracy. This technique spares as much surrounding healthy tissue as possible because individual beams can be of the lowest possible energy but they combine to produce the maximum intensity at their focal point at the tumor site. For γ-rays, radionuclides in use include cobalt-60, gold-198, and iodine-131. Gold-198 concentrates in the liver, and iodine-131 is used to treat thyroid cancers because iodine accumulates in this gland.

A significant proportion of tumor cells are hypoxic (i.e., have a low oxygen level) and are thus less sensitive to damage by irradiation, which works through the formation of DNA-damaging oxygen-free radicals. Therefore, prior to and during radiation therapy, oxygen is sometimes administered to sensitize the tumor cells. In addition, radiosensitizing drugs such as metronidazole (normally used as an anti-protozoal/antibacterial agent) have been coadministered experimentally prior to treatment in an attempt to improve therapeutic outcomes (Structure 1.2). However, such agents can also sensitize healthy tissues thus leading to no net improvement in therapeutic index.

Metronidazole (Flagyl^{TM})

STRUCTURE 1.2 Structure of the experimental radiosensitizing agent metronidazole.

Neutrons (which are particles rather than γ- or X-rays) are also used experimentally in cancer therapy (see Chapter 7). For example, a process known as *high linear energy transfer* has been developed to kill hypoxic cells by irradiating the tumor with neutrons that then decay to α-particles, the latter causing cell damage in an oxygen-independent manner. A more-sophisticated treatment known as *boron neutron capture therapy* involves administration of a boron-10 (^{10}B)–enriched delivery agent that is taken up by the tumor. The target area is then irradiated with low energy neutrons that are captured by the ^{10}B atoms. This leads to a reaction that produces α-particles (^{4}He) and lithium-7 (^{7}Li) ions that destroy the tumor tissue.

1.7.3 PHOTODYNAMIC THERAPY (PDT)

PDT involves the initial systemic administration of a photosensitizer such as the porphyrin derivative Photofrin™, which has a degree of selectivity for the tumor tissue (see Chapter 7). After the agent has localized, the tumor is irradiated with an intense light source of an appropriate wavelength (usually a laser), which excites the Photofrin. Upon decay to its ground state, available oxygen is then transformed into the singlet form (i.e., free radical), which is highly cytotoxic and damages the tumor cells. Some research has also shown that, by damaging endothelial cells, PDT can restrict blood flow to tumors.

Laser light sources are now well developed, and the use of flexible optical fibers means that tumors in body cavities such as the gastrointestinal (GI) tract, bladder, and esophagus can be easily reached. Furthermore, key-hole surgery techniques offer the possibility of reaching other organs. As other less-expensive nonlaser light sources become widely available and new types of photosensitizers are developed, the use of PDT is likely to escalate.

1.7.4 BIOLOGICAL RESPONSE MODIFYING AGENTS

Many so-called *biological response modifiers* (BRMs), or *biologicals,* are either in use or development. These include agents as diverse as antibodies, antibody-drug conjugates, interferons, interleukins, enzymes, vaccines, and other types of immune stimulants (see Chapter 8). For example, several tumor types, including some types of breast cancer, have been found to produce specific tumor antigens on their cell surfaces, and this has led to the development of monoclonal antibodies specific for these tumors (e.g., Herceptin™ [see Chapter 5 and Chapter 7]). Tumor-specific antibodies can also be used to selectively deliver a cytotoxic agent (e.g., Mylotarg™) or a radionuclide (see Chapter 7) to the tumor site. An antibody-enzyme conjugate designed to release an active form of a cytotoxic agent from a nontoxic prodrug selectively at the tumor site (i.e., antibody-directed enzyme prodrug therapy [ADEPT]; see Chapter 7) is presently being evaluated in clinical trials. Finally, research is ongoing into the development of vaccines that may either prevent tumor formation or modify the growth of established tumors (see Chapter 8). For example, one recent success is the development of an anti-HPV vaccine which, when given to young females, appears to virtually eradicate the risk of cervical cancer in later life.

1.7.5 CHEMOTHERAPY

Chemotherapy involves the use of low-molecular-weight drugs to selectively destroy a tumor or at least limit its growth. Nitrogen mustards were the first agents to be used clinically; their use resulted from the accidental discovery that the mustard gas used in World War II had antileukemic properties. Since then, important advances have been made in the development of new anticancer drugs. For example, cisplatin, which was also discovered serendipitously, provided a major advance in the treatment of testicular and ovarian carcinomas, and the more recent discovery of imatinib (Gleevec™) has led to very high response rates in chronic phase Philadelphia-chromosome-positive CML patients.

One advantage of chemotherapy is that, after intravenous administration, low-molecular-weight drugs distribute throughout most tissues of the body and so can kill tumor cells in protected areas (e.g., the brain) or those cells in the process of metastases. However, the disadvantages of many cytotoxic agents include unpleasant side effects, such as bone marrow suppression, GI tract lesions, hair loss, nausea, and the rapid development of clinical resistance. The side effects occur because cytotoxic agents, such as DNA-interactive drugs, tubulin inhibitors, and antimetabolites, act on both tumor cells (often triggering apoptosis) and healthy cells. Their mechanisms of action include a differentially faster uptake and action in the more rapidly dividing cancer cells. Alternative mechanisms include, in the case of DNA-interactive agents, the lack of ability of cancer cells to repair DNA adducts. The side effects reflect the fact that cells of the bone marrow, GI tract, and hair follicles divide at a faster rate than most healthy tissues. Some of the newer families of agents, such as the kinase inhibitors, are much more selective for cancer cells. They can be given orally and have far fewer toxic side effects. Chapter 2 through Chapter 6 describe the various classes of low-molecular-weight chemotherapeutic agents in current clinical use. Chapter 11 describes adjunct therapies sometimes used with them to counteract side effects such as nausea, and to enhance activity. Chapter 9 describes new research areas that are likely to lead to novel types of anticancer agents in the future.

1.8 DISCOVERY OF ANTICANCER DRUGS AND PRECLINICAL EVALUATION

Most anticancer agents in current use were discovered either by chance (e.g., cisplatin and the nitrogen mustards) or through screening programs (e.g., vinblastine and paclitaxel [Taxol™]). Only recently has a more detailed knowledge of the fundamental biochemical differences between normal and tumor cells allowed a truly rational approach to the drug design process (e.g., the kinase inhibitors Gleevec™ and gefitinib [Iressa™]). A combination of the power of rational drug design and modern screening techniques has been realized with the current trend toward the use of both diverse and focused compound libraries to provide large numbers of molecules for *in vitro* and cellular screening against specific cancer-related targets.

New lead agents are nearly always evaluated initially in *in vitro* tumor cell lines. This only measures the *cytotoxicity* (i.e., cell-killing ability) of an agent and provides no indication of whether it is likely to have useful antitumor activity *in vivo*. However, by studying panels of different tumor cell types, one can establish whether an agent has selective cytotoxicity toward a particular tumor type, which may then suggest suitable *in vivo* experiments. Initial *in vivo* experiments may involve observation of the effect of systemically administered novel agents on tumor cells growing in porous fibers inserted subcutaneously or intraperitoneally in mice or rats (i.e., the *Hollow Fiber* assay pioneered by the National Cancer Institute [NCI]). However, *Human Tumor Xenograft* assays are the most widely used animal model in which human tumor fragments are transplanted into immunosuppressed rodents (usually mice, but less commonly rats). The effect on tumor growth and extension of life span is then observed after administration of the novel agent, using different schedules. These experiments also provide an initial indication of the toxicity profile of the agent.

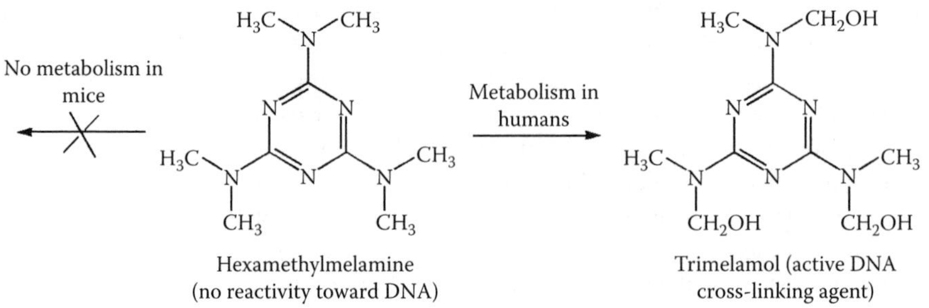

SCHEME 1.2 The metabolism of hexamethylmelamine to biologically active trimelamol in humans but not in mice.

Although useful as a tool in anticancer drug development, these animal models do not reliably relate to equivalent tumors in humans, as numerous differences exist including the integrity of the blood supply to the transplanted tumor and general biochemical species differences including metabolism. This means that, sometimes, drugs that are active in humans show no effect in animal models. For example, hexamethylmelamine [2,4,6-tris(dimethylamino)-1,3,5-triazine] was the lead compound for the melamine class of antitumor agents that reached clinical evaluation (Scheme 1.2). This agent has some (albeit limited) antitumor activity in humans (against bronchial, ovarian, and breast cancers) because it is metabolized to carbinolamine species (e.g., trimelamol), which can form interstrand cross-links with DNA. However, the parent compound exhibits only minimal activity when tested in rodent models, as mice fail to carry out the crucial oxidative metabolic step. Despite these problems, most drugs in clinical use today (with the exception of some hormonal agents) were introduced as a result of activity demonstrated in animal models.

The NCI has carried out a screening cascade of the type described above since the 1960s. Compounds sent to the NCI from academic and industrial sources worldwide are initially screened in a 60-cell line panel (or initially a 3-cell line precursor panel), and compounds with interesting activity are progressed through hollow fiber and then human tumor xenograft assays to establish efficacy. The NCI has now established a substantial database of information and uses a computer algorithm known as "COMPARE" to establish whether the characteristics of a novel agent evaluated in the 60-cell line panel are similar to any existing families of agents or individual compounds, thus predicting the likely mechanism of action.

1.9 ACCESSIBILITY OF DRUGS TO TUMOR CELLS

The accessibility of anticancer drugs to tumor cells in the body varies greatly. While leukemia cells are fully exposed to drugs in the bloodstream, most solid tumors have a less reliable blood supply. Small, early stage tumors can be reasonably well supplied and are, therefore, more susceptible to drug action. Larger tumors, on the other hand, often have poor capillary access, particularly in their centers, which can be hypoxic and even necrotic. The degree of accessibility of a chemotherapeutic agent is therefore one of a number of reasons for the greater sensitivity of small primary and early metastatic tumors to chemotherapy and highlights the importance of early diagnosis and treatment. It is noteworthy that brain tumors are particularly resistant to chemotherapy because few drugs are capable of crossing the blood–brain barrier. Although beyond the scope of this book, it is important to note that favorable pharmacokinetic characteristics of an anticancer agent (including all ADME considerations [Absorption, Distribution, Metabolism, and Excretion]) are crucial for a drug to be available to enter the tumor.

1.10 ACHIEVING SELECTIVE TOXICITY

Many of the anticancer agents presently used, including the antimetabolites, DNA-interactive agents, and tubulin inhibitors, are cytotoxic agents. Although the reasons for their selective toxicity are not fully understood, they are thought to have a greater effect on tumor cells because these cells usually divide more rapidly than healthy cells. However, cells of the bone marrow, GI tract, and hair follicles also divide rapidly, which explains the consistent pattern of side effects accompanying chemotherapy that are dose-limiting in practice.

The development of more-effective chemotherapeutic agents is critically dependent upon the discovery of exploitable biochemical differences between normal and tumor cells. Such differences should allow a more rational approach to drug design rather than relying on the empirical manner in which many of the present-day drugs have been discovered and developed. Examples of drugs resulting from such a fundamental change in the discovery process are presently limited, but the kinase inhibitors Gleevec™ and Iressa™ are good examples of the trend in this direction (see Chapter 5). An older example is the discovery that some lymphoid malignancies are dependent on an exogenous supply of asparagine, whereas healthy cells can synthesize their own. This led to the clinically useful agent asparaginase (Ewinase™) (see Chapter 8).

As a result, much current research in cancer drug discovery is aimed at identifying genes and gene products (i.e., proteins, enzymes, receptors) either unique to, or overexpressed in, tumor cells. The kinase family of signaling enzymes represents a good example of a set of targets up-regulated in certain tumor cell types, and many types of kinase inhibitors are in development. Imatinib (Gleevec) is a good example of a new generation of tyrosine kinase–targeted drugs; it inhibits a specific fusion protein (BCR-ABL) produced by a genetic sequence (the Philadelphia fusion sequence) found uniquely in CML tumor cells. It is hoped that complete selectivity may one day be achieved through targeting individual genes, thus directly exploiting the genetic differences between tumor cells and healthy cells. This approach is known as *antigene technology*, using gene-targeting agents that can block transcription (see Chapter 9).

1.11 LIMITING THE TOXICITY OF CHEMOTHERAPEUTIC AGENTS

With few exceptions — such as the cumulative toxicities associated with Adriamycin (cardiac), bleomycin (pulmonary), and cisplatin (renal) — the most common side effects of cytotoxic agents such as bone marrow suppression, GI tract lesions, and nausea and vomiting are usually reversible within 2 to 3 weeks. For example, mucositis (associated with actinomycin D, Adriamycin, bleomycin, methotrexate, 5-fluorouracil, and daunorubicin) is reversible over a period of 5 to 10 days, reflecting the rapid recovery of normal tissues. Nonetheless, all of these side effects are distressing for patients and, in some cases, are so severe that potentially beneficial treatments have to be halted.

Therefore, in order to maximize the therapeutic index (a ratio of the effective/toxic doses), a number of approaches have been developed to limit the toxicity of chemotherapeutic agents. These include dose scheduling and the coordination of drug types with cell cycle, the coadministration of adjuvants, the design of novel prodrugs and dosage forms, and the use of physical devices such as the cold cap. Recent studies have also focused on the effect of circadian rhythms, and a growing body of evidence suggests that the efficacy and toxicity of some agents can be influenced by their time of administration during a 24-hour cycle. Finally, the study of pharmacogenomic markers is a rapidly growing research area; in the future, it may be possible to screen out patients who are most likely to develop serious side effects to a particular drug.

1.11.1 DOSE SCHEDULING AND CELL CYCLE

Competition between cells within a tumor favors those that continuously progress through the cell cycle. Tumor cells generally proliferate and differentiate at a faster rate than the immediately surrounding normal cells, and this leads to disease progression. However, healthy rapidly dividing cells are also affected by chemotherapy, which gives rise to the recognized side effects, such as myelosuppression, alopecia, dermatitis, and GI, liver, and kidney toxicities.

The cell cycle (Scheme 1.3) is a target for many chemotherapeutic agents, particularly cytotoxic agents. It has proved possible to limit bone marrow toxicity by exploiting the cell kinetic differences between normal and tumor stem cells. Stem cells constitute the smallest, yet most important, compartment in a proliferating system. They are capable of an indefinite number of divisions and are responsible for maintaining the integrity and survival of a cell population. However, only 20% of bone marrow stem cells are usually in an active cycle at any one time, the remaining 80% being in the resting phase (G_0). This small dividing marrow stem cell population may be significantly reduced in size by chemotherapy, thus leading

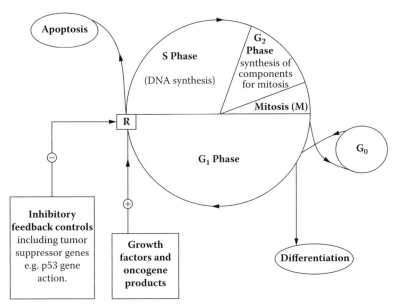

SCHEME 1.3 The cell cycle. Class 1 drugs act at specific phases of the cycle (usually S or M); Class 2 drugs act at any phase of the cycle. (R = Restriction/Check Point.)

to the observed toxicity of bone marrow suppression. However, within 3 to 4 days, the remaining stem cells can move from G_0 into active cycle. Therefore, toxicity can be reduced by administering very high doses of drugs for 24- to 36-hour periods but allowing coordinated interdispersed recovery periods for the bone marrow to regenerate.

Studies of the cell kinetic patterns of tumor growth have led to a classification of cytotoxic agents based on their ability to reduce the stem cell population of normal bone marrow and lymphoma cells in mice. The first class, called *Class 1,* consists of cell-cycle specific agents that kill cells in only one phase of the cycle. Drugs acting in *S-phase* (the period of DNA synthesis) include 6-mercaptopurine, cytosine arabinoside, and methotrexate; those active in M phase (during mitosis) include vinblastine and vincristine. In the case of Class 1 agents, an increased dose kills no more bone marrow stem cells than would be killed by the initial dose. The other category, called *Class 2,* consists of non-cell-cycle-specific agents can still be cells in all phases of the cell cycle (although some of these agents are still more active in a given phase of the cycle). Examples include cyclophosphamide, melphalan, chlorambucil, cisplatin, carmustine, lomustine, 5-fluorouracil, actinomycin D, and daunorubicin. An increased dose of Class 2 agents increases the number of bone marrow stem cells killed. Another strategy to enhance efficacy while minimizing toxicity is to utilize mixtures of Class 1 and Class 2 agents.

More recently, reports have appeared of the possible effect of circadian rhythms on the efficacy and toxicity of anticancer agents. Although this work is at an early stage, the evidence so far suggests that the benefit of some drugs can be maximized by administering at certain times of the day in synchronization with a patient's rhythms.

1.11.2 USE OF ADJUVANTS

There are a number of examples of the coadministration of other agents to minimize the toxicity of anticancer drugs (particularly cytotoxics). For example, folic acid is administered with methotrexate as a "rescue therapy" (see Chapter 2), and pretreatment of a patient with steroids (e.g., prednisolone or dexamethasone) can help reduce the side effects of some agents (e.g., the edema associated with Taxol [see Chapter 11]). In addition, some chemotherapeutic agents induce side effects such as nausea and vomiting. Drugs that are prone to this (e.g., cisplatin) can be coadministered with antinausea agents. A significant advance was made in antinausea treatments with the development of the 5-hydroxytryptamine type 3 antagonists, which are described in Chapter 11.

1.11.3 NOVEL FORMULATIONS AND PRODRUGS

Sometimes, placing a drug in a novel formulation can prevent certain side effects. For example, doxorubicin is renowned for its cardiotoxicity and for causing tissue necrosis at the injection site. Liposomal formulations, such as Caelyx™ and Myocet™ (see Chapter 3), are claimed to reduce the incidence of these side effects. It is postulated that because the drug becomes sequestered in liposomes, the amount of free drug available in the blood to cause cardiotoxicity or affect the veins at the injection site is minimized. Furthermore, it is thought that liposomal particles may be taken up with a degree of selectivity by a tumor where they degrade to deliver the drug in higher concentrations than would occur by administering the free agent.

Another strategy to reduce toxicity is to develop novel prodrugs that only release active drug at the tumor site. Alternatively, the drugs can be coupled to antibodies for selective delivery to tumors. These and a number of other approaches to increase efficacy and reduce toxicity are discussed in more detail in Chapter 7.

1.11.4 THE COLD CAP

Hair loss is a serious concern for cancer patients being treated with certain cytotoxin agents and, until recently, has been addressed with the provision of custom wigs for patients significantly affected. However, alternative strategies are now available. For example, a Swedish company called Dignitana supplies a cap (the DigniCap™) that, when worn on the head during chemotherapy sessions, cools the vessels that supply blood to the hair follicles, thus restricting blood flow and consequent exposure to the drug (Figure 1.2). This significantly reduces hair loss during the treatment period.

1.11.5 PHARMACOGENOMIC MARKERS OF TOXICITY

Currently, much research is underway to identify pharmacogenomic markers of both efficacy and toxicity for anticancer therapies. This knowledge would allow patients to be screened to predict the risk of developing serious side effects to a particular drug. One example where this is already possible is with 5-fluorouridine. A small percentage of people (3%-5%) are deficient in the enzyme dihydropyrimidine dehydrogenase (DPD), which is important for metabolizing the agent. Due to a buildup

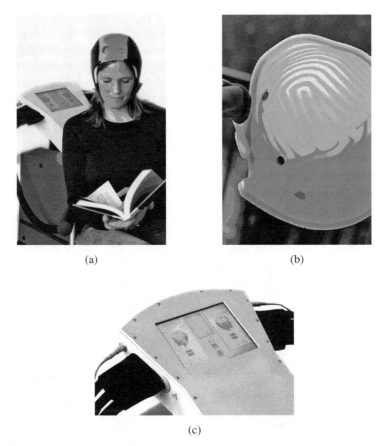

(a)　(b)

(c)

FIGURE 1.2 The DigniCap™ device which, when in place on the head (a) cools the scalp, reduces blood flow to the hair follicles and thus reduces hair loss during and after chemotherapy. The DigniCap works by circulating a cooling fluid through embedded cooling channels (b) under the control of a software-driven control station (c).

of the drug during chemotherapy at what would be a normal dose for most patients, these individuals experience severe vomiting, diarrhea, mouth sores, and other dangerous side effects which, in some cases, can lead to death. Screening for DPD status can prevent patients from suffering these side effects and alternative treatments can be instigated (see Chapter 10). It is anticipated that many more pharmacogenomic markers of both efficacy and toxicity will be discovered in the future, and their use in clinical practice is likely to become widespread.

1.12 OVERVIEW OF MECHANISMS OF ACTION OF CHEMOTHERAPEUTIC AGENTS

The anticancer agents either in clinical use today or in development work through a wide variety of different mechanisms in order to interfere with tumor cell growth, motility or survival, or with angiogenesis. By far, the largest and most diverse group of agents is those that interact with DNA (see Chapter 3) (Scheme 1.4). Some do this by blocking the synthesis of DNA (e.g., 6-mercaptopurine), while others become incorporated into DNA and then interfere with its function (e.g., 6-thioguanine). However, others interact directly with DNA by mechanisms including intrastrand (e.g., cisplatin) or interstrand (e.g., nitrogen mustards) cross-linking, intercalation between base pairs (e.g., mitoxantrone), or interaction in the minor (e.g., ET-387 and SJG-136) or major (e.g., temozolomide [Temodal™]) grooves. Other drugs interact with DNA (by one of the above mechanisms) and then cause strand scission (i.e., breakage or cleavage) at the binding site.

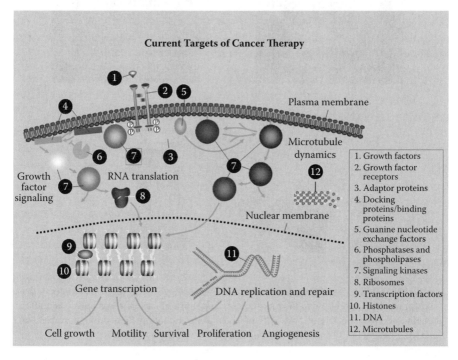

SCHEME 1.4 Summary of the various mechanisms of action of anticancer agents.

Antimetabolites (see Chapter 2) work by interfering with key biosynthetic pathways (e.g., methotrexate), and antitubulin agents (e.g., vinblastine and Taxol) interfere with cell division by interacting with the microtubule components of the spindle apparatus (see Chapter 4). The recently developed signal transduction inhibitors (see Chapter 5) are a rapidly growing group of agents that work by selectively blocking pathways involved in, for example, growth factor signaling (e.g., Gleevec and Iressa). The vascular targeting agents (see Chapter 7) work by either blocking new capillary growth (e.g., the antiangiogenic agents) or by reducing the capacity of existing tumor vasculature (e.g., the combretastatins). Other drugs are classed as hormonal agents (see Chapter 6), and examples of their actions include blocking the estrogen receptor (e.g., tamoxifen) or steroidal biosynthesis via aromatase inhibition (e.g., ana-strozole). There is also a broad family of agents known as the *biologicals* (see

Chapter 8) that includes such agents as antibodies, interleukins, immune response modifiers (interferons), and vaccines that are usually produced through biological processes such as cell culture and genetic engineering. Some agents, such as pro-drugs, have been developed for use in more sophisticated cancer cell–targeting strategies, such as ADEPT and GDEPT (see Chapter 7). Finally, agents being developed against new targets, such as RNA or protein-protein interactions (e.g., p53/HDM2), and those that are still at the research stage are described in Chapter 9.

1.13 DRUG RESISTANCE

The development of drug resistance is one of the most significant problems encountered in cancer chemotherapy, since up to 50% of tumors have either *de novo* drug resistance or otherwise develop resistance to anticancer drugs after initial treatment. Therefore, preliminary selective cytotoxicity toward a tumor can sometimes be followed by a rapid recovery in the rate of tumor growth, and the degree of resistance may increase after each subsequent administration to the point where the chemotherapeutic agent becomes completely ineffective.

In addition to the *induction* of resistance by chemotherapeutic agents, it should be noted that *selection* also plays a key role. Within all tumor populations, an intrinsic variation in gene, and hence protein, expression levels exists. Once a tumor is exposed to chemotherapy, the subclones with the most resistant genotype or phenotype are effectively selected because the more sensitive subclones are killed. The surviving tumor cells go on to divide and thus repopulate the tumor mass. Subsequent rounds of therapy then result in a diminished response because of this drug-induced skewing of the tumor population. In most human malignancies, the development of drug resistance is likely to be the result of a complex mixture of induction and selection, and it is the complex nature of this problem that explains the difficulties encountered in trying to overcome drug resistance.

Drug resistance has been observed in most drug-sensitive tumor types and for most classes of drugs. It can occur through a number of different mechanisms, the most important of which are outlined below:

- *Reduced intracellular drug concentrations.* This may result from enhanced drug efflux or decreased cellular penetration. Examples of drugs for which resistance may develop through this mechanism include the vinca alkaloids, dactinomycin, the anthracyclines, and the epi-podophyllotoxins.
- *Enhanced drug inactivation.* The antimetabolites, bleomycin, and the alkylating agents are examples of drug classes that can develop resistance through this mechanism. For the alkylating agents, glutathione transferase production may increase, which catalyzes their covalent interaction with glutathione rather than DNA.
- *Decreased activation of drug.* Examples of drugs that are affected by this mechanism include certain antimetabolites that must be enzymatically converted to nucleotides in order to exert their effects.

- *Modified production of a receptor or enzyme (i.e., gene amplification).* A good example of a drug that develops resistance through this mechanism is methotrexate. Amplification of dihydrofolate reductase can occur in resistant tumors so that increasingly higher concentrations of drug are required for a similar effect. A cell may also attempt to reduce the amount of a target receptor to decrease the effect of the drug.
- *Reduced affinity of receptor or enzyme for the drug.* Drug classes prone to this mechanism include the antimetabolites, the tyrosine kinase inhibitors, and hydroxyurea. For example, despite the promise of the latest generations of kinase inhibitors, such as Gleevec and Iressa, cancer cells can mutate the relevant kinases by one or more amino acids, thus causing a substantial reduction in drug-binding affinity which can lead to clinical resistance.
- *Increased repair of drug-induced DNA damage.* The alkylating agents are an example of a drug class that can develop resistance through this mechanism (although other mechanisms may also be operative). In this case, the cell has enzymes that detect DNA damage, such as alkylation or cross-linking, usually through a change in conformation of the helix around the lesion. Then, a number of specialist DNA repair enzymes repair the damage by, for example, excising the damaged DNA base pairs and resynthesizing the missing segment.
- *Reduction in activity of a mechanism-critical enzyme.* One of the best-known examples in this category is topoisomerase II. A decrease in activity of this enzyme is important for resistance to drugs such as the epi-podophyllotoxins, m-AMSA, and doxorubicin, which work by forming a ternary complex with topoisomerase II and DNA that leads to strand cleavage.
- *Multidrug resistance (MDR).* The discovery of the multidrug resistance gene (MDR1) has led to an understanding of how tumors can become resistant to several, often unrelated, drugs with different mechanisms of action simultaneously. MDR1, which is now considered to be one of the most important instigators of resistance to cancer chemotherapy, encodes an adenosine triphosphate–dependent efflux pump transmembrane protein known as *p-glycoprotein* that can become amplified in drug-resistant tumors. The resistance is generated through stepwise selection involving up-regulation of the p-glycoprotein that removes the drugs from the cell. Thus, resistant cells have reduced intracellular drug levels compared to drug-sensitive parental ones. The overproduction of p-glycoprotein has been correlated with the extent of drug resistance in a number of human cancers. In some cases, MDR can be successfully reversed using such drugs as cyclosporine; calcium channel blockers, such as verapamil; and tamoxifen.

One way to avoid the problem of resistance is to use a combination of different cytotoxic agents. However, there is also significant interest in developing strategies and agents to down-regulate pathways of resistance that may be coadministered with chemotherapeutic agents to maximize their efficacy.

PaTrin-2 (Patrin™, Lomeguatrib^BP)

STRUCTURE 1.3 Structure of the O6-alkylguanine-DNA-alkyltransferase inhibitor Patrin™.

One example, PaTrin-2 (Patrin™, Lomeguatrib[BP]), inhibits the methyl transferase enzyme O6-alkylguanine-DNA-alkyltransferase (ATase) that removes the methyl groups added to DNA by temozolomide (Structure 1.3). Patrin is already being investigated in the clinic in combination with temozolomide (see Chapter 3 and Chapter 12). Similarly, there is much interest in developing drugs that can inhibit the action of MDR1 as a means to enhance the effectiveness of existing chemotherapeutic agents. There is also growing interest in developing inhibitors against the recently discovered breast cancer resistance protein (BCRP or ABCG2). This protein, an ABC transporter and marker of stem and progenitor cells, confers a strong survival advantage to breast cancer cells under hypoxic conditions. Blocking BCRP function in BCRP+/+ progenitor cells markedly reduces survival under hypoxic conditions, hence the interest in developing drugs to inhibit this pathway.

1.14 COMBINATION CHEMOTHERAPY

Attempts to treat tumors with single agents are often disappointing. A single drug usually kills the most sensitive population of cells in a tumor, leaving a resistant fraction unharmed and still dividing. Therefore, each time the same agent is readministered, the tumor becomes more resistant and the treatment less efficacious. This led, in the early 1960s, to the first use of a combination of drugs for treating testicular tumors. Due to its success, this "cocktail" approach was then rapidly extended to other tumor types and is still commonly used in clinical practice today.

As a rule, each drug included in a particular combination should be active as a single agent and have different toxic (dose-limiting) side effects compared to the others. Multiple drug therapy also enables the simultaneous attack of different biological targets, thus enhancing the effectiveness of treatment. An example of the

TABLE 1.1
Efficacy Improvement Gained by Combining Agents
For the Treatment of Acute Lymphoblastic Leukemia

Drugs Used	Complete Remission (%)
Methotrexate (M)	22
Mercaptopurine (MP)	27
Prednisone (P)	63
Vincristine (V)	57
Daunorubicin (D)	38
P, V	94
P, V, M, MP	94
P, V, D	100

improvement in efficacy gained by using combinations of agents of different mechanistic classes compared to use of the same agents alone is illustrated in Table 1.1.

A variety of combination schedules using different drugs and dosage regimes is now available and accepted as superior to single drug therapies. Many cancers are already disseminated at the time of clinical diagnosis, and so combination chemotherapy may be commenced concurrently with local treatment (e.g., surgery, radiotherapy) to maximize patient benefits. Early diagnosis is always advantageous because micrometastases associated with the primary tumor are often very sensitive to combination chemotherapy since they have a good blood supply that facilitates drug access. They are also less likely than older tumors to develop drug resistance.

1.15 USE OF ADJUVANTS

In cancer chemotherapy, it is sometimes necessary to coadminister other agents that can either enhance the activity of an anticancer drug or counteract side effects resulting from it. An example of the former is coadministration of Patrin with temozolomide to enhance efficacy by slowing removal of the methylated DNA adducts formed (see Chapter 3). In terms of adjuvants to reduce side effects, antiemetics are commonly administered to counteract the nausea associated with many chemotherapeutic agents (see Chapter 13). Drugs that cause myelosuppression (common with cytotoxic agents) can lead to an increased risk of infection, and so antibiotic or antifungal therapy may be required. Finally, steroids such as prednisolone or dexamethasone are coadministered with some anticancer agents to reduce the severity of side effects. For example, patients are often pretreated with steroids before Taxol is administered to reduce the occurrence of edema.

1.16 INFERTILITY FOLLOWING CANCER TREATMENTS

Both chemotherapy and radiotherapy treatments can cause sterility in males and females of childbearing age. This is a particular problem with DNA-interactive agents such as cisplatin, which damage the genome of germ line cells in the testes and ovaries. One solution to this problem in young male patients is sperm storage prior to treatment. A similar solution for women was not realized until 2004, when Belgian researchers showed that it was possible to remove and store ovarian tissue from a patient prior to both radiotherapy and chemotherapy. After seven years of storage, the ovarian tissue was grafted back on to the patient's ovaries and she was able to conceive normally. It is anticipated that this procedure will become more widely available in the future.

FURTHER READING

Baguley, B.C., and Kerr, D.J., eds. *Anticancer Drug Development*. San Diego, CA: Academic Press, 2002.

Bishop, J.M., and Weinberg, R.A., eds. *Molecular Oncology*. New York: Scientific American, 1996.

Browne, M.J., and Thurlby, P.L., eds. *Genomes, Molecular Biology and Drug Discovery*. London: Academic Press, 1996.

Culver, K.W., et al. "Gene Therapy of Solid Tumors," *Brit. Med. Bull.*, 51:192-204, 1995.

Dalgleish, A.G. "Viruses and Cancer," *Brit. Med. Bull.*, 47:21-46, 1994.

Hanahan, D., and Weinberg, R.A. "The Hallmarks of Cancer," *Cell*, 100:57-70, 2000.

Hochhauser, D., and Harris, A.L. "Drug Resistance," *Brit. Med. Bull.*, 47:178-196, 1991.

"Intelligent Drug Design," *Nature*, 384:1-26, 1996.

Larson, E.R., and Fischer, P.H. "New Approaches to Antitumor Therapy," *Annu. Rep. Med. Chem.*, 24:121-128, 1989.

Macdonald, F., and Ford, C.H.J. *Molecular Biology of Cancer*. Oxford: BIOS Scientific Publishers, 1997.

Malcolm, A., ed. *The Cancer Handbook*, vol. 1. London: Nature Publishing Group, 2002.

Pratt, W.B. *The Anticancer Drugs*. Oxford: Oxford University Press, 1994.

Silverman, R.B. *The Organic Chemistry of Drug Design and Drug Action*, Second Edition. London: Academic Press, 2004.

Summerhayes, M., and Daniels, S. *Practical Chemotherapy: A Multidisciplinary Guide*. London: Radcliffe Medical Press, 2003.

Teicher, B.A., ed. *Anticancer Drug Development: Preclinical Screening, Clinical Trials, and Approval*. Totowa, NJ: Humana Press, 2003.

Thurston, D.E., "The Chemotherapy of Cancer" in Smith, H.J. (ed.): Smith and Williams' Introduction to the Principles of Drug Design and Action, 4th Edition, CRC Press, Boca Raton, FL, 411–522, 2005.

Vousden, K.H., and Farrell, P.J. "Viruses and Human Cancer," *Brit. Med. Bull.*, 50:560-581, 1994.

"What You Need to Know About Cancer," *Sci. Am. (Special Issue)*, 275:4-167, 1996.

Workman, P., ed. "New Approaches in Cancer Pharmacology," in *Drug Design and Development*. London: Springer-Verlag, 1992.

Yarnold, J.R., et al., eds. *Molecular Biology for Oncologists*, 2nd ed. London: Chapman & Hall, 1996.

2 Antimetabolites

2.1 INTRODUCTION

Antimetabolites work by blocking crucial metabolic pathways essential for cell growth. Their selectivity is thought to be related to the fact that tumor cells grow faster than normal cell populations, with the exception of the bone marrow, hair follicles, and parts of the gastrointestinal (GI) tract, and their effects on the latter leads to the well-known side effects. Thus, they are also known as *antiproliferative* agents. Although this growth differential can be significant in some leukemias, older solid tumors usually have only a small fraction of cells in active growth and so only partially respond. Most drugs of this type in clinical use, such as the antifolates and the purine and pyrimidine antimetabolites, work at the molecular level by interfering with deoxyribonucleic acid (DNA) synthesis. Other types of agents in this family include a specific thymidylate synthase inhibitor (raltitrexed), and adenosine deaminase (pentostatin) and ribonucleotide reductase (hydroxycarbamide) inhibitors.

2.2 DHFR INHIBITORS (ANTIFOLATES)

2.2.1 METHOTREXATE

Tetrahydrofolic acid is produced by the action of the enzyme dihydrofolate reductase (DHFR) on dihydrofolic acid and is required for the synthesis of thymine, which becomes incorporated into DNA (Scheme 2.1). In the late 1940s, slight modification of the structure of folic acid produced the lead antimetabolite aminopterin, which is now used as a rodenticide. Methotrexate, which was shown to be more selective, followed in the 1950s (also see Scheme 2.1). It binds more strongly to the active site of DHFR than the natural substrate by a factor of 10^4 due to the presence of an amino rather than a hydroxyl moiety at the 4-position, which increases the basic strength of the pyrimidine ring.

The most basic center in the methotrexate molecule is at the N1 and adjacent C2-NH_2 position, as confirmed by ^{13}C-NMR measurements at C2. Examination of the drug-enzyme complex by X-ray diffraction has shown that the pyrimidine ring is situated in a lipophilic cavity with the cation of N1/C2-NH_2 binding to an aspartate-26 anion of the enzyme. Other binding points revealed by X-ray include hydrogen bonding between C4-NH_2 and the carbonyl groups of both Leu-4 and Ala-97, and ionic interactions between the α-COOH of the glutamate residue and the basic side chain of Arg-57. The *p*-aminobenzoyl residue lies in a pocket formed on one side by the lipophilic side chains of Leu-27 and Phe-30 and, on the other side, by Phe-49, Pro-50, and Leu-54. A neighboring pocket lined by Leu-4, Ala-6, Leu-27, Phe-30, and Ala-97 accommodates the pteridine ring. The nicotinamide (NADPH)

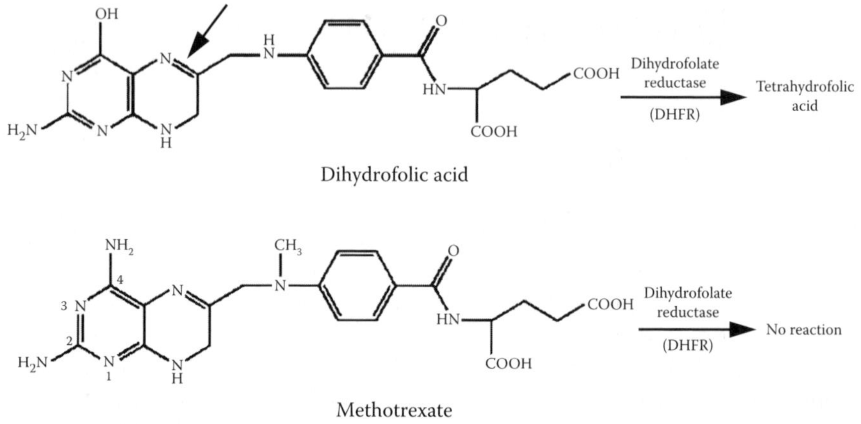

Dihydrofolic acid

Methotrexate

SCHEME 2.1 Reduction of dihydrofolic acid to tetrahydrofolic acid (bond to be reduced indicated by arrow) by DHFR. Methotrexate inhibits DHFR so that no tetrahydrofolic acid is produced.

portion of the fully extended coenzyme lies sufficiently close to the pteridine ring to facilitate transfer of a hydride anion from the pyridine nucleus to the C6-position.

Surprisingly, methotrexate occupies the reverse position at the active site of the enzyme compared to the substrate. Although the *p*-aminobenzoyl and glutamate portions of both are identically bound, with dihydrofolate, N-1 is unbound, C2-NH$_2$ and C4-OH bind only to water molecules, N3 is hydrogen bonded to Asp-26, N5 is unbound, and N8 interacts with Leu-4 via van der Waals forces. This structure results in the substrate being more loosely bound, a consequence of the differences in position and strength of the most basic centers in the substrate and inhibitor molecules.

Methotrexate is used as maintenance therapy for childhood acute lymphoblastic leukemia, where it can be given intrathecally for central nervous system prophylaxis. It is also useful in choriocarcinoma, non-Hodgkin's lymphoma, and a number of solid tumors. Side effects include myelosuppression, mucositis, and GI ulceration with potential damage to the kidneys and liver that may require careful monitoring. Resistance is also a problem, with the tumor cells eventually increasing production of DHFR.

In high-dose intermittent schedules, the adverse effects on bone marrow can be relieved by the periodic administration of the calcium salt of N5-formyltetrahydrofolic acid (folinic acid or Leucovorin™), which enables blockade of tetrahydrofolic acid production to be bypassed (i.e., folinic acid "rescue therapy"; see Structure 2.1).

Many derivatives of methotrexate have been synthesized in an attempt to reduce its toxicity; however, it is still the major DHFR inhibitor in clinical use. Analogs containing a fluorine atom have also been synthesized so that their interaction with the DHFR enzyme can be studied by Nuclear Magnetic Resonance (NMR) both *in vitro* and *in vivo*.

STRUCTURE 2.1 Structure of Leucovorin used in "rescue therapy" after treatment with methotrexate.

2.3 PURINE ANTIMETABOLITES

Purine antimetabolites inhibit a later stage in DNA synthesis than do DHFR inhibitors (see Structure 2.2). Their major problem is a lack of selective toxicity because purines are involved in many other cellular processes in addition to nucleic acid synthesis. Mercaptopurine (Puri-Nethol™; 6-MP) is used almost exclusively as maintenance therapy for acute leukemias (Structure 2.2). The free-base form is converted by sensitive tumor cells into the ribonucleotide 6-mercaptopurin-9-yl (MPRP), which results from interaction of the compound with 5-phosphoribosyl transferase. Resistance to 6-MP usually arises due to loss of production of this enzyme within tumor cells.

6-Mercaptopurine (Puri-Nethol™, 6-MP)   6-Tioguanine (Lanvis™)

STRUCTURE 2.2 The purine antimetabolites 6-mercaptopurine and 6-tioguanine.

Although MPRP inhibits several enzymatic pathways in the biosynthesis of purine nucleotides, including the conversion of inosine-5′-phosphate to adenosine-5′-phosphate, the main inhibitory action appears to occur at an earlier stage, when 5′-phosphoribosylpyrophosphate is converted into phosphoribosylamine by phosphoribosylpyrophosphate amido-transferase.

Allopurinol is contraindicated during treatment because it interferes with the metabolism of 6-MP by inhibiting xanthine oxidase mediated degradation of 6-MP to thiouric acid, which can lead to renal damage. Interestingly, the immunosuppressant agent azathioprine (Imuran™) is metabolized to 6-mercaptopurine.

Another cytotoxic drug used for treating myeloblastic leukemia, tioguanine (previously thioguanine or 6-thioguanine), is metabolized to the 9-(1′-ribosyl-5′-phosphate) by tumor cells (Structure 2.2). However, in contrast to MPRP, this intermediate does not inhibit an enzyme but is further phosphorylated to the triphosphate and then incorporated into DNA as a "false" nucleic acid. The main side effect of bone marrow suppression is caused by rapid assimilation of 6-thioguanine into the genome of bone marrow cells. The drug is used orally to induce remission in acute myeloid leukemia.

Fludarabine Phosphate (Fludara™)   Cladribine (Leustat™)

STRUCTURE 2.3 The purine antimetabolites fludarabine (Fludara) and cladribine (Leustat).

Fludarabine (Fludara™) and cladribine (Leustat™) are more recently introduced agents of this family that retain the purine nucleus but have sugar-like moieties already attached (see Structure 2.3). Fludarabine is recommended for patients with B-cell chronic lymphocytic leukemia (CLL) after initial treatment with an alkylating agent has failed. In addition to myelosuppression, fludarabine can also cause immunosuppression. Cladribine is an effective but potentially toxic drug given by intravenous infusion for the first-line treatment of hairy cell leukemia and the second-line treatment of CLL in patients who have failed on standard regimens of alkylating agents. Its usefulness is limited by both myelosuppression and neurotoxicity.

2.4 PYRIMIDINE ANTIMETABOLITES

Cytarabine (Cytosar™, ARA-C) and 5-fluorouracil (5-FU) are the two prototypic pyrimidine antimetabolites that work by interfering with pyrimidine synthesis (see Structure 2.4). Cytarabine is still one of the most effective single agents available for treating acute myeloblastic leukemia, the major side effect being myelosuppression. A disadvantage of cytarabine therapy arises from its rapid hepatic deamination (i.e., short half-life) by cytosine deaminase to give an inactive uracil derivative. These agents work by inhibiting the synthesis of DNA in S-Phase and by blocking the movement of cells through the G1/S part of the cell cycle.

Cytarabine (Cytosar^TM)   5-Fluorouracil

STRUCTURE 2.4 The pyrimidine antimetabolites cytarabine (Cytosar) and 5-fluorouracil (5-FU).

The short half-life of cytarabine can be counteracted by using continuous infusion methods of administration, although the agent can also be administered subcutaneously and intrathecally. The rapid deamination has led to a quest for deaminase-resistant agents or pyrimidine nucleoside deaminase inhibitors that can be coadministered. Of the many halogenated analogs investigated, only fluoro-derivatives have any appreciable antitumor activity.

5-Fluorouracil (5-FU) is used for the treatment of breast tumors and cancers of the GI tract, including advanced colorectal cancer (Structure 2.4). It is also highly effective as a 5% cream (Efudex^TM) for treating certain skin cancers. The drug's main side effects include myelosuppression and mucositis. It is initially metabolized to the 2′-deoxyribonucleotide, 5-fluoro-2′-deoxyuridylic acid (FUdRP), which is a potent inhibitor of thymidylate synthetase. The latter causes the transfer of a methyl group from the coenzyme methylenetetrahydrofolic acid to deoxyuridylic acid, which is converted to thymidylic acid and incorporated into DNA. 5-FU has been shown to have an affinity for thymidylate synthetase several thousand times greater than that of the natural substrate. This remarkable property is associated with the fluorine atom whose van der Waals radius compares favorably with that of hydrogen, although the bond strength is considerably greater. Additionally, the high electronegativity of fluorine affects the electron distribution, conferring a lower pK_a on the molecule compared to uracil. These two features combine to enable FUdRP to fit into the active site of the enzyme extremely well although the fluorine cannot be removed, thus effectively inhibiting the enzyme. Further studies have suggested that a nucleophilic sulphydryl group at the active site forms a covalent bond to FUdRP, leading to a "dead end" adduct of the enzyme, coenzyme, and 5-FU. Structure-activity studies have shown that the increased size but lower electronegativity of other types of halogen atoms reduce activity. It has been postulated that the high selectivity of 5-FU, especially in skin treatments, may reflect the fact that certain types of cancer cells lack the relevant enzymes to degrade it.

Gemcitabine (Gemzar™)   Capecitabine (Xeloda™)   Tegafur (Uftoral ™)

STRUCTURE 2.5 The pyrimidine antimetabolites gemcitabine (Gemzar), capecitabine (Xeloda), and tegafur (Uftoral).

More-recently introduced antimetabolites in this family include gemcitabine (Gemzar™), which is used intravenously with cisplatin for metastatic non-small-cell lung, pancreatic, and bladder cancers (Structure 2.5). It is generally well tolerated but can cause GI disturbances, renal impairment, pulmonary toxicity, and influenza-like symptoms. Intracellularly, gemcitabine is metabolized to biologically active diphosphate and triphosphate nucleosides that are responsible for inhibiting DNA synthesis and inducing apoptosis. Capecitabine (Xeloda™) is metabolized to 5-FU and is useful as oral monotherapy for metastatic colorectal cancer. Similarly, tegafur (Uftoral™) is a prodrug of 5-fluorouracil given orally in combination with uracil which inhibits degradation of the 5-FU. It is used together with calcium folinate to treat metastatic colorectal cancer. (See Structure 2.5.)

2.5 THYMIDYLATE SYNTHASE INHIBITION

Although working through the same mechanism as fluorouracil by inhibiting thymidylate synthase, raltitrexed (Tomudex™) is a recently introduced potent inhibitor that represents a new structural class (see Structure 2.6).

Raltitrexed (Tomudex™)

STRUCTURE 2.6 The thymidylate synthase inhibitor raltitrexed (Tomudex™).

Presently approved in Canada, it is given intravenously for palliation of advanced colorectal cancer in cases where 5-FU and folinic acid cannot be used. It is generally well tolerated but can cause myelosuppression and GI toxicity.

STRUCTURE 2.7 The adenosine deaminase inhibitor pentostatin (Nipent™).

2.6 ADENOSINE DEAMINASE INHIBITION

The one adenosine deaminase inhibitor in clinical use is pentostatin (Nipent™) (see Structure 2.7). Pentostatin was initially isolated from *Streptomyces antibioticus* in the mid-1970s, and a full synthesis was reported in 1982.

Administered intravenously on alternate weeks, pentostatin is highly active in hairy cell leukemia and is capable of inducing prolonged remissions. It is potentially toxic, causing myelosuppression, immunosuppression, and a number of other side effects that can be severe.

2.7 RIBONUCLEOTIDE REDUCTASE INHIBITION

Only one ribonucleotide reductase inhibitor is clinically used. Hydroxycarbamide (Hydrea™), also known as hydroxyurea, is an orally active agent first synthesized in the 1860s (see Structure 2.8). Its antitumor activity, which was discovered much later, is thought to be due to inhibition of ribonucleotide reductase resulting in blockade (and thus synchronization) of the cell cycle at the G_1-S interface. This causes a depletion of the deoxynucleoside triphosphate pool and blocks DNA synthesis and repair.

Hydroxycarbamide (Hydroxyurea; Hydrea™)

STRUCTURE 2.8 Structure of hydroxycarbamide (Hydroxyurea; Hydrea).

Hydroxycarbamide is used mainly in the treatment of chronic myeloid leukemia, often in combination with other drugs, but has also been used (off-label) for the treatment of melanoma. It is occasionally used for polycythemia (the usual treatment is venesection). Interestingly, because it induces fetal hemoglobin production, hydroxyurea is also used to treat sickle cell anemia. The most common toxic effects are myelosuppression, nausea, and skin reactions.

FURTHER READING

Blanke, C.D., et al. "Trimetrexate: Review and Current Clinical Experience in Advanced Colorectal Cancer," *Sem. Oncol.,* 5, 18:S18-57-S18-63, 1997.

Budde, L.S., and Hanna, N.H. "Antimetabolites in the Management of Non-Small Cell Lung Cancer," *Curr. Treat. Options in Oncol.,* 6:83-93, 2005.

Kaye, S.B. "New Antimetabolites in Cancer Chemotherapy and Their Clinical Impact," *Brit. J. Cancer,* 78: Supp. 3, 1-7, 1998.

Peters, G.J., et al. "Basis for Effective Combination Cancer Chemotherapy with Antimetabolites," *Pharm. & Therap.,* 87(2-3):227-253, 2000.

3 DNA-Interactive Agents

3.1 INTRODUCTION

A large number of anticancer drugs exert their effect by interacting with deoxyribo-nucleic acid (DNA). These agents are presently the most widely used group in clinical practice. In broad terms, their mechanism of action involves interfering with DNA processing, thus leading to cell death, usually through invoking apoptosis. Although some of the more simple agents in this class (e.g., the nitrogen mustards) cause relatively nonspecific DNA damage (in this case, interstrand cross-linking) at numerous sites in the genome, some of the newer experimental DNA-interactive agents (e.g., SJG-136) are much more specific, recognizing and binding to specific DNA sequences.

Agents of this type interact with the DNA double helix through a variety of mechanisms. Some drugs intercalate (i.e., insert) between the base pairs of DNA, whereas others alkylate DNA bases in either the minor or major grooves (see Figure 3.1). Some agents cross-link the DNA strands together in either an *intrastrand* or *interstrand* manner in either the minor or major grooves, and yet other agents exert their effect by binding to the helix and then cleaving the DNA strands.

As with other families of anticancer drugs, the observed selective toxicity toward cancer cells may arise solely from the difference in growth rate of populations of cancer cells compared to normal cells, which also explains their main side effects of toxicity toward the bone marrow and cells of the gastrointestinal (GI) tract. Another possibility is that selectivity may arise through a reduced capacity of cancer

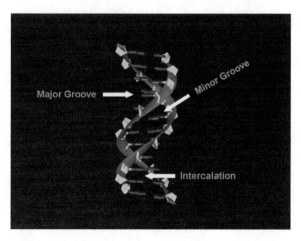

FIGURE 3.1 Structure of DNA showing the main sites of drug interaction either in the minor or major grooves or by intercalation between the base pairs.

cells to repair DNA lesions compared with normal cells, thus triggering apoptosis. There is also evidence that some DNA-interactive agents selectively target specific DNA regions (e.g., GC-rich sequences) that may be prevalent (e.g., "GC Islands") in parts of the genome, such as the promoter regions of some cancer-relevant genes in tumor cells. Some experimental agents are now being specifically designed to recognize distinct sequences of DNA.

As a class, the short-term side effects of DNA-interactive drugs include alopecia (hair loss), GI toxicity (often mucositis and diarrhea) and, more seriously, reversible bone marrow suppression (which is usually the dose-limiting toxicity). However, two problems are linked to their prolonged use. First, gametogenesis can be severely affected, and sperm storage is now recommended in young male patients prior to treatment. Second, prolonged use, particularly in combination with radiotherapy, can increase the risk of acute nonlymphocytic leukemia occurring later in life, presumably due to the DNA damage caused by these treatments.

This chapter describes several groups of DNA-interactive agents. They are categorized according to their mechanism of action.

3.2 ALKYLATING AGENTS

Dacarbazine and temozolomide are known as *methylating agents* because they work by methylating guanine bases within the major groove of DNA, predominantly at the O6-position. However, the experimental agent ET-743 *alkylates* the N2-position of guanine in the minor groove, thus forming a bulky adduct that blocks DNA processing.

3.2.1 METHYLATING AGENTS

3.2.1.1 Dacarbazine

Dacarbazine (DTIC-Dome™) was one of several triazenes originally evaluated as potential inhibitors of purine biosynthesis. Although it was found to have a wide spectrum of activity, ranging from malignant lymphomas to melanomas and sarcomas, it was later established that its mechanism of action was not associated with inhibition of purine biosynthesis. Instead, it was demonstrated that N-demethylation occurs *in vivo* to afford 5-aminoimidazole-4-carboxamide and a transient methyldiazonium ion (see Scheme 3.1). Through radiolabeling experiments,

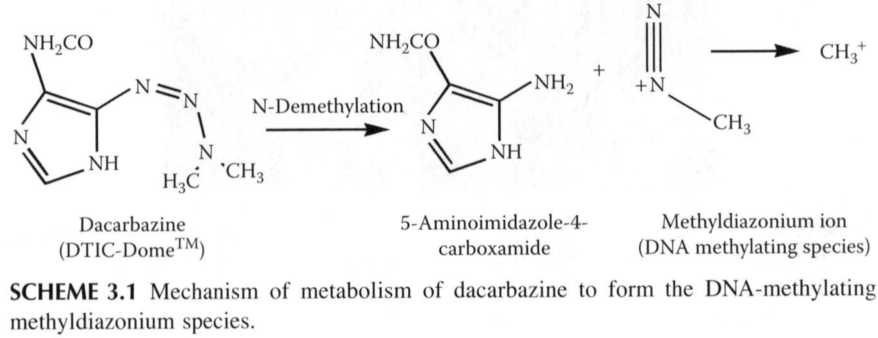

| Dacarbazine | 5-Aminoimidazole-4- | Methyldiazonium ion |
| (DTIC-Dome™) | carboxamide | (DNA methylating species) |

SCHEME 3.1 Mechanism of metabolism of dacarbazine to form the DNA-methylating methyldiazonium species.

it has been shown that the latter methylates DNA at guanine N7 positions in the minor groove.

Given intravenously, the irritant properties of dacarbazine preclude contact with skin and mucus membranes. Also, because triazenes are prone to photochemical decomposition, an intravenous infusion bag containing dacarbazine must be protected from light. The drug is used as a single agent to treat metastatic melanoma and in combination with other drugs for soft tissue sarcomas. It has also been used as a component of a combination therapy for Hodgkin's disease known as ABVD (doxorubicin [Adriamycin™], Bleomycin, Vinblastine, and Dacarbazine). The predominant side effects are myelosuppression and intense nausea and vomiting.

3.2.1.2 Temozolomide

Temozolomide (Temodal™) is a more-recently introduced alkylating agent (a methylating agent) licensed for the second-line treatment of malignant gliomas and also used experimentally for melanomas (Scheme 3.2). It is structurally related to dacarbazine but is more similar to the nitrosoureas in terms of its spectrum of activity and cross-resistance profile. As with dacarbazine and the nitrosoureas, temozolomide's major dose-limiting toxicity is bone marrow suppression. However, one significant advantage of this agent is that it possesses good oral bioavailability and distribution properties. In particular, it penetrates the central nervous system. Positron emission tomography (PET) scans with the labeled drug have shown direct evidence of tumor localization in the brain.

Like many other anticancer drugs in clinical use today, temozolomide is an example of a drug discovered through innovative synthetic chemistry linked to *in vitro* screening. In the 1980s, a wide range of analogs of imidazotetrazines were prepared and studied by Professor Malcolm Stevens' group at Aston University (Birmingham, U.K.), who were primarily interested in the chemistry of imidazotet-

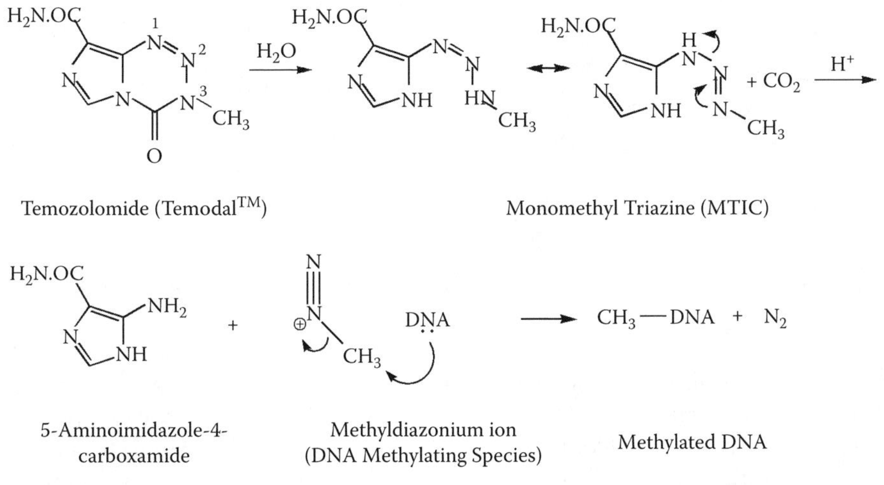

SCHEME 3.2 Mechanism of degradation of temozolomide to form the DNA-methylating methyl species.

razines. Screening programs identified two analogs (temozolomide and mitozolomide) with interesting biological activity, and it took an additional 15 years for temozolomide to reach the market in the late 1990s.

Like dacarbazine, temozolomide is a prodrug and serves to transport a methylating agent (the methyldiazonium ion) to guanine bases within the major groove of DNA. The mechanism of activation involves chemical hydrolytic cleavage (as opposed to the enzymatic cleavage that occurs with dacarbazine) of the tetrazinone ring at physiological pH to give the unstable monomethyl triazeno imidazole carboxamide (MTIC), which then undergoes further cleavage to liberate the stable 5-aminoimidazole-4-carboxamide and the highly reactive methyldiazonium methylating species. After methylating DNA (or otherwise decomposing by reacting with water to give methanol), the latter forms nitrogen gas (N_2). The small, stable molecules 5-aminoimidazole-4-carboxamide, carbon dioxide, and N_2 provide the driving force for the mechanism of action of temozolomide.

Early mechanistic studies involving deuterium labeling and nuclear magnetic resonance (NMR) indicated that the methyl hydrogens of the methyldiazonium species are freely exchangeable with those of aqueous solvent, suggesting that methyldiazonium has a sufficiently long lifetime in aqueous solution (estimated $t_{1/2}$ = 0.4 s) to reach its DNA target. Interestingly, the ethyl analog of temozolomide (N3-ethyl instead of N3-methyl) is completely unreactive toward DNA and devoid of any *in vitro* cytotoxicity or *in vivo* antitumor activity. The most likely explanation for this is that, assuming a similar activation pathway to temozolomide, ethyldiazonium ions are produced rather than methyldiazonium ions. Mechanistic studies (i.e., no exchange of protons with solvent by NMR) have shown that the ethyldiazonium ion is inherently less stable than methyldiazonium. It either reacts immediately with the first nucleophile it encounters (i.e., usually water) or eliminates to give ethane. Thus, there is no extended half-life as is the case with the methyldiazonium ion, and so no opportunity for the ion to reach the target DNA.

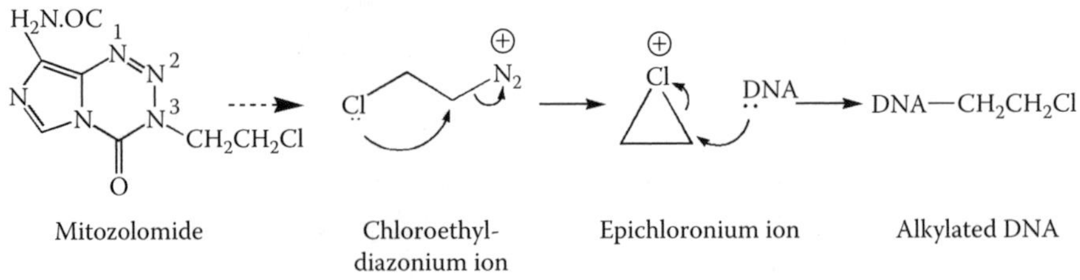

| Mitozolomide | Chloroethyl-diazonium ion | Epichloronium ion | Alkylated DNA |

SCHEME 3.3 Mechanism of degradation of mitozolomide to form the DNA-alkylating epichloronium species.

The experimental N3-chloroethyl (N3-CH_2CH_2-Cl) analog, known as mitozolomide, is also an efficient DNA alkylating agent (Scheme 3.3). In this case, fragmentation occurs to give a chloroethyl triazine species equivalent to the MTIC intermediate in the temozolomide breakdown pathway. This generates a chloroethyldiazonium species that cyclizes to the electrophilic epichloronium ion that initially monoalkylates DNA but can react further to form cross-links. However, in contrast to temozolomide, the resulting DNA lesions appear to be ineffective in producing an antitumor effect *in vivo*.

Studies have shown that the antitumor activity of temozolomide correlates with its accumulation in tumors where it methylates the O6- and N7-positions of guanine in the DNA of tumor cells, with the N7-adducts predominating. After patients have been treated with labeled temozolomide (^{11}C label at the methyl group), the relative abundance of covalently modified DNA in brain tumors relative to normal tissue can be observed by PET imaging. It has been postulated that this selectivity at the cellular level may be attributable to the slightly different pH environments of normal versus malignant tissues in the brain, coupled with differential capacities to repair the methylated lesions by O6-alkylguanine-DNA-alkyltransferase (ATase) or other repair processes. Also, methylation is favored in guanine-rich regions of DNA (for electrostatic reasons), and an alternative hypothesis is that in some cases, tumor cells may have more exposed guanine-rich regions than normal cells.

At a biochemical level, it is thought that temozolomide's activity depends, in part, on the less abundant O6-methyl adducts and on the operation of the mismatch repair enzymes (MMRs) that detect the O6-MeG-T wobble base-pairs formed during replication of drug-modified DNA. Excision of the erroneous base T from the daughter strand and replacement with the best fitting alternative, also a T, leads to a futile cycle of excision and insertion of T opposite the drug-modified G. The increased frequency of MMR-induced strand breaks at the site then triggers arrest of the cell cycle. In support of this mechanism, tumor cell lines low in MMR activity, such as colorectal, biliary tract, and endometrial carcinomas, are refractory to temozolomide. In the absence of effective MMR capability further replication would lead to insertion of an A base opposite the aberrant T, and so a G→A mutation would occur over two rounds of cell division. To protect against such events, another repair protein has evolved, ATase, that cleaves O6-modifications stoichiometrically to restore native guanine bases. This is known to be the principal mechanism of resistance to temozolomide. Therefore, the response of a tumor to temozolomide depends on the relative levels of expression of both MMR and ATase in the tumor cells which, as a result, may be responsive, actively resistant, or not susceptible as summarized in Table 3.1.

Strategies to overcome ATase-mediated resistance include divided dose scheduling, or preadministration of an ATase inhibitor, the prototype of which is O6-benzylguanine (Structure 3.1). Structure activity studies show that replacing the

TABLE 3.1
Response of Tumor Cells to Temozolomide According to their ATase and MMR Status

ATase	MMR	Phenotype
−	+	Responsive
+	+	Actively resistant
+/−	−	Not susceptible (passively resistant)

benzene ring of the O6-benzyl substituent with more-polar unsubstituted or substituted heterocyclic rings can lead to a better fit at the enzyme's active site and thus greater inhibitory activity. The bromothiophene analog PaTrin-2 (Patrin™, Lomeguatrib[BP]) has emerged as one of the most potent compounds of this class and with a good toxicity profile (Structure 3.1). It is presently being investigated in crossover clinical trials with temozolomide in both melanoma and colorectal cancers.

O6-Benzylguanine   PaTrin-2 (Patrin™, Lomeguatrib[BP])

STRUCTURE 3.1 Structures of the O6-alkyl-DNA alkyltranferase (ATase) inhibitors O6-benzylguanine and Patrin™.

Investigators are trying to ascertain whether it is possible to predict the likely efficacy of temozolomide in patients by establishing whether a link exists between levels of clinical activity and ATase gene polymorphisms. A genetic element to the interindividual variability in ATase expression levels is already evident from the observation that the two ATase alleles can be expressed at greatly different levels in the healthy nontumor lung tissues of different lung cancer patients and the fact that at least some of this variation maps close to or within the ATase locus. Studies are presently underway to establish whether this can be turned into a robust laboratory assay to predict responders.

In the case of MMR, it has also been suggested that a laboratory assay for MMR status could also be useful in identifying patients suitable for treatment with temozolomide. Interestingly, there is also a strategy for potentially overcoming a lack of MMR expression. For example, MMR is not expressed in some colorectal cancers and other types of tumors because the gene is silenced by methylation of the promoter. The triazine agent decitabine (Structure 3.2) can modify the process of gene methylation and can reverse promoter methylation thus reactivating genes, including silenced tumor suppressor genes. Combinations of Patrin and decitabine in MMR down-regulated human ovarian cancer cells have been shown to reverse both ATase and MMR resistance to temozolomide. The potential benefit of this strategy is now being evaluated in the clinic.

Another consideration is that methyl group delivery to the critical O6-position of guanine by temozolomide is a relatively inefficient process. First, it is thought that most of the methyldiazonium species produced are converted to methanol (by interaction with water) before reaching the target DNA of the cancer cells. Second,

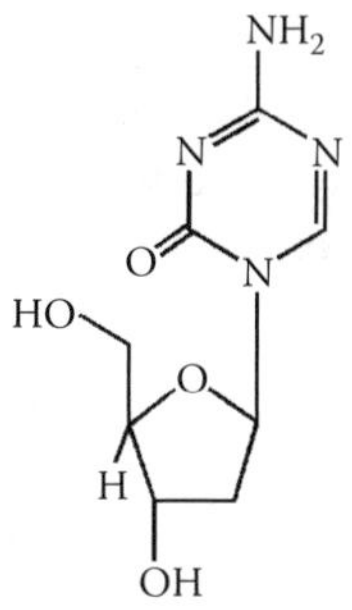

STRUCTURE 3.2 Structure of the DNA methylation modulator decitabine.

of the methyl groups successfully transferred to DNA, labeling experiments have demonstrated that only 5% locate to the O6-position of guanine, whereas approximately 70% attach to the N7-position. Therefore, another potentiation strategy is to coadminister agents that inhibit poly(ADP-ribose) polymerase (PARP), the enzyme that recognizes DNA damage (including N7-lesions) and tags strand breaks for later repair, thus allowing the relatively benign N7-methylation to become a cytotoxic event. Researchers are also attempting to adapt the structure of the group transferred from the N3-position of the imidazotetrazine moiety to the N7-position of guanine, so that the lesion may become more lethal. Using these approaches to avoid ATase-mediated resistance and the dependence on MMR for activity may expand the range of tumors susceptible to temozolomide-type agents.

3.2.1.3 Procarbazine

Procarbazine, N-(1-methylethyl)-4-[(2-methylhydrazino)methyl]benzamide, is a hydrazine derivative first synthesized as a monoamine oxidase inhibitor in the early 1960s (Scheme 3.4). It was only later discovered to have significant activity in lymphomas and carcinoma of the bronchus.

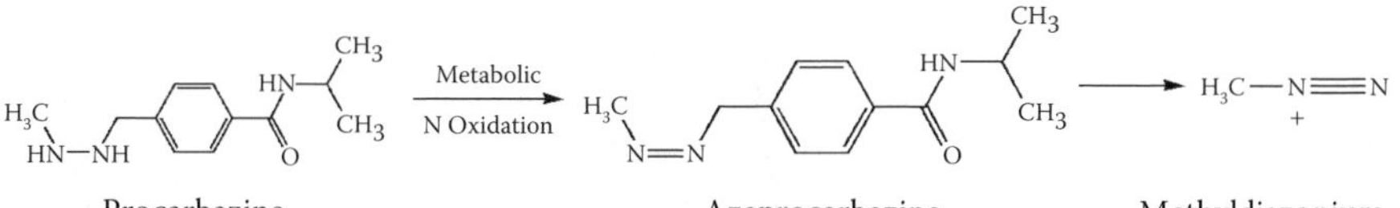

SCHEME 3.4 Mechanism of metabolism of procarbazine to form the DNA-methylating methyldiazonium species.

Its mechanism of action is thought to involve initial metabolic N-oxidation to an azaprocarbazine species, followed by subsequent rearrangement to produce either methyl diazonium or methyl radicals that act as DNA-methylating agents toward guanine residues.

Procarbazine is most often used in Hodgkin's disease, for example in "MOPP" combination therapy, which includes Mustine (chlormethine), Oncovin™

(vincristine), Procarbazine, and Prednisolone. It is administered orally and its toxic effects include nausea, myelosuppression, and a hypersensitivity rash that prevents further use of the drug. The mild monoamine-oxidase inhibitory effect of procarbazine does not require any dietary restriction; however, alcohol ingestion may cause a disulfiram-like reaction.

3.2.2 ECTEINASCIDIN-743

Ecteinascidin (ET–743, Yondelis™) is a novel DNA-binding agent derived from the marine tunicate *Ecteinascidia turbinata* that is presently being evaluated in the clinic (Structure 3.3). Rather than methylating DNA in the major groove, as is the case with dacarbazine and temozolomide, ET-743 alkylates the N2 of guanine in the minor groove, thus forming a bulky adduct that blocks DNA processing. It has significant *in vitro* activity against melanoma, breast, ovarian, colon, renal, and non-small-cell lung and prostate cell lines and is being evaluated in the clinic against soft tissue sarcoma (STS), which is one of the most difficult forms of cancer to treat. The majority of chemotherapeutic agents have only marginal activity against STS, with the most active agents (doxorubicin and ifosfamide) providing objective response rates of only about 20% as first-line treatments, and with their use limited by serious toxicities and resistance.

STRUCTURE 3.3 Structure of Ecteinascidin-743 (Yondelis™, ET-743). The electrophilic DNA-interactive carbinolamine [-NH-CH(OH)-] moiety is indicated with an arrow.

The drug has a unique mechanism of action that involves covalent binding to the N2-position of guanine within the minor groove of DNA, which causes the double helix to bend toward the major groove. This is a unique feature distinguishing ET-743 from all currently available DNA-binding agents, which usually perturb DNA by bending it toward their site of interaction rather than away from it. The ET-743 structure consists of three fused tetrahydroisoquinoline ring systems, two of which (subunits A and B) provide the framework for covalent interaction within the minor groove. The carbinolamine unit [-NH-CH(OH)-] formed from the nitrogen of the B subunit and the adjacent secondary alcohol is thought to be the electrophilic moiety responsible for alkylating the N2 of guanine, a mechanism identical to that used by the pyrrolobenzodiazepines (PBDs) (see below). The third tetrahydroisoquinoline system (subunit C) protrudes from the DNA duplex and interacts with adjacent nuclear proteins, contributing to the molecule's activity.

At a biochemical level, the cytotoxicity of ET-743 appears to be associated with the DNA repair pathways of cells. Both inhibition of cell cycle progression (leading to p53-independent apoptosis) and inhibition of transcription-coupled nucleotide excision repair (TC-NER) pathways have been demonstrated in *in vitro* studies. The TC-NER pathway involves recognition of DNA damage and recruitment of various nucleases at the site of DNA damage. At micromolar concentrations, ET-743 has been shown to trap these nucleases in a malfunctioning nuclease-(ET-743)-DNA adduct complex, thereby inducing irreparable single-strand breaks in the DNA. This process is supported by the fact that mammalian cell lines deficient in TC-NER show resistance to ET-743. *In vitro* exposure of human colon carcinoma cells to clinically relevant (i.e., low nanomolar) concentrations of ET-743 induces a strong perturbation of the cell cycle. Cell cycle arrest in G2 phase (resulting in p53-independent apoptosis) occurs after an initial delay of cell progression from G1 to G2 phase, and inhibition of DNA synthesis also occurs. Furthermore, there is evidence that nanomolar concentrations of ET-743 cause inhibition of the expression of genes involved in cellular proliferation (e.g., *c-jun*, *c-fos*) through promoter-specific interactions and interference with transcriptional activation. Finally, unlike other DNA-damaging drugs (e.g., doxorubicin) that cause rapid induction of expression of the multidrug resistance gene (MDR1) in human sarcoma cells, this agent selectively blocks transcriptional activation of MDR1 in these cells *in vitro*.

ET-743 is generally well tolerated by patients, with the most frequently reported side effects being noncumulative hematological and hepatic toxicities. Reversible and transient elevation of hepatic transaminases, nausea, vomiting, and asthenia are common but are seldom severe or treatment-limiting. Other side effects commonly associated with cytotoxic agents, such as mucositis, alopecia, cardiotoxicity, and neurotoxicities, are not observed.

3.2.3 PYRROLOBENZODIAZEPINE (PBD) MONOMERS

The pyrrolo[2,1-c][1,4]benzodiazepine (PBD) family of antitumor agents is based on the natural product anthramycin, which was the first member to be isolated from *Streptomyces refuineus* var. *thermotolerans* in the early 1960s (Structure 3.4). Other

STRUCTURE 3.4 Structure of the pyrrolobenzodiazepine anthramycin showing the N10-C11 carbinolamine, carbinolamine methyl ether, and imine forms.

well-known members of the family include tomaymycin, sibiromycin, and neothramycin.

The PBD structure consists of three fused rings (A, B, and C) with a chiral center at the C11a-position that provides the molecule with a three-dimensional shape perfectly matched for a snug fit within the DNA minor groove spanning three DNA base pairs. The molecules also contain an electrophilic carbinolamine moiety at the N10-C11 position (which can also exist in the equivalent methyl ether or imine forms); once in the minor groove, this alkylates the C2-amino group of a guanine through formation of an aminal linkage to C11 of the PBD (Scheme 3.5). Due to the snug fit of the PBD in the minor groove, very little distortion of the DNA helix occurs, as is the case with other alkylating agents (e.g., ET-743) and cross-linking agents (e.g., Cisplatin and the nitrogen mustards). For this reason, PBD-DNA adducts do not appear to attract the attention of the DNA repair proteins, which could be a significant clinical advantage in that the development of resistance through DNA repair may be avoided or delayed. Crucially, interaction of the PBD molecules with DNA is sequence-selective, with a preference for purine-guanine-purine sequences (with the central guanine covalently bound). This sequence preference can be explained by the length of the molecule and hydrogen bonding interactions

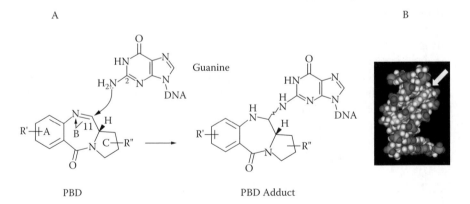

SCHEME 3.5 A. Mechanism of binding of a PBD to the C2-amino group of a guanine base within the DNA minor groove. B. Molecular model of a PBD binding in the DNA minor groove (indicated with arrow).

between parts of the PBD molecule and other DNA bases. For example, the N3-position of one flanking adenine forms a hydrogen bond to the N10-proton of the PBD. Most importantly, the PBD-DNA adducts are sufficiently robust to block endonuclease enzymes and also transcription, both in a sequence-dependent manner.

A number of PBD monomers were evaluated in the clinic in the 1960s and 1970s. Anthramycin itself was demonstrated to have antitumor activity but could not be developed due to a serious dose-limiting cardiotoxicity. This was later shown to be caused by the phenolic hydroxyl group at C9, which was being converted to quinone species that produced free radicals capable of damaging heart muscle. Other side effects included bone-marrow suppression and tissue necrosis at the injection site.

Synthetic routes became available in the 1990s that allowed relatively large quantities of PBD analogs to be produced without a C9-hydroxyl group, thus avoiding the cardiotoxicity problem. These developments allowed extensive structure activity relationship (SAR) studies to be carried out, and it is now known that C2-C3-unsaturation, along with the presence of unsaturated (and preferably conjugated) substituents at the C2-position, are important for maximizing potency. One such PBD monomer lacking a C9-hydroxyl but containing an optimized C2-substituent is presently being developed for clinical evaluation.

A related family of PBD compounds known as the *PBD dimers,* in which two PBD monomeric units are joined together through their A-rings to produce DNA interstrand cross-linking agents, has also been developed. One example — SJG-136 — is described elsewhere in this chapter.

3.3 CROSS-LINKING AGENTS

Cross-linking agents consist of two DNA-alkylating moieties separated by a chemical linker of varying length. By alkylating two nucleophilic functional groups on DNA bases positioned either on the same or opposite strands either directly adjacent or with a varying number of other base pairs in between, adducts known as *intrastrand* or *interstrand* cross-links, respectively, are formed (see Scheme 3.6). Both types of adducts interfere with DNA processing, including replication and transcription, and usually lead to cell death by triggering apoptosis. The cross-linked adducts often produce distortion of the DNA helix, and this change in three-dimensional shape is recognized by DNA repair enzymes. The adducts can be excised by these enzymes and the DNA repaired, which represents the predominant mechanism of clinical resistance for agents of this type.

The discovery that a DNA cross-linking agent may be used to treat cancer was made serendipitously during World War II, when on December 3, 1943, Allied military personnel were accidentally exposed to sulfur mustard gas [$S(CH_2CH_2Cl)_2$], a chemical weapon used in World War I. A U.S. warship, the USS *Liberty,* carrying a secret stockpile of 100 tons of this material (to be used in retaliation if chemical agents were used by the Axis forces), was subjected to an air attack in the harbor of Bari in Italy. Several hundred people were injured and many killed by exposure to the gas. Autopsy observations of individuals exposed to the gas revealed that profound lymphoid and myeloid suppression had occurred. Two pharmacists, Louis

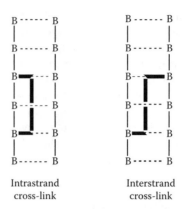

SCHEME 3.6   The two types of DNA cross-linked adducts: intrastrand and interstrand (B = DNA base, B------B = DNA base pair; adducts may span two or more base pairs).

Goodman and Alfred Gilman, recruited by the U.S. Department of Defense to investigate the pharmacology of chemical warfare agents, reasoned that mustard gas might be useful to treat lymphoma because it is a tumor of the lymphoid cells. They first set up an animal model, establishing lymphomas in mice, and then went on to demonstrate that they could treat them with mustard agents and provide some benefit. Realizing that sulfur mustard was too toxic to use in the treatment of the human disease, in collaboration with a thoracic surgeon, Gustav Linskog, they injected a mustard analog into a patient with non-Hodgkin's lymphoma and observed a dramatic reduction in the patient's tumor masses. Although this effect only lasted a few weeks, it was the first step to the realization that cancer could be treated with pharmacological agents. Further progress was made in the following years through medicinal chemistry studies carried out by Goodman and Gilman and others, demonstrating that replacement of the sulfur atom of sulphur mustard with a nitrogen and other substituents led to less-reactive and more clinically effective compounds. These analogs, now known as the *nitrogen mustards,* had a more acceptable toxicity profile and several members of the family are still in use today.

Despite their use in practice, the nitrogen mustards are highly reactive and relatively unselective at the DNA level forming multiple cross-linked sites across the genome, which are thought to lead to their high toxicity. Their lack of selectivity has driven the design of new experimental mustard analogs such as tallimustine (a nitrogen mustard conjugated to a netropsin analog), which has enhanced DNA-sequence recognition properties, binding in the minor groove, spanning five base pairs and recognizing a 5'-GAAAT sequence. It has also led to the design of nonmustard cross-linking agents such as SJG-136, which are much more selective, forming cross-linked adducts at specific DNA sequences. Through this approach, researchers hope to develop agents that are more selective for specific DNA sequences, thus reducing the frequency of cross-linked sites in genomic DNA and potentially reducing the associated toxicity.

The biochemical pathways that lead from cross-link formation to cell death still have not been completely elucidated, although it is known that if the genome of any cell becomes too damaged to repair, then p53 signals the cell to move into apoptosis. One hypothesis used to explain the cellular selectivity of cross-linking agents is that healthy cells are, on average, more proficient at repairing cross-linked adducts, whereas many tumor cells lose some or all DNA repair proficiency through their many genetic defects. For example, cisplatin is very effective clinically in certain testicular tumors, and these cells when obtained from biopsy and grown *in vitro*, can be shown to be highly resistant to DNA repair. An alternative hypothesis is that selective toxicity toward tumor cells occurs because they often divide more rapidly than most healthy tissues, which would also explain why toxic effects are observed in the few tissues known to contain rapidly dividing cells, such as the GI tract and hair follicles.

3.3.1 NITROGEN MUSTARDS

As described above, the nitrogen mustards were developed as bioisosteric derivatives of sulfur mustard gas that was first synthesized in 1886 but made its debut as a war gas in 1922. These analogs had an improved therapeutic index compared to sulfur mustard and were introduced into the clinic by Goodman in 1946. The first aliphatic example was mechlorethamine hydrochloride (now called *chlormethine* or *mustine hydrochloride*), which was shown to efficiently depress the white blood cell count at tolerable doses and was initially used for treating certain leukemias.

Despite much research on the medicinal chemistry and pharmacology of the nitrogen mustards, their precise mechanism of action at the molecular level and selectivity for tumor cells remain to be properly explained. These molecules "staple" the two strands of DNA together via covalent interactions with the N7-positions of guanine bases on opposites strands in the major groove, and it can be shown *in vitro* that these adducts block replication and inhibit enzymes such as ribonucleic acid (RNA) polymerase. It can also be demonstrated by electrophoresis-based experiments and molecular modeling that mustard adducts distort the helical structure of DNA at their binding site, and this effect can be transmitted through a number of base-pairs (i.e., a *teleometric* effect). Therefore, DNA processing may also be affected at a point distant from the adduct sites. Apart from the kinetic (i.e., killing faster-growing cells) or repair (i.e., cancer cells being repair-deficient compared to normal cells) explanations for the selectivity of the nitrogen mustards, another possibility is that their GC-selectivity may play a role. For example, it is known that some of the gene sequences associated with Burkitt's lymphoma are particularly GC-rich, and this disease is highly responsive to cyclophosphamide, which cross-links guanine bases.

In addition to MDR-related resistance, the activity of nitrogen mustards can be significantly reduced by cancer cells elevating their levels of glutathione. The highly nucleophilic glutathione forms adducts with mustards that are then no longer electrophilic and are unable to react with DNA. Alternatively, some cancer cells become resistant to nitrogen mustards by carrying out repair processes whereby the mustard adducts are excised and DNA from the damaged area resynthesized. To deal with this type of resistance, repair inhibitors that can be coadministered with mustards to enhance their clinical effectiveness are under development.

3.3.1.1 Aliphatic Nitrogen Mustards

3.3.1.1.1 Chlormethine (Mustine)

The only example of a simple aliphatic nitrogen mustard in clinical use is chlormethine (Mustine) (see Structure 3.5). The high chemical reactivity of the aliphatic nitrogen mustards toward DNA and also their vulnerability to attack by a wide range of nucleophiles is thought to account, in part, for the observed toxicities.

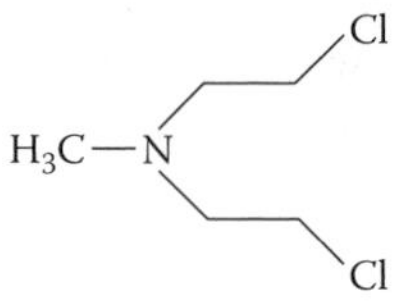

STRUCTURE 3.5   Structure of chlormethine (Mustine).

Under physiological conditions, chlormethine and related compounds undergo initial internal cyclization through elimination of chloride to form a cyclic aziridinium (ethyleneiminium) ion. This cyclization process involves intramolecular catalysis through neighboring group participation and results in a positive charge on the nitrogen, which is delocalized over the two adjacent carbon atoms. This cation, although relatively stable in aqueous biological fluids, is highly strained and reacts readily with any nucleophile. The clinically useful antitumor activity results from initial attack of an N7-atom of a guanine base in the major groove of DNA. This process is then repeated with the second chloroethyl group and a further guanine N7-atom located on the opposite DNA strand, giving rise to an interstrand cross-link that effectively locks the two strands of DNA together (see Scheme 3.7).

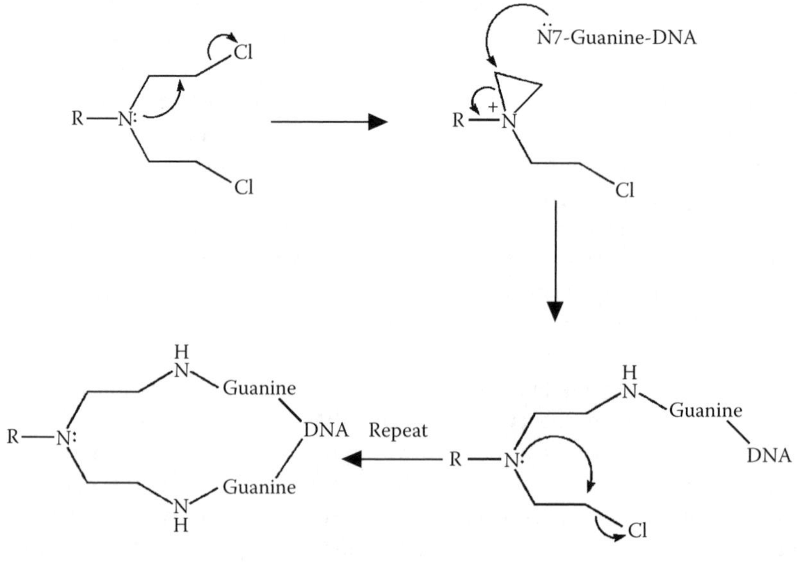

SCHEME 3.7   Mechanism of interstrand cross-linking of DNA by the aliphatic nitrogen mustards (R = alkyl).

This type of alkylation resembles an S_N2 process because the rate-controlling step is the bimolecular reaction between the cyclic aziridinium ion and the nucleophile, which involves simultaneous bond formation and breakage. The preceding step, involving iminium ion formation, is a fast unimolecular process more like an S_N1 reaction. However, in practice, it is difficult to draw a sharp distinction between the contributions of S_N1- and S_N2-type mechanisms.

Chlormethine is used in some chemotherapy regimens for the treatment of Hodgkin's disease, although severe vomiting can occur. Due to its chemical reactivity, it must be freshly prepared prior to administration, and then delivered via a fast-running intravenous infusion. Care must be taken with the infusion needle because local extravasation (i.e., leakage into surrounding tissue) causes severe tissue necrosis.

3.3.1.2 Aromatic Nitrogen Mustards

The aromatic nitrogen mustards were introduced in the 1950s and are milder alkylating agents than the aliphatic mustards. For these agents, the aromatic ring acts as an electron-sink, withdrawing electrons from the nitrogen atom and discouraging aziridinium ion formation. In comparison to aliphatic mustards, the central nitrogen atom is not sufficiently basic to form a cyclic aziridinium ion because the nitrogen electron pair is delocalized through interaction with the electrons of the aromatic ring. Therefore, alkylation most likely proceeds via an S_N1 mechanism, with carbocation formation (resulting from chloride ion ejection) providing the rate-determining step (Scheme 3.8).

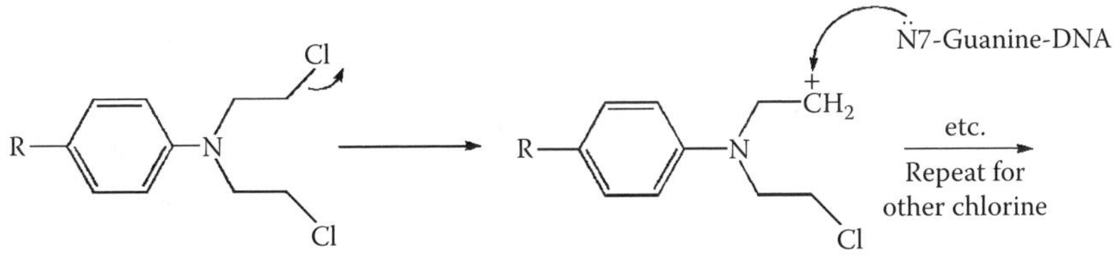

SCHEME 3.8 Mechanism of interstrand cross-linking of DNA by the aromatic nitrogen mustards (R = various functional groups).

Therefore, these analogs are sufficiently reduced in activity as electrophiles that they can reach their target DNA before being deactivated by reaction with collateral nucleophiles. This means that the aromatic mustards can be taken orally, which is a significant advantage.

3.3.1.2.1 Chlorambucil (Leukeran™)

During early SAR studies, researchers found that inclusion of a carboxylic acid group greatly improved the water solubility of aromatic mustard analogs. However, direct attachment of a carboxyl group to the aromatic ring (i.e., R = COOH in Scheme 3.8) reduced chemical and biological activity to an unsatisfactory level due to the additional electron-withdrawing effect. Therefore, the carboxyl group was electronically "insulated" from the aromatic ring by a number of methylene groups

STRUCTURE 3.6 Structure of chlorambucil (Leukeran™).

(three proving optimal), giving rise to chlorambucil (Leukeran), one of the slowest acting and least toxic nitrogen mustards (Structure 3.6).

Chlorambucil is useful in the treatment of ovarian cancer, Hodgkin's disease, indolent non-Hodgkin's lymphomas, and chronic lymphocytic leukemia (CLL). Its lower chemical reactivity is reflected in oral dosing and the relative lack of side effects, with the exception of reversible bone marrow suppression. Occasionally, patients develop widespread rashes, which can progress into serious dermatological conditions unless treatment is stopped.

3.3.1.2.2   Melphalan (Alkeran™)

Melphalan (L-phenylalanine mustard, Alkeran™), which also contains a carboxylic acid, was designed in the early 1950s to possess greater tumor selectivity based on the hypothesis that incorporation of an amino acid residue (i.e., phenylalanine) might facilitate selective uptake by tumor cells in which rapid protein synthesis is occurring (Structure 3.7). Because melphalan prepared from D-phenylalanine is much less active than that prepared from the L-form, it has been postulated that this agent may be taken-up into cells by the L-phenylalanine active transport mechanism.

STRUCTURE 3.7 Structure of melphalan (Alkeran™).

Melphalan is used to treat myeloma and, occasionally, solid tumors (e.g., breast and ovarian) and lymphomas. Like chlorambucil, melphalan is usually given orally but can also be given intravenously. Bone marrow toxicity, the most common side effect of melphalan, is delayed and so the drug is usually administered at 6-week intervals to allow for this.

3.3.1.3   Oxazaphosphorines

A further attempt to produce more-selective mustards was based on the hypothesis that some tumors may overproduce phosphoramidase enzymes and that these may be utilized to activate a prodrug form of a mustard. Cyclophosphamide (Endox-ana™), the most successful mustard to result from this work, has been safely used for many years. Ifosfamide (Mitoxana™) is a related analog, more recently intro-duced.

3.3.1.3.1 Cyclophosphamide (Endoxana™)

The design of this mustard prodrug, first synthesized in 1958, was based on the concept that the P=O group should decrease the availability of the nitrogen lone pair in an analogous manner to the phenyl ring of the aromatic mustards, thus reducing the electrophilicity of the molecule (Structure 3.8). Furthermore, it was postulated that the P=O group might be cleaved by phosphoramidase enzymes that were thought to be overexpressed by some tumor cells, thus releasing the nitrogen lone pair and restoring the electrophilicity of the molecule selectively at the tumor site.

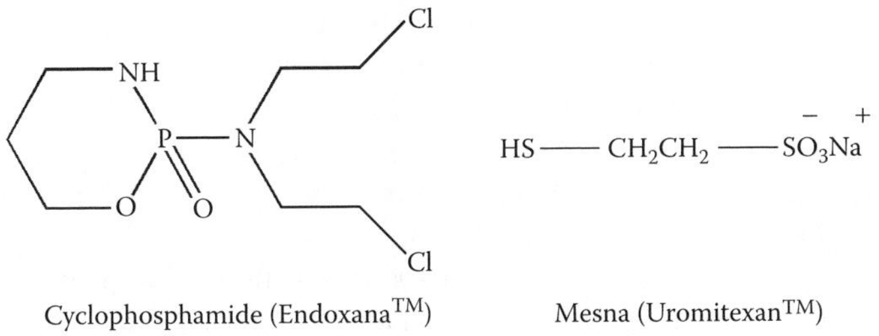

Cyclophosphamide (Endoxana™)   Mesna (Uromitexan™)

STRUCTURE 3.8 Structures of cyclophosphamide (Endoxana™) and Mesna (Uromitexan™).

Although at the time this appeared to be a rational design concept, it was later shown that activation *in vivo* is not due to enzyme-catalyzed hydrolysis of the P=O group but rather to initial oxidation at the C4-position of the oxazaphosphorine ring by liver microsomal enzymes. After 4-hydroxylation, the molecule then fragments to give phosphoramide mustard, which is thought to be the biologically active species, along with the highly electrophilic acrolein, which is toxic and can cause hemorrhagic cystitis. The further breakdown of phosphoramide mustard to normustine [$HN(CH_2CH_2Cl)_2$] has also been observed (Scheme 3.9).

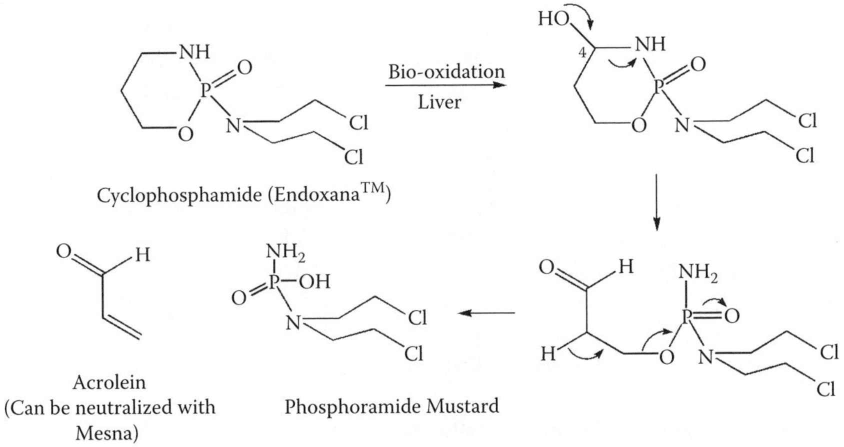

SCHEME 3.9 Mechanism of activation of cyclophosphamide (Endoxana™) to give the DNA cross-linking species phosphoramide mustard and the toxic by-product acrolein.

Cyclophosphamide has a broad spectrum of clinical activity and is widely used in the treatment of solid tumors, including carcinomas of the bronchus, breast, and ovary, and various sarcomas. It is also used to treat CLL and lymphomas. This agent is administered intravenously or orally but is not active until metabolized by the liver. The by-product acrolein is excreted in the urine and is a potent electrophile, reacting with nucleophiles on the surfaces of cells that line the bladder and causing the very serious but fortunately rare complication of hemorrhagic cystitis in suscep-tible patients. An increased fluid intake after administration of cyclophosphamide can help to avoid this problem. However, after high-dose therapy (more than 2 g intravenously), the adjuvant agent mesna (Uromitexan™) is routinely coadminis-tered (and also given afterwards) to neutralize the effects of acrolein. Mesna, which is sodium 2-mercaptoethanesulfonate, acts as a "sacrificial" nucleophile, reacting with acrolein via a Michael addition to form a water-soluble biologically inactive adduct that is safely eliminated in urine. Mesna is also given before and after oral therapy with cyclophosphamide (see Scheme 3.10).

In addition to the usual side effects associated with mustard agents, cyclophos-phamide also suppresses B-cell activity and antibody formation.

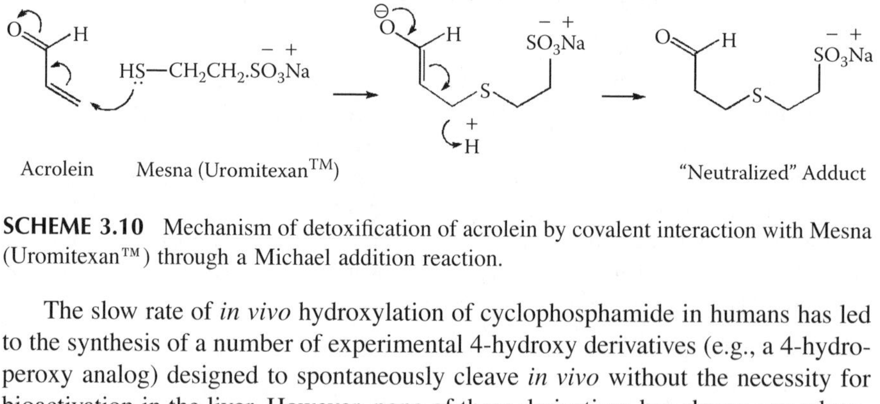

Acrolein   Mesna (Uromitexan™)   "Neutralized" Adduct

SCHEME 3.10   Mechanism of detoxification of acrolein by covalent interaction with Mesna (Uromitexan™) through a Michael addition reaction.

The slow rate of *in vivo* hydroxylation of cyclophosphamide in humans has led to the synthesis of a number of experimental 4-hydroxy derivatives (e.g., a 4-hydro-peroxy analog) designed to spontaneously cleave *in vivo* without the necessity for bioactivation in the liver. However, none of these derivatives has shown any advan-tage over cyclophosphamide, which remains an important drug in clinical practice today.

3.3.1.3.2 Ifosfamide (Mitoxana™)

Ifosfamide (Mitoxana™) is an analog of cyclophosphamide that is not technically a nitrogen mustard due to the translocation of one chloroethyl moiety to another position within the molecule (i.e., to a nitrogen attached to the P=O group) (see Structure 3.9). It can only be administered intravenously, and has a similar spectrum of activity to cyclophosphamide. Mesna is routinely coadministered with ifosfamide and also given afterward to reduce bladder and urethral toxicity.

3.3.1.4 Conjugated Nitrogen Mustards

Given the significant antitumor activity of the mustards, extensive medicinal chem-istry research throughout the years has been devoted to attaching them to other

STRUCTURE 3.9 Structure of ifosfamide (Mitoxana™).

therapeutic moieties either to achieve a combination effect or for the purposes of targeting the mustard to a particular organ or cell type. The most commercially successful of these approaches has been the combination of a mustard moiety and an estrogen to give estramustine, which appears to provide a hormonal response and an antimitotic effect (by reducing testosterone levels) rather than working through a DNA-cross-linking mechanism.

3.3.1.4.1 Estramustine Phosphate (Estracyt™)

Estramustine phosphate (Estracyt™) is a conjugate consisting of chlormethine (Mustine) chemically linked to an estrogen moiety. It is used orally in patients with metastatic prostate cancer as both our primary and secondary therapies. At one time, estrogens represented the primary treatment for metastatic prostate cancer; however, with the discovery of their serious cardiovascular side effects in the early 1970s, this treatment fell from favor. However, it is now accepted that estrogens have a multiplicity of actions and can be useful in the chemotherapy of prostate cancer. For example, estrogens increase the levels of sex hormone-binding globulin (thus preventing free testosterone from attaching to prostate cancer cells), lower the levels of luteinizing hormone (resulting in reduced testosterone levels), and block 5-α-reductase (thus inhibiting the conversion of testosterone to dihydrotestosterone). More recently, estrogens have been shown to have a direct cytotoxic effect on prostate cancer cells growing *in vitro*.

STRUCTURE 3.10 Structure of estramustine (Estracyt™).

Estramustine (Estracyt™), the combination of an estrogen joined to a nitrogen mustard through a carbamate linkage, was first synthesized in the early 1960s, although it was not used in practice until 1976. Although it is feasible that the intact molecule alkylates or cross-links DNA, the carbamate linker is thought to be cleaved by plasma or cellular carbamase enzymes with the released estrogen fragment being the biologically active intermediate.

A number of clinical studies have shown that estramustine must be given in high doses at the beginning of the treatment if useful estrogenic effects are to develop. In patients with poorly differentiated prostate cancer, estramustine at a dose level of 10 mg/kg is slightly superior to orchiectomy (surgical removal of the testicles).

3.3.2 AZIRIDINES

Rather than form aziridinium ions as reactive intermediates through an activation process (as with the mustards), thiotepa and related analogs, such as the experimental agents AZQ and BZQ, have an aziridine ring already incorporated into their structures (Structure 3.11). Ring-opening of the aziridines with nucleophiles is slower than that the fully charged aziridinium ions formed from mustards. However, depending on the pK_a of the aziridine nitrogen, significant protonation at physiological pH is likely, meaning that, in practice, the aziridinium ion may be the reactive species anyway.

Thiotepa

AZQ: R = HNCOOEt
BZQ: R = HNCH$_2$CH$_2$OH

STRUCTURE 3.11 Structures of thiotepa, AZQ, and BZQ.

Thiotepa was first synthesized in the mid-1950s and is used as an insect sterilant as well as an anticancer agent. It is usually used as an intracavitary drug (e.g., by intrapleural, intrabladder, or peritoneal infusion) for the treatment of bladder or ovarian cancers or malignant effusions. The drug can also be given parentally and is occasionally used to treat breast cancer by this route of administration.

A substituted benzoquinone ring has also been employed as an "anchor" for the aziridine groups in the experimental agents AZQ and BZQ. The aziridine moieties are deactivated by the withdrawal of electrons from the nitrogens into the quinone carbonyl groups via the six-membered ring. These molecules have been employed as experimental bioreductive prodrugs because reduction of the quinone ring to either the semiquinone or the hydroxyquinone species reverses the electron flow, thus increasing the basicity of the aziridine nitrogens and allowing them to become activated via protonation.

The cross-linking agent mitomycin C also contains an aziridine moiety and is discussed elsewhere in this chapter.

3.3.3 EPOXIDES

Epoxide functionalities are similar to aziridines in their propensity to alkylate DNA. Although some carcinogens, such as the aflatoxins, are known to alkylate DNA via an epoxide moiety, there are currently no drugs in clinical use that contain preformed epoxide moieties. However, treosulfan (L-threitol 1,4-bismethanesulfonate, Ovastat™) is a prodrug that converts nonenzymatically to L-diepoxybutane via the corresponding monoepoxide under physiological conditions (see Scheme 3.11).

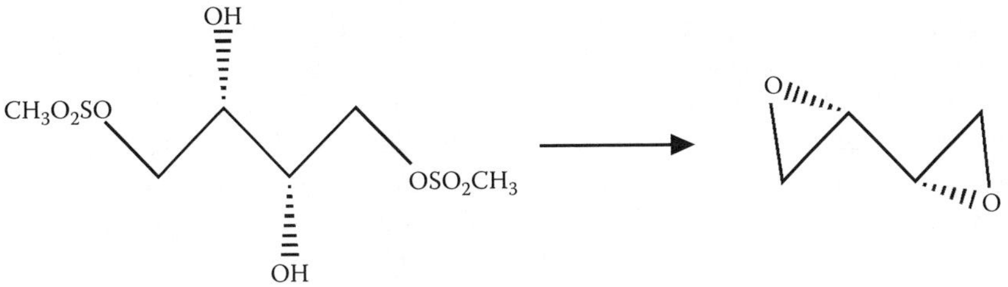

SCHEME 3.11 Conversion of treosulfan (Ovastat™) into the DNA-reactive L-Diepoxybutane.

In vitro studies have shown that this conversion is required for the alkylation and interstrand cross-linking of plasmid DNA, and it is assumed that it also occurs *in vivo*. Alkylation occurs at guanine bases with a sequence selectivity similar to other alkylating agents, such as the nitrogen mustards. In treosulfan-treated K562 cells growing *in vitro*, cross-links form slowly, reaching a peak after approximately 24 hours. However, incubation of K562 cells with preformed epoxides such as L-diepoxybutane itself provides faster and more efficient DNA cross-linking, supporting the proposed prodrug conversion step.

Treosulfan is mainly used to treat ovarian cancer and can be administered either orally or by intravenous or intraperitoneal injection. The major side effects are similar to those for the nitrogen mustards and include bone marrow suppression and skin toxicities. Nausea, vomiting, and hair loss occur less frequently.

3.3.4 METHANESULFONATES

Busulfan [1,4-di(methanesulfonyloxy)butane] is the best known example of an alkyl dimethanesulfonate with significant antitumor activity (see Structure 3.12). It is known to form DNA cross-links, with the methanesulfonyloxy moieties acting as leaving groups after attack by nucleophilic sites on DNA. From a mechanistic viewpoint, the methanesulfonate groups are presumed to participate as leaving groups in S_N2-type alkylation reactions.

STRUCTURE 3.12   Structure of busulfan (Myleran™).

SAR studies have revealed that unsaturated analogs of known stereochemistry, such as the corresponding butyne and trans-butene derivatives are inactive, whereas cis-butene derivatives retain activity. Interestingly, the activities of the saturated busulfan and the cis-butene-analog depend on their structural flexibility and three-dimensional shape, respectively, which allows them to form cyclic derivatives by 1,4-bisalkylation of suitable nucleophilic groups. For example, 1,4-di(7-gua-nyl)butane has been identified as a product of the reaction between busulfan and DNA, suggesting that this agent acts as an interstrand cross-linking agent in a similar manner to the nitrogen mustards. However, studies of the structure of urinary metab-olites suggest that cysteine residues in certain proteins are also alkylated.

Busulfan is administered orally and causes significantly less nausea and vomiting than other DNA cross-linking agents, and is thus more acceptable to patients. It is used almost exclusively for the treatment of chronic myeloid leukemia, in which it is highly effective and can keep patients almost symptom-free for long periods. Unfortunately, it causes excessive myelosuppression that can result in irreversible bone marrow aplasia and so requires careful monitoring. Hyperpigmentation of the skin is another common side effect and, more rarely, pulmonary fibrosis can occur.

3.3.5 NITROSOUREAS

The nitrosoureas alkylate DNA and lead to both mono-adducts and interstrand cross-links at a number of different sites. The study of a large number of nitrosourea analogs has established the structural unit for optimal activity as the 2-chloroet-hyl-N-nitrosoureido moiety, and the likely mechanism of both monoalkylation and cross-linking is shown in Scheme 3.12. Although these molecules possess chloro-ethyl fragments, their activity is not associated with aziridinium ion formation as in the mustards because the corresponding nitrogen atom is part of a urea structure and so the electron pair on the nitrogen is not available to participate in a cyclization reaction. Instead, it is thought that the alkylation of nucleic acids proceeds via generation of a chloroethyl carbonium ion. The alkyl isocyanate fragment also formed is thought to carbamoylate the amino groups of proteins.

The most significant property of the nitrosoureas is their activity toward cancer cells in the brain and cerebrospinal fluid, the so-called "sanctuary sites." This is due to the relatively high lipophilicity of these agents compared to other drugs (e.g., nitrogen mustards). However, they are also active in malignant lymphomas and carcinomas of the breast, bronchus, and colon. Carcinoma of the GI tract, which is

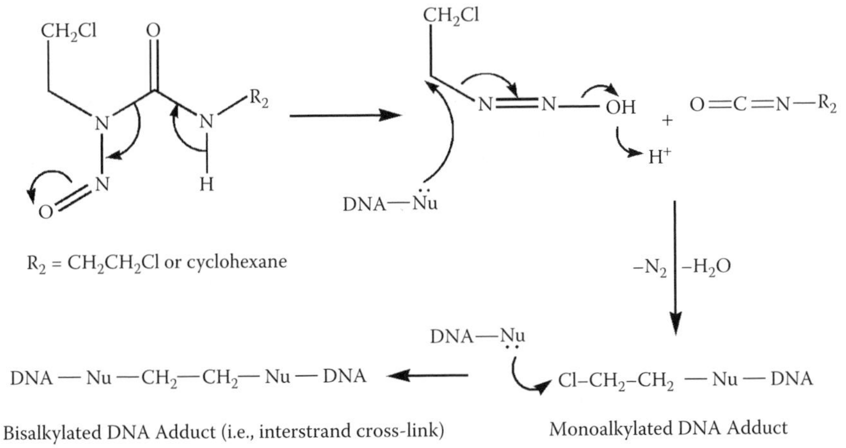

SCHEME 3.12 Mechanism of action of the nitrosoureas showing the formation of mono- and bisalkylated (i.e., cross-linked) adducts.

notably intractable to drug treatment, also responds to nitrosoureas. Unfortunately, these agents cause severe bone marrow toxicity, which is usually dose-limiting. Carmustine and lomustine are two examples of clinically useful nitrosoureas.

3.3.5.1 Lomustine

Lomustine [N-(2-chloroethyl)-N'-cyclohexyl-N-nitrosourea (CCNU)] (CeeNU™) is a nitrosourea analog with a high degree of lipid solubility (see Structure 3.13). Administered orally, it is mainly used for the treatment of certain solid tumors and also Hodgkin's disease. Because this agent produces delayed bone marrow toxicity, it should be administered at 4- to 6-week intervals. Lomustine frequently causes moderately severe nausea and vomiting, and permanent bone marrow damage can results from prolonged exposure.

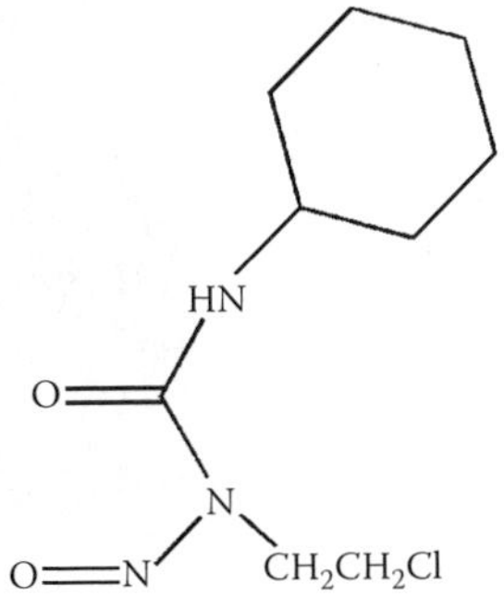

STRUCTURE 3.13 Structure of lomustine [CNNU] (CeeNU™).

3.3.5.2 Carmustine

Carmustine [N,N'-Bis(2-chloroethyl)-N-nitrosourea] (BiCNU™) has a similar activity and toxicity profile to lomustine but is usually administered intravenously (see Structure 3.14). It is most frequently used to treat brain tumors, lymphomas, and myelomas. A major problem with carmustine is that delayed pulmonary fibrosis and cumulative renal damage can occur.

$$
\begin{array}{c}
\text{HN} \diagup \text{CH}_2\text{CH}_2\text{Cl} \\
\text{O} = \text{C} \\
\text{N} \diagdown \text{CH}_2\text{CH}_2\text{Cl} \\
\text{O} = \text{N}
\end{array}
$$

STRUCTURE 3.14 Structure of carmustine (BiCNU™).

Recently, a new formulation of carmustine with no systemic toxicity has been developed for the local treatment of brain tumors. The carmustine is formulated into a slow release "wafer" dosage form (Gliadel Wafer™; polifeprosan 20 with carmustine) that is implanted directly into the resection cavity left after surgical removal of the tumor (Figure 3.2). Recent Phase II studies of Gliadel in adult patients with newly diagnosed, high-grade malignant glioma have demonstrated acceptable toxicities when the wafer is combined with daily oral temozolomide and standard radiotherapy (followed by up to 18 cycles of oral monthly temozolomide). These encouraging results suggest that Gliadel may become a standard treatment in the future.

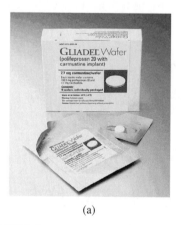
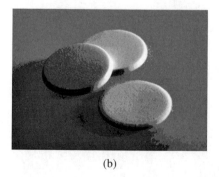

(a)   (b)

FIGURE 3.2 (a) Gliadel Wafer™ for the local delivery of carmustine to brain tumors; (b) detail of dosage forms which contain polifeprosan 20 and carmustine (www.guilfordpharm.com).

3.3.6 PLATINUM COMPLEXES

The discovery and development of the platinum complexes, which include cisplatin, carboplatin (Paraplatin™), (Structure 3.15) and oxaliplatin (Eloxatin™), is often quoted as one of the great success stories of cancer chemotherapy due to the pronounced activity of cisplatin, particularly in testicular and ovarian cancers. As with many other clinically useful drugs, the first member of the family, cisplatin, was discovered by serendipity rather than by design.

Cisplatin   Carboplatin (Paraplatin™)   Oxaliplatin (Eloxatin™)

STRUCTURE 3.15 Structures of cisplatin, carboplatin (Paraplatin™), and oxaliplatin (Eloxatin™).

Cisplatin, also known as Peyrone's salt or Peyrone's chloride, is a coordination complex originally prepared and reported in 1845. In the 1960s, Rosenberg and co-workers observed that passing an alternating electric current through platinum electrodes in an electric cell containing *Escherichia coli* led to arrest of cell division without killing the cells. Continued growth without division led to unusually elongated cells with a spindle-like appearance (Figure 3.3). The cause of the cytostatic effect was eventually traced to platinum complexes formed electrolytically at concentrations of only 10 parts per million in the presence of ammonium salts and light. Eventually, *cis*-diamminedichloroplatinum (cisplatin) was identified as one of the most active complexes in the mixture and, given its ability to block cell division while still allowing cell growth, Rosenberg recognized its potential in cancer

(a)   (b)

FIGURE 3.3 Photomicrographs showing the effect of passing an alternating electric current through platinum electrodes in an electric cell containing *E. coli*: (a) No current — normal cell division and growth; (b) current passed — arrest of cell division without killing and continued growth lead to unusually elongated cells with a spindle-like appearance.

chemotherapy. He then went on to show that cisplatin possessed significant cyto-toxicity toward cancer cells growing *in vitro* and also had antitumor activity *in vivo*. As a result, cisplatin was introduced into practice in the U.K. in 1979. Mechanistic studies showed that it works by interacting in the major groove of DNA, where it forms intrastrand cross-links.

The related analogs, carboplatin and oxaliplatin, were developed later in an attempt to reduce the problematic side effects of cisplatin. The platinum complexes are all administered intravenously, although an orally active experimental analog (JM-216) has been studied in the clinic but not commercialized. One disadvantage of the metal complexes is their high cost due to the platinum content.

3.3.6.1 Cisplatin

Cisplatin produces intrastrand cross-links in the major groove of DNA with prefer-ential interaction between (guanine N7)-(guanine N7), (guanine N7)-(adenine N7) (in both cases with the bases adjacent to one another) and (guanine N7)-X-(guanine N7) (with one base "X" in between the alkylated guanines) (Structure 3.16). Based on techniques such as gel electrophoresis, high-field NMR, X-ray crystallography and atomic force microscopy, these intrastrand cross-links can be shown to kink the DNA at adduct sites, a phenomenon that is recognized by DNA repair enzymes. Interestingly, the configurational isomer *trans*-platin cannot interact so efficiently with DNA due to its shape, and this is reflected in significantly lower antitumor activity.

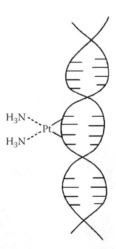

STRUCTURE 3.16 Schematic diagram of the cisplatin-DNA adduct showing the predomi-nant intrastrand cross-link which kinks the DNA.

Cisplatin is of significant value in patients with metastatic germ cell cancers (i.e., seminomas and teratomas). For example, clinical trials of cisplatin in combi-nation with vinblastine have produced complete remission in 59% of patients with testicular cancer and 30% complete remission in ovarian carcinoma patients.

The main mechanism of resistance to cisplatin is through repair of the DNA adducts. It can be shown in cisplatin-resistant cell lines that the adducts are rapidly repaired, and it is thought that the DNA repair surveillance enzymes recognize the distortion around the adduct site. One possible explanation for the significant activity of cisplatin in testicular cancer is that the germ cells have a limited ability to repair cisplatin adducts compared to somatic cells. Other tumors that respond to cisplatin include squamous cell carcinoma of the head and neck, bladder carcinoma, refractory choriocarcinoma, and lung, upper GI, and ovarian cancers. However, carboplatin is now preferred for ovarian cancer.

Cisplatin has a number of dose-limiting side effects, including nephrotoxicity (monitoring of renal function is essential), ototoxicity, peripheral neuropathy, hypomagnesemia, and myelosuppression. It requires intensive intravenous hydration during treatment, and progress may be complicated by severe nausea and vomiting. However, despite these problems, cisplatin is increasingly given in day care settings.

3.3.6.2 Carboplatin

Carboplatin (Palaplatin™) is an analog of cisplatin that incorporates a cyclobutyl-substituted hexa dilactone ring (Structure 3.15). It is better tolerated than cisplatin in terms of nausea and vomiting, GI toxicity, nephrotoxicity, neurotoxicity, and ototoxicity, although myelosuppression is more pronounced. It is widely used in the treatment of advanced ovarian cancer and lung cancer (particularly the small-cell type). The dose is determined according to renal function rather than body surface area, and it is often used on an outpatient basis.

3.3.6.3 Oxaliplatin

Oxaliplatin (Eloxatin™), which contains cyclohexyl and pentadilactone rings, is licensed for the treatment of metastatic colorectal cancer in combination with fluorouracil and folinic acid (Structure 3.15). Neurotoxic side effects, which include sensory peripheral neuropathy, are dose limiting. Other side effects include GI disturbances, ototoxicity, and myelosuppression. Renal function is normally monitored during treatment.

3.3.7 CARBINOLAMINES

The carbinolamine group [-NR-CH(OH)-] (R = alkyl or aryl) is an electrophilic moiety found in a number of synthetic and naturally occurring compounds that can form a covalent bond with various nucleophilic sites on DNA bases. Trimelamol, which contains three carbinolamine moieties, is an experimental agent developed from the clinically active hexamethylmelamine and the closely related pentamethylmelamine, which do not contain carbinolamine groups themselves but obtain them during oxidative metabolism. Trimelamol was designed to have the carbinolamine moieties already in place. Phase II trials carried out in the early 1990s showed that trimelamol is active in refractory ovarian cancer and is less emetic and neurotoxic than pentamethylmelamine.

STRUCTURE 3.17 Structures of hexamethylmelamine, pentamethylmelamine, and trimelamol. Trimelamol can lose three water molecules to form the DNA-reactive tri-iminium form.

The precise mode of action of trimelamol has not yet been established. It is known to be a reasonably potent DNA interstrand cross-linking agent, although the exact sites of alkylation or cross-linking have not been established. It is known that the carbinolamine moieties dehydrate to afford iminium ions, and it is these highly electrophilic species that most likely alkylate DNA bases (see Structure 3.17). In practice, there is likely to be a pH-dependent equilibrium mixture of the carbinolamine and imine forms in aqueous solution.

One reason why trimelamol has not been commercialized is its poor solubility in a number of physiologically compatible solvents, thus making it difficult to formulate. It also suffers from stability problems in that amine-containing degradation products are known to couple with trimelamol itself to form dimeric and higher order polymers that are extremely insoluble and can precipitate *in vivo*, potentially causing emboli. Analogs of trimelamol that partially overcome these problems have been developed, and so an improved version of trimelamol may be studied in the future.

PBD monomer (DNA-alkylating) and dimer (DNA cross-linking) agents also possess carbinolamine moieties or the equivalent. However, due to their unique mechanism of action (i.e., DNA sequence recognition), they are discussed elsewhere in this chapter.

3.3.8 CYCLOPROPANES

Cyclopropane moieties are the homocarbon analogs of aziridines and epoxides. They are also electrophilic and capable of alkylating DNA. Part of the driving force for such an alkylation reaction is the release of strain within the three-membered ring. The experimental agent bizelesin is the only known example of a DNA interstrand cross-linking agent that forms covalent bonds with DNA bases through cyclopropane moieties. However, as this agent is also an example of a sequence-selective cross-linking agent, it is discussed elsewhere in this chapter.

3.3.9 MITOMYCIN-C

Mitomycin-C (Mitomycin-C Kyowa™) is a member of a group of naturally occurring antitumor antibiotics produced by *Streptomyces caespitosus (griseovinaceseus)* and was first isolated in 1958. It is unusually rich in chemical functional groups, and the components of the molecule essential for its mode of action are the quinone, aziridine, and carbamate moieties. Its mechanism of action involves a bioreductive step, and so mitomycin is regarded as a "bioreductive agent" (see Chapter 7). However, it is discussed in this chapter due to its DNA cross-linking properties.

The detailed mechanism of activation of mitomycin-C is complex (see Scheme 3.13). It is thought that initial reduction of the quinone moiety (one-electron reduction yields a semiquinone, while a two-electron reduction provides the hydroquinone) transforms the heterocyclic nitrogen from a conjugated amido to an amino form, thus making it more electron rich and facilitating elimination of the ring junction methoxy group. Tautomerization of the resulting iminium ion and loss of the carbamate group then creates an electrophilic center that is susceptible to nucleophilic attack by a DNA base. Nucleophilic attack of the aziridine moiety by a

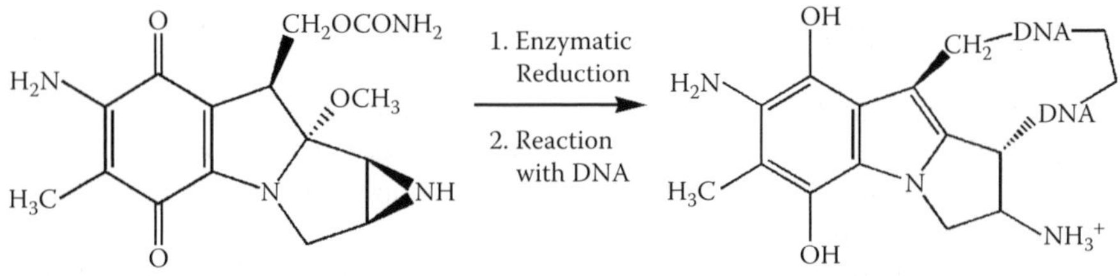

SCHEME 3.13 Bioreduction of mitomycin-C (Mitomycin-C Kyowa™) followed by DNA cross-linking.

nucleophile on the opposite strand of DNA also occurs, leading to an interstrand cross-link. The predominant adducts appear to form between two guanine-N2 groups within the minor groove of DNA.

The most important feature of mitomycin is the bioreductive "trigger" that is required before DNA cross-linking can occur. It is known that the centers of some tumors, particularly old ones of large size, are hypoxic due to a poor blood supply. Therefore, the bioreductive conditions that exist at the centers of these tumors are thought to account, in part, for the tumor selectivity of mitomycin. This concept of bioreductive activation continues to attract interest in the research community, and a number of other agents have been designed based on this mechanism of action (see Chapter 7).

Intravenous mitomycin is used to treat upper GI and breast cancers, and administration by bladder instillation is used to treat superficial bladder tumors. Adverse events include delayed bone marrow toxicity; therefore, the drug is usually given at 6-week intervals. Permanent bone marrow damage can result from prolonged use, and renal damage and lung fibrosis can also occur.

3.3.10 Sequence-Selective DNA Cross-Linking Agents

Most of the cross-linking agents described in the previous sections are relatively small in size and rarely span more than one or two DNA base pairs. In the last decade, there has been an effort to develop agents that can span longer stretches of DNA (e.g., six base pairs or more), recognizing specific sequences of base pairs within the length spanned. The rationale for this goal is that, for agents with lower sequence selectivity, a large number of adducts are formed across the genome, many of which may contribute to the toxic side effects (e.g., myelosuppression) observed with these agents. If longer, more sequence-selective agents can be developed, then fewer adducts will form across the genome, with a potential reduction in the occurrence and intensity of side effects and a likely improvement in therapeutic index. The ultimate design objective is an agent that recognizes and cross-links a DNA sequence that occurs uniquely in tumor cells (e.g., mutated genes) and not in healthy cells. Two families of agents have so far been developed based on this principle: the PBD dimers such as SJG-136, which cross-link guanine bases on opposite DNA strands via their N2 positions, and the CBI dimers, which cross-link adenine bases on opposite strands via their N3 positions.

3.3.10.1   PBD Dimers (SJG-136)

The experimental agent SJG-136 (NCI-694501, BN2629) is a PBD dimer that consists of two monomeric PBD units joined through their C8/C8'-positions via a propyldioxy linker (see Structure 3.18).

STRUCTURE 3.18   Structure of the sequence-selective minor groove interstrand cross-linking agent SJG-136.

The two imine moieties (i.e., N10-C11/N10'-C11') at the top of each seven-membered ring bind covalently to the N2-positions of guanines on opposite DNA strands, thus forming an interstrand cross-link in the minor groove. The molecule spans six DNA base pairs and occupies half a complete turn of the DNA helix. It has a preference for binding to purine-GATC-pyrimidine sequences while recognizing the central GATC sequence. Recognition of the central AT base pairs occurs through hydrogen bonding interactions between the N3-position of adenine and the N10-proton of the PBD unit on either strand (Figure 3.4). Due to the S stereochemistry at the C11a/C11a' positions of the molecule, it is perfectly shaped to fit into and follow the contours of the minor groove with little of the molecule exposed outside of the groove. Unlike other DNA cross-linking agents, this causes virtually no distortion of the DNA helical structure, thus avoiding recognition of the adducts by repair enzymes. *In vitro* experiments have confirmed this by demonstrating that SJG-136 adducts are repaired very slowly.

SJG-136 has shown significant activity in the NCI 60 cell line screen and other cell lines, and in follow-up hollow fiber and human tumor xenograft assays in both mice and rats, in which it provides a dose-dependent antitumor effect in a number of tumor models and is curative in many. In particular, because it forms nondistorting DNA adducts that are difficult for cells to repair, SJG-136 retains full potency in a number of drug-resistant cell lines and has significant antitumor activity in cisplatin-resistant xenograft models, suggesting its potential use against drug-resistant cancers. This agent is presently being evaluated in Phase I clinical trials.

3.3.10.2   Cyclopropanepyrroloindole (CPI) Dimer (Bizelesin)

The experimental agent bizelesin is similar to SJG-136 in binding in the minor groove of DNA and spanning approximately six base pairs. However, it cross-links two adenines on opposite strands via their N3 positions, as opposed to the G-G cross-link formed by SJG-136. In the case of bizelesin, the alkylating moieties are cyclopropane rings that form *in situ*. Thus, bizelesin itself is a prodrug that undergoes

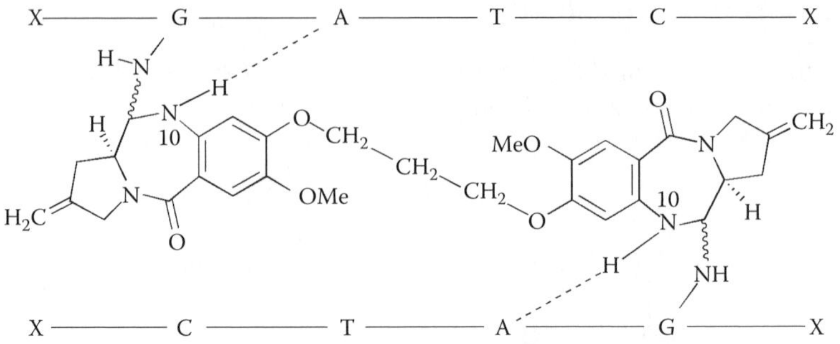

(a)

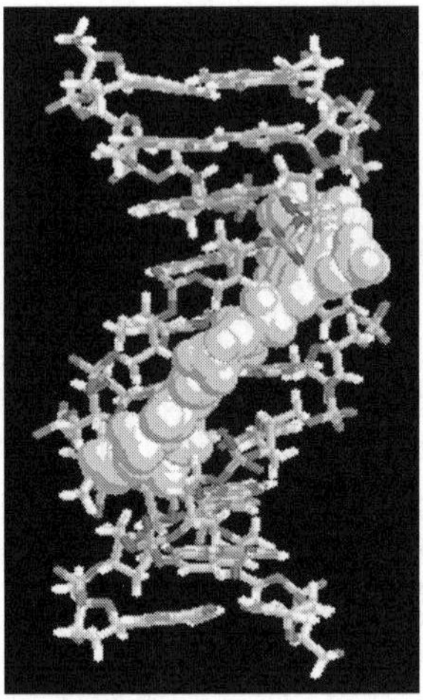

(b)

FIGURE 3.4 (a) Schematic diagram of the SJG-136/DNA adduct showing the molecule bound to its preferred GATC sequence through an interstrand cross-link between the guanines on opposite strands. The central AT base pairs are recognized by hydrogen bonds between the adenine N3 and PBD N10-H atoms; (b) molecular model showing how SJG-136 fits snugly in the minor groove of DNA with no distortion of the helical structure and little of the molecule exposed beyond the boundaries of the groove.

chemical transformation into the electrophilic dicyclopropane species. As shown in Scheme 3.14, this involves spontaneous formation of the quinone and cyclopropane moieties through loss of chloride ions.

Cyclopropane moieties are the homocarbon analogs of aziridines and epoxides. They are electrophilic and thus capable of alkylating DNA. However, in the case of, once formed, bizelesin, each cyclopropane moiety, once formed, is attached to the *para*-position of a dihydroquinone and so is further activated. Attack of a cyclopropane ring by an adenine N3 is driven by the energy released in re-aromatizing the dihydroquinone to a phenol within each unit. The central indole units also recognize AT base pairs but noncovalently, and so overall bizelesin recognizes a run of AT base pairs.

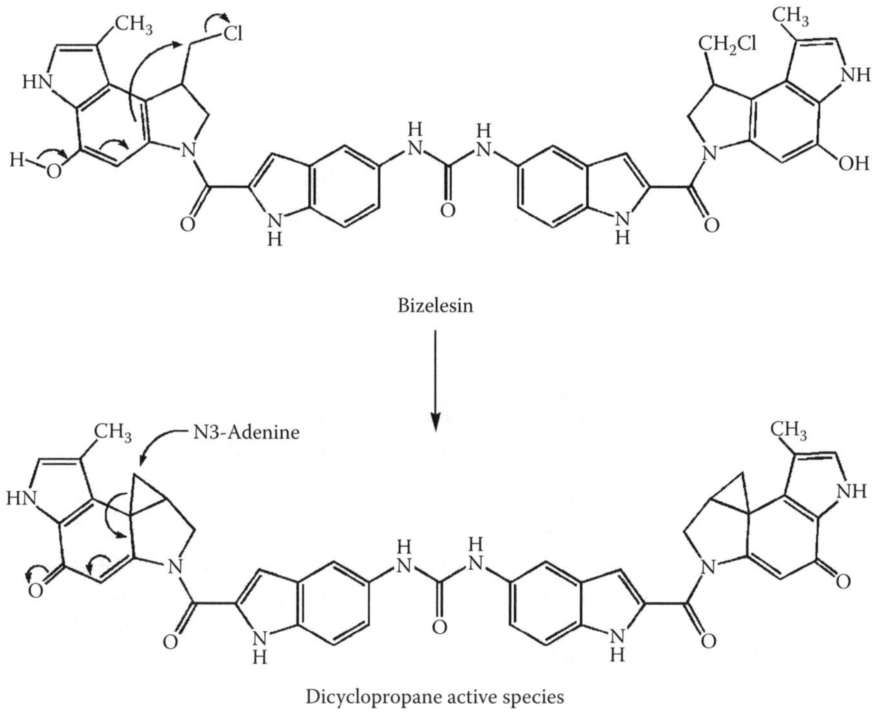

Bizelesin

Dicyclopropane active species

SCHEME 3.14 Activation of bizelesin to form the DNA-reactive dicyclopropane species.

Unfortunately, in early clinical trials, bizelesin was shown to cause severe myelosuppression and so was not progressed any further. A number of related experimental agents, such as adozelesin, have been investigated but all have shown similar toxicity profiles. Research continues on sequence-selective cross-linking agents such as SJS-136 and Bizelesin and related analogs in an effort to fully understand their SAR and clinical potential.

3.4 INTERCALATING AGENTS

The DNA intercalating agents represent one of the most widely used groups of antitumor agents. They are flat in shape, consisting of three or four fused aromatic rings, and their mechanism of action involves insertion between the base pairs of DNA, perpendicular to the axis of the helix. Once in position, they are held in place by interactions, including hydrogen bonding and van der Waals forces. In addition, many intercalators have side chains rich in hydrogen-bonding functionalities (e.g., the amino sugar of doxorubicin (Structure 3.19) or the pentapeptide rings of dactinomycin [Structure 3.22]) that position themselves in the DNA minor or major grooves, where they further stabilize the adduct by forming hydrogen bonds and other noncovalent interactions. Some intercalators with arrays of functional groups at either end of the molecule protrude into both the minor and major grooves and are sometimes referred to as *threading agents*.

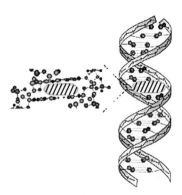

SCHEME 3.15 Model showing an intercalating agent (hatched) inserted between two base pairs of DNA, and an expanded view of the two base pairs involved. Insertion of intercalating agent has the effect of lengthening the DNA helix, which can be detected by viscosity or electrophoresis measurements.

Intercalation can be detected in naked DNA by an increase in helix length which can be measured as an enhancement in viscosity using sedimentation values or as a change in mobility of DNA fragments during electrophoresis (see Scheme 3.15). Adducts of intercalating agents and DNA can also be studied by such techniques as NMR and X-ray crystallography.

A number of different modes of action have been ascribed to intercalating agents. For example, it can be demonstrated that, *in vitro*, some intercalators and threading agents block transcription and interfere with other DNA processing enzymes. Many intercalating agents have a preference for GC-rich sequences and, as with some alkylating agents, this may also play a part in their mechanism of action (i.e., the presence of "GC-Islands" in promoter regions of genes). Some intercalators are known to "trap" complexes between topoisomerase enzymes and DNA, thus leading to strand cleavage. Others are known to chelate metal ions and produce DNA-cleaving free radicals or to interact with cell membranes.

The largest family of intercalating agents in clinical use is the anthracyclines, members of which contain four fused, aromatic rings and include the naturally occurring antibiotics doxorubicin, daunorubicin, and aclarubicin (Structure 3.19), and the related semisynthetic analogs epirubicin and idarubicin (Structure 3.20). Members of the anthracene family possess three aromatic rings, with mitoxantrone (also known as mitozantrone) being the main agent in common use (Structure 3.21). The third group is the phenoxazine family, members of which contain three fused, six-membered rings but with the central ring containing oxygen and nitrogen heteroatoms. The best known member of this group is dactinomycin, which also contains two cyclic peptide side chains that stabilize the drug-DNA adduct by interacting in the minor groove of DNA (Structure 3.22).

Finally, it should be noted that many intercalating agents function as radiomimetics. Therefore, radiotherapy should be avoided during drug treatment in order to avoid enhanced toxicity.

3.4.1 ANTHRACYCLINES

The anthracyclines (sometimes known as the *anthraquinones*) are a group of anti-tumor antibiotics, the first example of which was isolated from *Streptomyces peucetius*. It is the best-known family of intercalating agents, and members consist of a planar anthraquinone nucleus attached to an amino-containing sugar. Although doxorubicin, daunorubicin, and aclarubicin are natural products (Structure 3.19), semisynthetic analogs such as epirubicin and idarubicin have also been developed (Structure 3.20).

Several different mechanisms have been suggested to explain the biological activity of the anthracyclines, and controversy still exists about the relative importance of each. The first centers on the fact that the planar ring system inserts between two DNA base pairs perpendicular to the long axis of the double helix, with the amino sugar conferring stability on the adduct through hydrogen bonding interactions with the sugar phosphate backbone. It is known from *in vitro* experiments that the adduct formed is sufficiently stable to interfere with DNA processing, including transcription. Second, the anthracyclines are known to form complexes (i.e., ternary complexes) with topoisomerase enzymes and DNA, which can lead to strand breaks. Third, binding to cell membranes has been observed, and this may alter membrane fluidity and ion transport as well as disturb various biochemical equilibria in the cell. Lastly, generation of semiquinone species can lead to free radical or hydroxy radical production, which may cause DNA and other types of cellular damage. Radical formation may be mediated by chelation of divalent cations such as calcium and ferrous ions by the phenolic and quinone functionalities, and this is thought to be responsible for the cardiotoxicity observed with the anthracyclines.

3.4.1.1 Doxorubicin

Doxorubicin (Caelyx™, Myocet™) was first extracted from *Streptomyces peucetius* and is presently one of the most successful and widely used anticancer drugs because of its broad spectrum of activity (Structure 3.19). It is used to treat acute leukemias, lymphomas, and a variety of solid tumors, including carcinomas of the breast, lung, thyroid, and ovary as well as soft-tissue carcinomas. The drug is given by injection into a fast running infusion, usually at 21-day intervals, although care must be taken to avoid local extravasation, which can cause severe tissue necrosis. Frequently encountered adverse events include nausea and vomiting, myelosuppression, mucositis, and alopecia. Doxorubicin is mostly excreted by the biliary tract, and an elevated bilirubin level is considered a marker for reducing the dose. Higher cumulative doses are associated with cardiomyopathy, and potentially fatal heart failure can occur. Thus, cardiac monitoring is sometimes used to assist in safely limiting the total dosage. Liposomal formulations of doxorubicin are also available and are claimed to reduce the risk of cardiotoxicity and local tissue necrosis. For example, Caelyx™ is licensed for advanced acquired immunodeficiency syndrome (AIDS)–related Kaposi's sarcoma and for advanced ovarian cancer when platinum-based chemotherapy has failed. A similar product, Myocet™, is licensed for use with cyclophosphamide for metastatic breast cancer. Doxorubicin can also be given by bladder instillation for superficial bladder tumors.

STRUCTURE 3.19 Structures of the naturally occurring anthraquinone antitumor antibiotics, doxorubicin (Caelyx™ or Myocet™), daunarubicin (DaunoXome™) and aclarubicin.

3.4.1.2 Daunorubicin

Daunorubicin was originally found in the fermentation broth of *Streptomyces peucetius* and is an important agent in the treatment of acute lymphocytic and myelocytic leukemias (Structure 3.19). Administered intravenously, its general properties are similar to those of doxorubicin. An intravenous liposomal formulation (DaunoXome™) is licensed for advanced AIDS-related Kaposi's sarcoma.

3.4.1.3 Aclarubicin

Aclarubicin is produced by *Streptomyces galilaeus*, and its general properties are similar to those of doxorubicin (Structure 3.19). It is used for acute nonlymphocytic leukemia in patients who have relapsed or are resistant or refractory to first line chemotherapy.

Epirubicin (Pharmorubin™)   Idarubicin (Zavedos™)

STRUCTURE 3.20 Structures of the semi-synthetic anthraquinone antitumor antibiotics, epirubicin (Pharmorubin™) and Idarubicin (Zavedos™).

3.4.1.4 Epirubicin

Epirubicin (Pharmorubicin™) is a semisynthetic analog of doxorubicin differing only in the stereochemistry of the C4-hydroxy group of the sugar moiety (Structure 3.20). Clinical trial data show that it is similar in efficacy to doxorubicin for the treatment of breast cancer; however, it is necessary to carefully monitor the maximum cumulative dose to help avoid cardiotoxicity. In a similar manner to doxorubicin, it is administered either intravenously or by bladder instillation.

3.4.1.5 Idarubicin

Idarubicin (Zavedos™) is a semisynthetic analog of daunorubicin that differs only in the absence of a methoxy group in its A-ring (Structure 3.20). It has general properties similar to doxorubicin but is the only anthracycline that can be given orally as well as intravenously. Idarubicin is used in advanced breast cancer after failure of first-line chemotherapy (not including anthracyclines) and also in acute nonlymphocytic leukemia.

3.4.2 ANTHRACENES

Agents in the anthracene family are based on the anthracene nucleus and have three fused rings rather than four (as in the anthracyclines). Mitoxantrone (also known as mitozantrone) is the most important anthracene agent used in the clinic (Structure 3.21).

3.4.2.1 Mitoxantrone

Mitoxantrone (Novantrone™, Onkotrone™) is rich in oxygen- and nitrogen-containing substituents and side chains that allow extensive stabilization of the intercalated adduct by hydrogen-bonding interactions. In particular, it has two identical side chains containing both amino and hydroxyl functionalities (Structure 3.21). The molecule is known to have a preference for binding to GC-rich sequences and, as with the anthracyclines, evidence shows that DNA is cleaved, although the mechanism is not thought to be linked to the generation of reactive oxygen species.

STRUCTURE 3.21   Structure of Mitoxantrone (Novantrone™ or Onkotrone™).

Mitoxantrone is used for the treatment of metastatic breast cancer and is also licensed for the treatment of adult nonlymphocytic leukemia and non-Hodgkin's lymphoma. It is administered intravenously and is generally well tolerated, although dose-related cardiotoxicity and myelosuppression are serious possible side effects. Although cardiotoxicity is less prominent than with the anthracyclines, cardiac examinations and monitoring are still recommended after a certain cumulative dose has been reached.

3.4.3 PHENOXAZINES

Members of the phenoxazine family are similar to the anthracenes in that they contain three fused six-membered rings but differ in that the central ring contains oxygen and nitrogen heteroatoms. The best known member is dactinomycin, which contains two cyclic peptide side chains that stabilize the drug-DNA adduct by interacting with functional groups in the minor groove (see Structure 3.22).

3.4.3.1 Dactinomycin

Dactinomycin (Cosmegen Lyovac™), which is known as a chromopeptide antibiotic, was isolated from *Streptomyces parvulus* in the 1950s and was initially developed as a potent bacteriostatic agent, although it was later found to be too toxic for general use. The antitumor activity of dactinomycin did not emerge until 10 years later, when it was tried with great success in the treatment of Wilm's tumor (a kidney tumor in children) and a type of uterine cancer.

STRUCTURE 3.22   Structure of dactinomycin (Cosmegan Lyovac™).

The molecule consists of a tricyclic phenoxazin-3-one chromophore with two identical cyclic pentapeptide side chains. The mechanism of action, which is concentration-dependent, involves either blockade of DNA synthesis or inhibition of DNA-directed RNA synthesis, thus preventing RNA chain elongation. Adduct formation may also lead to single-strand DNA breaks in a manner similar to that of doxorubicin, either through radical formation or by interaction with topoisomerase. X-ray crystallography studies have shown that the phenoxazine ring intercalates preferentially between GC base pairs where it can interact with the N2-amino groups of guanines. The cyclic peptide moieties then position themselves in the minor groove and participate in extensive hydrogen bonding and hydrophobic interactions with functional groups in the floor and walls of the groove, thus providing significant stabilization of the adduct. The stability of the adduct is crucial for the blockage of RNA polymerase.

Given intravenously, dactinomycin is mainly used to treat pediatric cancers. For example, it is used in the treatment of gestational choriocarcinoma and Wilm's tumor in combination with methotrexate or vincristine, respectively. It is also used in some testicular sarcomas and in AIDS-related Kaposi's sarcoma. The side effects of dactinomycin are similar to those of doxorubicin except that cardiac toxicity is less prominent. Tumor resistance to dactinomycin is common, possibly due to reduced uptake and also active transport of the drug out of tumor cells.

3.5 TOPOISOMERASE INHIBITORS

DNA topoisomerases are a family of enzymes responsible for the cleavage, anneal-ing, and topological state (e.g., supercoiling) of DNA. In particular, these enzymes bring about the unwinding and winding of the supercoiled DNA that comprises the chromosomes. If the DNA cannot be unwound, transcription of the message cannot occur and cell death results because the corresponding proteins cannot be synthesized (see Scheme 3.16).

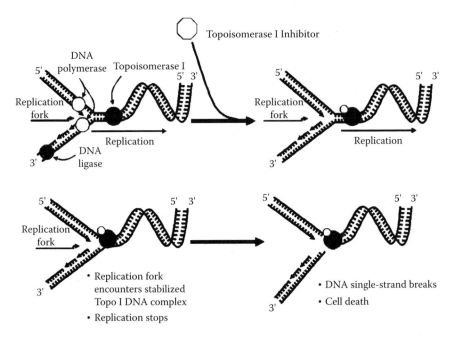

SCHEME 3.16 Mechanism of the formation of DNA breaks and resulting cell death induced by Topoisomerase I inhibitors.

There are two types of topoisomerase enzymes, known as I and II. Topoisomerase I (Topo I) enzymes are capable of removing negative supercoils in DNA without leaving damaging nicks. They work by breaking only one strand of DNA, and then by attaching the free phosphate residue of the broken strand to a tyrosine residue of the enzyme. The complex then rotates, relieving the supercoiled tension of the DNA, and the two ends are resealed. Topoisomerase II (Topo II) enzymes cleave both strands of double-stranded DNA simultaneously, passing a complete duplex strand through the cut, followed by resealing of both strands.

The mode of action of intercalators such as the anthracyclines, anthracenes, and acridines (e.g., amsacrine) is thought to be partly associated with inhibition of the topoisomerase enzymes. However, other families of agents, such as the campto-thecins and ellipticines, have been discovered that specifically bind to and inhibit topoisomerase enzymes.

3.5.1 TOPOISOMERASE I INHIBITORS

Topo I is an essential DNA topology-controlling enzyme that works by transiently breaking one DNA duplex and passing the second strand through the break followed by resealing. This is an absolute requirement of many nuclear processes, including replication, transcription, and recombination. Topo I inhibitors work by keeping the chromosomes wound tight so that the cell cannot make proteins and cell death results. Because some cancer cells grow and reproduce at a much faster rate than normal cells, they are thought to be more vulnerable to topoisomerase inhibition, which is one possible explanation for their selective toxicity.

STRUCTURE 3.23   Structure of camptothecin.

The lead structure for Topo I inhibitors is the natural product camptothecin, which is a cytotoxic quinoline-based alkaloid with a unique five-ring system extracted from the barks of the Chinese *camptotheca* tree (e.g., *Camptotheca acuminate*) and the Asian *nothapodytes* tree. Camptothecin and related compounds cause cancer cell death by inhibiting the enzyme DNA Topo I by a unique mechanism that involves stabilization of a covalent reaction intermediate known as the *cleavable complex*.

The use of camptothecin in the clinic was found to be limited due to poor water solubility and a number of serious side effects. However, a number of derivatives of camptothecin were produced to address these issues and are now in use in the treatment of small-cell lung, colon, and breast cancers as well as some melanomas and leukemias. Topotecan was approved by the U.S. FDA in 1996 for the treatment of advanced ovarian cancers resistant to other chemotherapeutic agents. The injectable irinotecan hydrochloride was also approved in 1996 as a treatment for metastatic cancer of the colon or rectum, although it is normally only prescribed for colorectal cancer patients who have not responded to standard treatment with fluorouracil.

The major side effects of camptothecin-based drugs include severe nausea and diarrhea and lowered leukocyte (white blood cell) counts. Bone marrow damage may also occur with these drugs.

3.5.1.1 Topotecan

The solubility problems of camptothecin were solved by the addition of hydroxyl and dimethylaminomethyl substituents to the benzenoid ring to give topotecan (Hycamtin™), which has the same mechanism of action as camptothecin in inhibiting Topo I (see Structure 3.24). It binds to the Topo I-DNA adduct to form a ternary complex that prevents re-ligation of the DNA strand, thus leading to double-strand DNA breaks and cell death. At neutral pH, topotecan, unlike irinotecan, is found predominantly in the inactive carboxylate form (i.e., resulting from opening of the lactone ring). As a result, topotecan has different toxicity and antitumor profiles compared to irinotecan. The agent is cell-cycle specific for the S-phase and also sensitizes cells toward radiation.

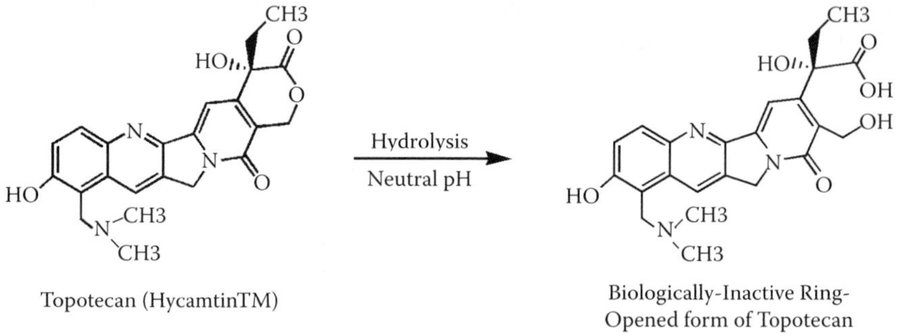

Topotecan (HycamtinTM)

Biologically-Inactive Ring-Opened form of Topotecan

STRUCTURE 3.24 Structure of topotecan (Hycamtin™).

Topotecan is given by intravenous infusion for the treatment of metastatic ovarian cancer when first line or subsequent therapy has failed. In addition to dose-limiting myelosuppression, side effects include GI disturbances such as delayed diarrhea, asthenia, alopecia, and anorexia.

3.5.1.2 Irinotecan

Irinotecan (Campto™ or Camptosar™), also known as CPT-11, represents a further attempt to improve the water solubility of camptothecin by the addition of a di-piperidine carbamate functionality to the benzenoid ring. Irinotecan is hydrolyzed *in vivo* to 7-ethyl-10-hydroxycamptothecin (SN-38), an active metabolite (see Structure 3.25). Therefore, irinotecan may be considered to be a prodrug. SN-38 is approximately 200- to 2000-fold more cytotoxic than irinotecan; however, despite

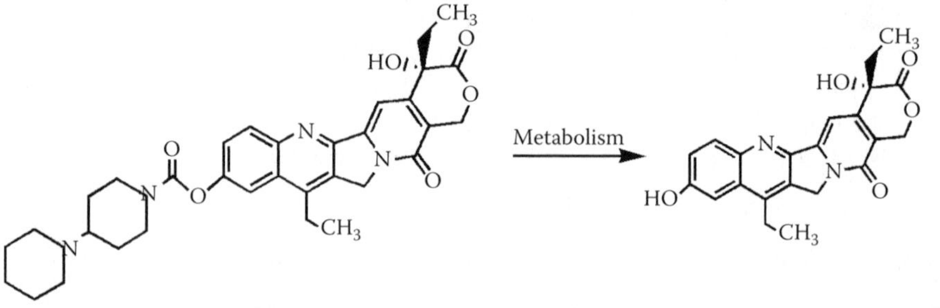

Irinotecan (Campto™ or Camptosar™)

SN-38 (Irinotecan metabolite)

STRUCTURE 3.25 Structure of irinotecan (Camptosar™) and its main metabolite SN-38.

its own potential as an anticancer agent, it has been hampered by poor solubility in most pharmaceutically acceptable solvents. Therefore, attempts have been made to develop a liposomal preparation of SN-38 to capitalize on its potency while solving the formulation problems.

Both SN-38 and irinotecan become attached to the Topo I–DNA complex, which prevents re-ligation of the DNA strand. This results in double-strand DNA breaks and cell death. However, as with all topsisomerase inhibitors, it is still unclear how this mechanism contributes to the selective antitumor activity of irinotecan or SN-38 in humans. Irinotecan, like topotecan, is specific for the S-phase of the cell cycle.

Irinotecan is licensed for metastatic colorectal cancer in combination with fluorouracil and folinic acid, or as monotherapy when fluorouracil-containing treatments have failed. It is given by intravenous infusion, and the side effects are the same as those for topotecan.

3.5.2 Topoisomerase II Inhibitors

Human cells express two genetically distinct isoforms of Topo II, known as Topo IIα and Topo IIβ. The lead structure for drugs such as etoposide and teniposide that inhibit these enzymes is podophyllotoxin, a plant alkaloid isolated from the American mandrake rhizome (see Structure 3.26).

STRUCTURE 3.26 Structure of podophyllotoxin.

Other agents such as the ellipticines and semisynthetic derivatives such as podophyllic acid ethyl hydrazide have been synthesized and studied but none have been commercialized. Amsacrine is also a known topsisomerase II inhibitor. As a class, topo II inhibitors appear to have activity in certain cancers, including testicular cancer, oat-cell carcinoma of the bronchus, malignant teratomas, and various leukemias and lymphomas.

3.5.2.1 Etoposide

Etoposide (Etopophos™ or Vepesid™) is a semisynthetic glucoside of epipodophyllotoxin used to treat small-cell bronchial carcinoma, for which it is claimed to be one of the most effective agents known (Structure 3.27). Etoposide works by inhibiting

STRUCTURE 3.27 Structure of etoposide (Etopophos™ or Vepesid™).

the ability of Topo IIα and IIβ to reseal cleaved DNA duplexes. Therefore, normally reversible DNA strand breaks are converted into lethal breaks by processes such as transcription and replication.

Etoposide is usually administered by slow intravenous infusion but can also be given orally, with the dose being twice that of the intravenous dose. When formulated as etoposide phosphate, it can be given by intravenous infusion or injection. Etoposide is usually administered daily for 3 to 5 days, with courses not being repeated more frequently than every 21 days. It is used to treat testicular cancer, small-cell carcinoma of the bronchus, and some lymphomas. Toxic effects of the drug include nausea and vomiting, myelosuppression, and alopecia.

3.5.2.2 Teniposide

Teniposide (Vumon™) is an analog of etoposide that differs in structure only in that the methyl substituent on the sugar moiety is exchanged for a thiophene ring (see Structure 3.28). It has broadly similar clinical activity to etoposide.

STRUCTURE 3.28 Structure of teniposide (Vumon™).

3.5.2.3 Ellipticene

The experimental agent ellipticine is a plant alkaloid isolated from *Ochrosia elliptica* that exerts its antitumor action through intercalation with DNA and inhibition of the topo II enzyme (see Structure 3.29). In *in vitro* cytotoxicity studies, ellipticine has significant activity in nasopharyngeal carcinoma cell lines. Despite much research on the ellipticines, no successful clinical candidate based on the ellipticine structure has yet emerged.

STRUCTURE 3.29   Structure of ellipticene.

3.5.2.4 Amsacrine

Amsacrine (Amsidine™) was first synthesized in 1974 and entered clinical trials in 1977. It has an acridine-based structure (see Structure 3.30). The mechanism of action is not completely understood, but it is known to intercalate with DNA, inhibit topo II, and cause double-strand DNA breaks. Its cytotoxicity appears to be most pronounced when topoisomerase levels are at a maximum during the S-phase of the cell cycle.

STRUCTURE 3.30   Structure of amsacrine (Amsidine™).

Clinically, amsacrine has an activity and toxicity profile similar to doxorubicin. It is administered intravenously and is used in the treatment of advanced ovarian carcinomas, myelogenous leukemias, and lymphomas. Side effects include myelo-suppression and mucositis. Electrolytes are normally monitored during treatment because fatal arrhythmias have occurred in association with hypokalemia.

3.6 DNA-CLEAVING AGENTS

DNA-cleaving agents work by binding to the DNA helix, usually in a sequence-selective manner, and then cleaving the double strand through radical production.

The best known example of this class is the natural product bleomycin, which is used clinically. The experimental enediyne agent calicheamicin has proved too toxic to be used in its own right as an antitumor agent; however, it has been successfully used as the cytotoxic "warhead" in the antibody-drug conjugate gemtuzumab ozogamicin (Mylotarg™) (see Chapter 7).

3.6.1 BLEOMYCINS

The bleomycins are a group of closely related natural products that exert their antitumor activity by binding to DNA in a sequence-selective manner and then causing strand cleavage. The pharmaceutical preparation known as *bleomycin sulfate* consists of a mixture of the glycopeptide bases (e.g., A_2, A_2I, B_{1-4}, etc.), with A2 as the predominant component. The mixture is obtained from *Streptomyces verticillus*, and the individual molecular weights are in the region of 1300. Despite the size and complexity of the molecule, particularly with regard to the number of chiral centers, the first total synthesis of bleomycin was reported in 1982.

Bleomycin molecules have three distinct regions that are believed to contribute toward their mechanism of action. First, the heterocyclic bithiazole moiety (right in Structure 3.31) is thought to intercalate with DNA. Electrostatic attraction of the highly charged sulfonium ion (A_2, $R = NH(CH_2)_3S^+Me_2$) to the phosphate residues in DNA then stabilizes the adduct. Once bound, the second domain (top left in Structure 3.31), which consists of a β-hydroxyhistidine, a β-aminoalanine, and a pyrimidine, forms an iron (II) complex that interacts with oxygen to generate free radicals, leading to single- and double-strand breaks. Currently, it is not clear if the

STRUCTURE 3.31   General structure of the bleomycins.

activation of this complex is self-initiating or the result of enzyme catalysis. The third region of bleomycin (bottom left in Structure 3.31) is glycopeptidic in nature and, although having no direct action of its own, may contribute to drug uptake by tumor cells or provide additional stabilizing hydrogen bonding interactions with DNA or associated histone proteins.

Bleomycin tends to accumulate in squamous cells and is therefore suitable for treatment of tumors of the head, neck, and genitalia, although is has also been used in the treatment of Hodgkin's disease and testicular carcinomas. Unlike most anti-cancer drugs, it is only slightly myelosuppressive, and dose-limiting toxicity is confined to the skin, mucosa, and lungs.

It is given intramuscularly or intravenously to treat squamous cell carcinoma, metastatic germ cell cancer and, in some drug regimens, non-Hodgkin's lymphoma. Bleomycin causes little bone marrow suppression but instead produces a serious progressive pulmonary fibrosis in 5% to 15% of patients. This adverse effect is dose related, and patients who have received prolonged and extensive treatment with bleomycin may be at risk of respiratory failure under some conditions. Another common adverse event is a dermatological toxicity that manifests as an increased pigmentation, particularly affecting the flexures. In addition, subcutaneous sclerotic plaques may occur. As a result, ulceration in these areas and also pigmentation of the nails may occur. Mucositis is another relatively common side effect, and an association with Raynaud's phenomenon has been reported. Finally, intravenous hydrocortisone is sometimes used to counter hypersensitivity reactions, which manifest as chills and fevers commonly occurring a few hours after administration. It is noteworthy that enzymes in most tissues rapidly deactivate bleomycin, probably as a result of deamination or peptidase activity.

3.6.2 ENEDIYNES

The enediynes are a family of approximately 15 to 20 antitumor antibiotics produced by *Micromonospora echinospora* ssp. *calichensis*, a bacterium isolated from chalky soil or caliche clay, a type of soil found in Texas. Naming of the individual compounds is based both on thin-layer chromatography mobility using Greek letters with subscripts and on the halogen substitution pattern that is indicted by the superscript. Some members of the enediyne family are also known as *esperamycins*.

These molecules are unique in structure, containing two triple bonds in close proximity (i.e., separated by an ethylene unit) as part of a larger ring (see Structure 3.32). They bind in the minor groove of DNA and initiate double-stranded DNA cleavage via a radical abstraction process that involves a unique thiol-mediated cyclization (the Bergman cyclization). During this process, the triple bonds rearrange to form an aromatic ring that provides the driving force for the reaction. This causes a proton to be abstracted from a sugar in the DNA backbone, thus leading to strand cleavage. DNA damage of this type is usually lethal because it is not repairable by the cell and this explains the significant *in vitro* cytotoxicity of these compounds.

Although the enediynes are exquisitely potent *in vitro*, they do not appear to be tumor-cell selective *in vivo*, and so no analogs have yet progressed to the clinic. However, a calicheamicin analog has been used as the cytotoxic "warhead" in the

STRUCTURE 3.32   Structure of calicheamicin γ_1.

antibody-drug conjugate gemtuzumab ozogamicin (Mylotarg™), as described in Chapter 7. The antibody portion of Mylotarg binds specifically to the CD33 antigen found on the surface of leukemic myeloblasts and immature normal cells of myelomonocytic lineage, and the calicheamicin then kills the cells by the mechanism described above.

FURTHER READING

Gilman, A. "The Biological Actions and Therapeutic Applications of the β-Chloroethyl Amines and Sulfides," *Science,* 103:409-436, 1946.

Gilman, A. "The Initial Clinical Trial of Nitrogen Mustard," *Am. J. Surg.,* 105:574-78, 1963.

Goodman, L.S., et al. "Nitrogen Mustard Therapy. Use of Methyl-bis(β-chloroethyl)amine Hydrochloride and Tris(β-chloroethyl)amine Hydrochloride for Hodgkin's Disease, Lymphosarcoma, Leukemia, and Certain Allied and Miscellaneous Disorders," *JAMA,* 105:475-76, 1946.

Lee, M.D., et al. "Calicheamicins: Discovery, Structure, Chemistry, and Interaction with DNA," *Accounts Chem. Res.,* 24:235-43, 1991.

Neidle, S.J., and Waring, M.J., eds. *Molecular Aspects of Anticancer Drug-DNA Interactions.* London: Macmillan Press, 1993.

Papac, R.J. "Origins of Cancer Therapy," *Yale J. Bio. Med.,* 74:391-98, 2001.

Pullman, B. "Sequence Specificity in the Binding of Anti-Tumour Anthracyclines to DNA," *AntiCancer Drug Des.,* 7:95-105, 1991.

Rosenberg, B., van Camp, L., and Krigas, T. "Inhibition of Cell Division in *Escherichia coli* by Electrolysis Products From a Platinum Electrode." *Nature (London),* 205, 698–699, 1965.

Rosenberg, B., van Camp, L., Trosko, J., and Mansour, V. "Platinum Compounds: A New Class of Potent Antitumour Agents." *Nature (London),* 222, 385–386, 1969.

Thurston, D.E. "Nucleic Acid Targeting: Therapeutic Strategies for the 21st Century," *Brit. J. Cancer,* 80 (Supp. 1): 65-85, 1999.

4 Antitubulin Agents

4.1 INTRODUCTION

Plants and trees have been an invaluable source of medicinal compounds throughout the centuries, including numerous anticancer agents such as the vinca alkaloids and the taxanes. These two families of agents work by interfering with microtubule dynamics (i.e., spindle formation or disassembly) in the cell but by different mechanisms. Microtubules are polymers of tubulin that are integral components of the mitotic spindle, and microtubule assembly (through polymerization) and disassembly (through depolymerization) are in dynamic equilibrium. Agents that disturb this equilibrium in either direction can block mitosis (division of the nucleus) and lead to cell death (see Scheme 4.1).

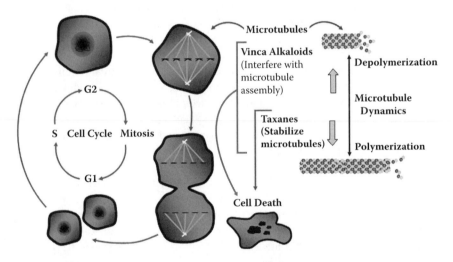

SCHEME 4.1 Microtubule dynamics as a target for cancer chemotherapy.

The prototype compound of this class is colchicine, a major alkaloid of *Colchicum autumnale* that was first extracted and isolated in pure form in 1928, although it can now be synthesized (Structure 4.1). Colchicine blocks or suppresses cell division by inhibiting mitosis. Specifically, it inhibits the development of spindles as the nuclei divide. Normally, cells would use the spindle fibers to line up their chromosomes, before making copies of them and dividing into two new cells with each daughter cell inheriting a single set of chromosomes. However, with colchicine present, it was observed that spindle fibers do not form, and so cells cannot properly relocate their chromosomes. Some or all of the chromosomes may be copied anyway,

STRUCTURE 4.1 Structure of colchicine.

but the cell cannot parcel them out to daughter cells and so division does not take place.

Because most types of cancer cells divide more rapidly than normal cells, they appear to be more susceptible to mitotic inhibitors, such as colchicine and the vinca and taxol alkaloids. However, colchicine was found to have only a narrow range of effectiveness as an anticancer agent in humans and so was not extensively developed, although it is still occasionally used in veterinary medicine to treat tumors in some animals. The main U.S. FDA-approved clinical use of colchicine today is for the treatment of gout (e.g., ColBenemid™), in which case it is thought to work by binding to tubulin in white cells (leukocytes) and inhibiting their migration into inflamed areas, thus causing a reduction in pain and inflammation. Colchicine is also used as an antimitotic agent to treat cell cultures in cancer research (see Structure 4.2).

The vinca alkaloids were discovered serendipitously in the 1960s when the pharmaceutical company Eli Lilly discovered their antitumor properties while screening plant extracts for antidiabetics drug leads. The company was particularly interested in leaf extracts of the subtropical plant *Catharanthus roseus* (the Madagascar periwinkle) because, in medical folklore, this plant was said to be useful for diabetes. However, when researchers attempted to confirm the antidiabetic characteristics of the extracts, their studies led instead to the isolation of two complex indole alkaloids, vinblastine and vincristine, which are still used today in the cancer clinic to treat a number of different tumor types. These alkaloids have a similar mechanism of action to colchicine in binding to tubulin and interfering with microtubule assembly, thus causing damage to the mitotic spindle apparatus and preventing chromosomes from traveling out to form daughter cell nuclei.

The taxanes (see Structure 4.3) were discovered by Wani and Wall through a plant screening program with the National Cancer Institute (NCI) in the 1960s. A historical marker in the form of a brass plaque fixed to a 2-ton stone was unveiled in 2002 to honor the collection of the original sample of *Taxus brevifolia* (the Pacific yew) (see Figure 4.1). The plaque is situated in La Wis Wis Campground in the Gifford Pinchot National Forest near Packwood, WA, approximately 7 miles from the location where the first sample was collected approximately 40 years ago by a group of botanists

> ### Discovery of Taxol
>
> Near this location on August 21, 1962, Arthur
> Barclay and a team of botanists from the U.S.
> Department of Agriculture collected bark of the
> Pacific Yew, *Taxus brevifolia* Nutt. Drs. Monroe Wall
> and Mansukh Wani, of the Research Triangle
> Institute, North Carolina, under contract to the U.S.
> National Cancer Institute, isolated Taxol from that
> sample. Since 1990 Taxol has been the drug of
> choice for treatment of ovarian cancer and is widely
> used in the treatment of breast cancer.
> Presented in 2002 by the USDA Forest Service,
> National Cancer Institute and the American Society
> of Pharmacognosy on the 40th anniversary of the
> collection.

FIGURE 4.1 The text of a historical marker located in the Gifford Pinchot National Forest (Packwood, WA) to commemorate the collection, 40 years ago, of the original sample of Pacific yew (*Taxus brevifolia*), which subsequently led to the discovery of paclitaxel (Taxol™).

led by Arthur Barclay (of the U.S. Department of Agriculture). Unlike the vinca alkaloids, the taxanes work by promoting the assembly of microtubules and inhibiting the tubulin disassembly process (i.e., a microtubule stabilization effect), thus interfering with cell division.

4.2 VINCA ALKALOIDS

The two alkaloids vinblastine and vincristine are minor constituents of the Madagascar periwinkle (*Vinca rosea*) (see Figure 4.2). The periwinkle is the common name for herbs of the dogbane family, which make up the genus *Vinca* of the family Apocynaceae. The lesser periwinkle, a native of many parts of Europe, grows in

FIGURE 4.2 The periwinkle (*Vinca rosea*), which produces the anticancer agents vinblastine and vincristine.

woods and thickets and is classified as *Vinca minor*. The greater periwinkle, which has much larger flowers and ovate to heart-shaped leaves, is a native of southern Europe and is classified as *Vinca major*. The periwinkle contains dozens of other alkaloids in addition to vinblastine and vincristine.

In the initial *in vivo* screens carried out by Eli Lilly, the plant extracts were shown to reduce white blood cell counts, which prompted an investigation of their anticancer properties, particularly in the leukemias. The isolated alkaloids, vinblastine and vincristine and the related semisynthetic vindesine and vinorelbine (Structure 4.2) are now used to treat certain solid tumors (mainly lung and breast), lymphomas, and acute leukemias. All of these agents are given by intravenous administration, and side effects include neurotoxicity, myelosuppression, and alopecia.

	R	R_1	R_2
Vinblastine (Velbe™)	Me	OMe	COMe
Vincristine (Oncovin™)	CHO	OMe	COMe
Vindesine (Eldesine™)	Me	NH_2	H

Vinorelbine (Navelbine™)

STRUCTURE 4.2 Structures of the vinca alkaloids.

The naturally occurring vinca alkaloids have complex but similar chemical structures. Vincristine is more widely used than vinblastine, but the plant produces the latter in approximately 100-fold greater amounts. Fortunately, vinblastine can be converted to vincristine by a simple chemical step involving oxidation of a methyl to a formyl group. Because these alkaloids have proved so useful in therapy, efforts have been made to design new analogs with reduced toxicity (mainly attempting to reduce neurotoxicity), which has resulted in the semisynthetic analogs vindesine and vinorelbine. Research has also been carried out into cell culture techniques and the use of immobilized plant enzymes as a means to produce the alkaloids more efficiently. Compounds representing the two halves of the dimeric vinca molecules (the A and B subunits; Structure 4.2) occur in much higher proportions in the plant extract, and linking these at the appropriate positions and with the correct stereochemistry have now become feasible.

The vinca alkaloids are classed as *cell-cycle-specific* because they block mitosis by causing cell-cycle arrest at metaphase. Their cytotoxic effects result from binding of the alkaloids to the microtubules. Microtubules are long tubular structures of about 25 nm in diameter that form the major component of the mitotic spindle apparatus responsible for the movement of chromosomes during cell division. These structures were first characterized in the cytoplasm over 25 years ago and comprise two main proteins, the α and β tubulins (M_r c 55 000), which form the microtubule scaffolding upon which many of the dynamic internal processes in living cells, including cell division, depend. Binding of the vinca alkaloids to the tubulins interferes with microtubule assembly, causing damage to the mitotic spindle apparatus and preventing chromosomes from traveling out to form daughter cell nuclei. This is similar to the action of colchicine but is different from the action of paclitaxel, which interferes with cell division by preventing the spindles from being broken down. There is some evidence that the vinca alkaloids may also block DNA and RNA synthesis with a degree of selectivity towards tumor cells. Overall, the basis for the tumor cell selectivity of these agents is not fully understood although it is assumed that, because of their mechanism of action, rapidly dividing cancer cells are more vulnerable than the nondividing cells of healthy tissues.

Myelosuppression and neurological toxicities are the major adverse effects associated with the vinca anticancer agents. Although all members of the family cause some neurological side effects, which normally manifest initially as autonomic or peripheral neuropathy, vincristine is the most affected and neurotoxicity is the dose-limiting toxicity for this agent. Conversely, vincristine causes the least myelosuppression of the four alkaloids, which is interesting from a structure activity relationship (SAR) standpoint given the similarity of chemical structures. The neurological effects may include loss of deep tendon reflexes, peripheral paresthesia, abdominal pain, constipation, and motor weakness, with the latter being taken as an indication that treatment should be discontinued. In general, most patients make a slow but complete recovery from these nervous system problems. The neurological toxicities are much less pronounced in the case of vinblastine, vindesine, and vinorelbine, which all have myelosuppression as their dose-limiting toxicity. Reversible alopecia commonly results from treatment with all members of the vinca family. Finally, it is important that administration techniques are carefully managed; all of the compounds are capable of causing severe local irritation at the injection site, so extravasation must be avoided.

4.2.1 VINBLASTINE

The isolation of vinblastine (Velbe™) from *Vinca rosea* and its structural identification were reported in 1959 and 1960, respectively, and a synthesis starting from the catharanthine and vindoline units was reported in 1979 (see Structure 4.2). As a sulfate salt, the agent is included on a weekly basis in several drug regimens for treating lymphocytic and histiocytic lymphomas, Hodgkin's disease, advanced disseminated breast carcinoma, Letterer-Siwe disease, Kaposi's sarcoma, choriocarcinoma, and advanced testicular carcinoma.

In addition to the acknowledged antitubulin activity of vinblastine, some research suggests that the agent is also able to inhibit the metabolism of glutamic acid, in particular the pathways leading to the Krebs cycle and then to urea formation. It is not known whether this contributes to the anticancer properties of vinblastine.

4.2.2 VINCRISTINE

The isolation and structural elucidation of vincristine (Oncovin™) were reported in 1961 and 1964, respectively (see Structure 4.2). Despite its similarity in structure to vinblastine, vincristine sulfate has a different spectrum of both antitumor activity and side effects. Notably, it is more neurotoxic than vinblastine, although it causes significantly less myelosuppression. Several drug combinations include vincristine for the treatment of acute lymphoblastic and myeloblastic leukemias, Hodgkin's disease, Wilms' tumor, rhabdomyosarcoma, neuroblastoma, retinoblastoma, soft tissue sarcomas, and disseminated cancer of the breast, testes, ovaries, and cervix. The relatively low bone marrow toxicity renders it suitable for combination with drugs that cause more significant bone marrow suppression.

4.2.3 VINDESINE

Vindesine sulfate (Eldisine™) is a semisynthetic product derived from vinblastine, and so its toxicity and side effects are similar to those of the parent compound (see Structure 4.2). It was first reported in 1974 and is mainly used to treat melanoma and lung cancers and, in combination with other drugs, to treat uterine cancers.

4.2.4 VINORELBINE

Vinorelbine (Navelbine™) was discovered serendipitously when it was observed that, when vinblastine is placed in strongly acid conditions, dehydration of the tertiary alcohol group in the piperidine ring of the A subunit occurs to introduce a double bond (see Structure 4.2). In the U.K., vinorelbine is presently recommended as one of the options for second-line or subsequent treatment of advanced breast cancer in cases where anthracycline-based therapies have failed or are unsuitable. However, it has also been used in Phase II clinical trials as a treatment for ovarian cancer, and has shown some activity in combination with cisplatin in non-small-cell lung tumors. Side effects include diarrhea, nausea, and hair loss, although vinorelbine appears to be significantly less neurotoxic than vindesine.

4.3 THE TAXANES

Paclitaxel (Taxol™) is a highly complex tetracyclic diterpene found in the needles and bark of *Taxus brevifolia*, the Pacific yew tree (see Structure 4.3). The cytotoxic nature of extracts of *Taxus brevifolia* was first demonstrated in 1964 through a screening program coordinated by the NCI. Pure paclitaxel was isolated in 1966 and its structure published in 1971. However, it did not appear in clinical practice until the 1990s, over 30 years after its discovery. Docetaxel (Taxotere™) is a more recently introduced semisynthetic analog with similar therapeutic and toxicological properties.

FIGURE 4.3 Paclitaxel is a highly complex tetracyclic diterpene found in the bark and needles of the Pacific (*Taxus brevifolia*) and English yew tree (shown).

The mechanism of action of these agents involves an effect on the microtubules but in a completely different way to the vinca alkaloids. In a cell, an equilibrium exists between the microtubules and tubulin dimers, and the assembly and disassembly of dimers is governed by cell requirements. Paclitaxel is known to promote microtubule assembly by stabilizing the microtubule complex and inhibiting their depolymerization to free tubulin, thus shifting the equilibrium in favor of the polymeric form of tubulin and reducing the critical concentration of the nonpolymerized form. This interferes with the ability of the chromosomes to separate during cell division. There are also some reports of paclitaxel acting as an immunomodulator, activating macrophages to produce interleukin-1 and tumor necrosis factor.

The barks from a large number of yew trees would be required to provide a single course of treatment of paclitaxel for an ovarian cancer patient, and the problems associated with producing sufficient quantities of the drug account, in part, for its delay in being introduced to clinical practice. Fortunately, semisynthesis is now possible by extracting baccatin (paclitaxel without the ester fragment; see Structure 4.3) in large amounts from the leaves (a renewable resource) of a related species, *Taxus baccata* (see Figure 4.4). The ester side chain can be made synthetically and then joined to baccatin to provide paclitaxel (see Structure 4.3). Similarly, docetaxel can be prepared from the natural precursor 10-deacetylbaccatin III (i.e., R_1 = H instead of $COCH_3$) extracted from the needles of *Taxus baccata* or *Taxus brevifolia* (the European or Pacific yew trees, respectively), through attachment of the relevant synthesized ester fragment [i.e., R = $CO.O.C(CH_3)_3$]. Although both

Synthetic Ester Fragment		Baccatin	

	R	R_1
Paclitaxel (Taxol™)	CO.Ph	CO.CH$_3$
Docetaxel (Taxotere™)	CO.O.C(CH$_3$)$_3$	H

STRUCTURE 4.3 Structures of paclitaxel (Taxol™) and docetaxel (Taxotere™).

paclitaxel and docetaxel are produced commercially by semisynthesis, the first total synthesis of paclitaxel was reported in early 1994.

Research is still ongoing into the development of new analogs of paclitaxel. One reason for this is that paclitaxel has relatively poor water solubility. A second reason is that the taxanes lack activity in some cancers and are susceptible to the development of resistance mediated by p-glycoprotein expression and by mutations in genes that encode tubulin.

FIGURE 4.4 Leaves harvested from *Taxus baccata* ready for the extraction of baccatin.

4.3.1 PACLITAXEL

Given by intravenous infusion, there is increasing evidence that paclitaxel (Taxol™) in combination with cisplatin or carboplatin is the treatment of choice for ovarian cancer. This combination is also used for women whose ovarian cancer is metastatic and considered inoperable, and in cases in which standard platinum therapies have failed. It is also used for non-small-cell lung cancer for which no further options (including surgery or radiotherapy) remain, and for metastatic breast cancer in which the usual anthracycline therapy has failed. Routine premedication with a corticosteroid, an antihistamine, and a histamine 2(H2)-receptor antagonist is recommended to prevent severe hypersensitivity reactions. More commonly, only bradycardia or asymptomatic hypotension occur. Other possible side effects include myelosuppression, peripheral neuropathy, and cardiac conduction defects with arrhythmias. Paclitaxel can also cause alopecia and muscle pain as well as mild to moderate nausea and vomiting.

Recently, scientists from the Institute for Bioprocessing and Analytical Measurement Techniques in Germany have discovered a more cost-effective procedure for producing paclitaxel. After isolating 10-deacetylbaccatin III from the leaves of the Pacific yew tree (*Taxus brevifolia*), they use a genetically modified strain of *Escherichia coli* to convert it to baccatin II, a precursor of paclitaxel.

4.3.2 DOCETAXEL

Docetaxel (Taxotere™) is licensed for initial treatment of advanced breast cancer in combination with doxorubicin, or alone where adjuvant cytotoxic chemotherapy has failed. It is also used for advanced or metastatic non-small-cell lung cancer in cases where first-line therapy has failed. Its side effects are similar to paclitaxel but persistent fluid retention (generally leg edema) can be resistant to treatment. Hypersensitivity reactions also occur, and prophylactic dexamethasone is recommended to reduce both fluid retention and the possibility of hypersensitivity reactions.

FURTHER READING

Goodman, J., and Walsh, V. *The Story of Taxol: Nature and Politics in the Pursuit of an Anti-Cancer Drug*. Cambridge, U.K.: Cambridge University Press, 2001.

Noble, R.L. "The Discovery of the Vinca Alkaloids — Chemotherapeutic Agents against Cancer," *Biochem. Cell Biology*, 68(12):1344-51, 1990.

Suffness, M. "Taxol: From Discovery to Therapeutic Use," *Ann. Rep. Medic. Chem.*, 27:305–14, 1993.

Suffness, M. "Taxol: Science and Applications," in *Pharmacology and Toxicology: Basic and Clinical Aspects*. Edited by Hollinger, M.A. Boca Raton, FL: CRC Press, 1995.

5 Molecularly Targeted Agents

5.1 INTRODUCTION

Conventional anticancer drugs, such as deoxyribonucleic acid (DNA)–alkylating and cross-linking agents, antimetabolites, topoisomerase inhibitors, and antitubulin agents, have been traditionally focused on targeting DNA processing and cell division. Although these drugs can be very efficacious, their lack of selectivity for tumor cells versus normal cells usually leads to serious side effects, such as bone marrow suppression and gastrointestinal (GI), cardiac, hepatic, and renal toxicities, which limit their use. In an attempt to avoid these unpleasant side effects, a new class of anticancer drugs known as *molecularly targeted agents* is being developed that works by targeting a biochemical pathway or protein that is unique to or up-regulated in cancer cells. Such agents are typically less toxic than drugs in the older classes and can be given orally with the objective of treating cancer as a chronic disease in the same way that conditions such as diabetes and heart disease are treated with long-term oral therapies.

Much of the pioneering work in this area has been carried out by targeting signaling pathways. Cells use a wide variety of intracellular and intercellular mechanisms to signal for processes, including growth, apoptosis, and intracellular protein degradation. More recently, molecular and genetic approaches to understanding cell biology have uncovered entirely new signaling networks that regulate cellular activities such as proliferation and survival. Many of these networks are now known to be radically altered in cancer cells, and these alterations have a genetic basis caused by somatic mutations. Due to up-regulation and a greater dependence on some of these pathways in tumor cells (sometimes referred to as *oncogene addiction* when a specific gene is involved), inhibition can lead to an anticancer effect. For this reason, drugs of this class are sometimes known as *signal transduction inhibitors* or *secondary messenger inhibitors.*

Research in the protein kinase area has recently led to the development of imatinib (Gleevec™), which is widely regarded as the first molecularly targeted anticancer drug and a landmark advance for patients with chronic myelogenous leukemia (CML). Taken orally, imatinib has few side effects and can keep patients in remission for long periods. A second kinase inhibitor, trastuzumab (Herceptin™), has also been successfully commercialized. It is a potent inhibitor of the human epidermal growth factor receptor 2 (HER2)/neu signaling pathway and provides significant benefits to breast cancer patients. A further example in this area is the development of the epidermal growth factor receptor (EGFR) inhibitor erlotinib

(Tarceva™), which is benefiting some cancer patients with non-small-cell lung cancer (NSCLC). A number of other kinase inhibitors potentially useful in treating different tumor types are presently in development, and this area is now a highly competitive one for a large number of pharmaceutical companies. Another research area described in this chapter is the development of farnesyltransferase (FTase) inhibitors (FTIs). It is thought that up to 30% of tumors develop from a mutated *Ras* oncogene, and this has pointed the way to FTIs designed to affect the pathway associated with *Ras*. Although not yet commercially available, inhibitors of D-type cyclins and their corresponding kinase counterparts such as CDKs 4 and 6 are being developed and are also described below. The ubiquitin-proteasome pathway is the main route for the degradation of proteins in cells, and bortezomib (Velcade™) has been developed to interfere with this pathway in tumor cells. Finally, mammalian target of rapamycin (mTOR) is a cellular enzyme that plays a key role in cell growth and proliferation as part of the mTOR signaling pathway and the development of inhibitors of this pathway is also described. It is worth noting that *molecular targeting* is one of the fastest growing areas of research in cancer chemotherapy, and there is currently a significant research effort to identify new pathways to target.

5.2   KINASE INHIBITORS

These agents act by inhibiting the actions of protein kinases that modulate the signaling systems necessary for cell division, growth, survival, and migration. Three such agents, imatinib (Gleevec™), trastuzumab (Herceptin™), and erlotinib (Tarceva™) have led the field in this area, and many new agents of this type are poised for commercialization.

5.2.1   Classification of Protein Kinases

Protein kinases are enzymes within cells that act by attaching phosphate groups to the amino acid residues of various proteins. This process of phosphorylation serves two primary roles: as a molecular on-off switch to trigger a cascade of cellular events and as a "connector" that binds proteins to each other. Therefore, protein kinases play a primary role in the complex signaling systems that transfer information between and within cells. They are generally classified into two groups:

A. Based on their specificity for target amino acids:
   - *Serine-* or *threonine-specific kinases* — Catalyze the phosphorylation of groups on serine and threonine residues
   - *Tyrosine-specific kinases* — Catalyze the phosphorylation of tyrosine residues
   - *Mixed function kinases* — Catalyze serine, threonine, or tyrosine phosphorylation
B. Based on their structure and cellular localization:
   - *Receptor kinases* — Defined by a hydrophobic transmembrane domain that passes through the plasma membrane, an extracellular ligand-binding domain, and a cytoplasmically located kinase domain. The extracellular

ligand-binding domain is typically glycosylated and conveys ligand specificity. The three main families are erbB (HER), platelet-derived growth factor receptor (PDGFR), and vascular endothelial growth factor receptor (VEGFR). Other well-known receptors include IGF-1R, KIT, and mutant FLt-3 receptors. (Scheme 5.1 and Scheme 5.2)

– *Nonreceptor kinases* — Unlike the receptor kinases, these agents have no transmembrane or extracellular domains and may be associated with the cytoplasmic surfaces by membrane localization via a lipid modification that anchors them to the phospholipid bilayer or by noncovalent binding to a membrane receptor. Examples include the ABL, JAK, FAK, and SRC kinases.

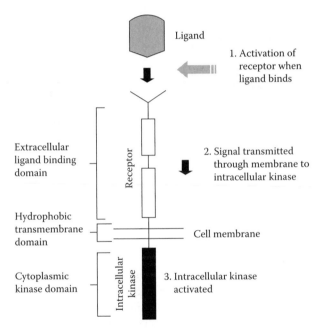

SCHEME 5.1 Structure of a receptor tyrosine kinase showing how it passes through the cellular membrane to allow a signal to be transmitted to its intracellular domain when a ligand binds to its extracellular domain.

5.2.2 FUNCTIONS OF PROTEIN KINASES

The function of protein kinases is to provide a mechanism for transmitting information from a factor outside a cell to the interior of the cell without the initiating factor having to cross the cell membrane. For example, growth factors and polypeptide hormones in the extracellular milieu can exert a regulatory effect on cells by activating specific gene transcription in the nucleus of target cells without passing through their cytoplasmic membranes. By this mechanism, protein kinases help regulate cellular functions such as proliferation, cell-matrix adhesion, cell-cell adhesion, movement, apoptosis control, transcription, and membrane transport (see Scheme 5.1 and Scheme 5.2).

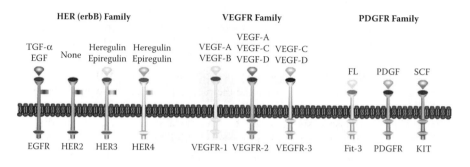

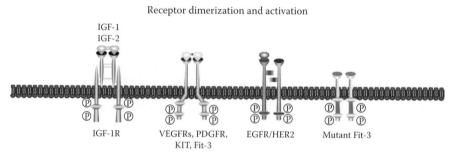

SCHEME 5.2 Receptor tyrosine kinase families and their activation.

5.2.3 Mechanism of Signal Transfer

The transfer of a signal from an extracellular factor through the membrane via a kinase usually occurs by one of two mechanisms. In the first, the receptor tyrosine kinase (RTK) is solely responsible for the transfer of signals across the membrane (Figure 5.3). Ligand binding to the RTK induces receptor dimerization or oligomerization, leading to interactions between adjacent cytoplasmic domains with consequent activation of the kinase. Activation of a nonreceptor kinase is similarly induced in response to the appropriate extracellular signal; however, dimerization may or may not be necessary for activation. In either case, activation of the cytoplasmic kinase domain is the key step in transferring the signal across the membrane. The activated kinase then initiates a cascade of phosphorylation reactions resulting in the activation of other proteins, as well as the production of secondary messenger molecules that transmit the signal initiated by the extracellular factor to the nucleus.

5.2.4 Regulation of Kinase Activity by Drugs

In order to control the effects of kinase activation on cellular processes using drugs, a number of mechanisms can be envisaged. First, use could be made of the opposing effects of different protein kinases. For example, activation rather than inhibition of some kinases can lead to cellular growth inhibition or cell death (e.g., activation of transforming growth factor-β [TGF-β] receptor signal transmission can modulate

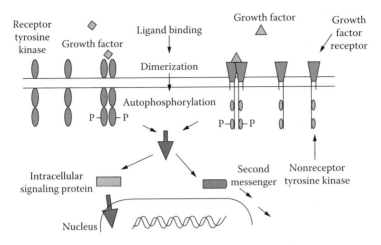

SCHEME 5.3 Mechanism of RTK signaling through dimerization after ligand binding.

cell cycle arrest). Second, activation of protein phosphatases that remove phosphates from kinase residues could switch off kinase signaling. Third, reducing or blocking the activity of enzymes involved in the formation of protein ligands or activating specific protease enzymes that degrade the ligands should modify signaling. Finally, the availability of protein kinases and their relative localization within cells could be manipulated.

5.2.5 Role of Protein Kinases in Cancer

Protein kinases can play major etiologic roles in the initiation of malignancy and may contribute to the uncontrolled proliferation of cancer cells, tumor progression, and the development of metastatic disease. It is thought that cancer cells depend almost entirely on signaling by protein kinases for their continued proliferation, whereas normal cells rarely invoke these pathways. Protein kinases can be altered in two main ways in tumor cells as described below.

5.2.5.1 Mutations in Protein Kinases

A number of examples in which specific kinases have become mutated in cancer cells are known. For example, a mutation of the protein kinase ABL (i.e., BCR-ABL) is the etiologic agent in CML. The cytoplasmic tyrosine kinase BCR-ABL, which is constitutively active, is present in 15% to 30% of cases of adult acute lymphoblastic leukemia (ALL) and virtually all cases of CML. This mutation has been put to good use in the design of imatinib (Gleevec™). A second example can be found in patients with multiple endocrine neoplasia (type 2), in which mutations in RET tyrosine kinase may be responsible for development of the disease. Finally, EGFR mutations with enhanced kinase activity have been detected in several human tumor types.

5.2.5.2 Overexpression of Protein Kinases

Expression of EGFR and its associated primary ligands epidermal growth factor (EGF) and transforming growth factor α (TGF-α) has been studied in several human malignancies with coexpression of EGFR and EGF observed to have both prognostic significance and a possible role in the pathogenesis of several human cancers. Specifically, overexpression of EGFR and EGF in several tumor types significantly reduces patient prognosis. For example, members of the EGFR kinase family (EGFR, ErbB-2, HER2/neu, ErbB-3, and ErbB-4) are known to be overexpressed in some types of breast tumors. The HER2/neu RTK has been found to be amplified up to 100 times in the tumor cells of approximately 30% of cancer patients with invasive breast disease, and its presence is also associated with poor prognosis. Similarly, overexpression of PDGF and PDGFR has been reported in meningioma, melanoma, and neuroendocrine cancers as well as tumors of the ovary, pancreas, stomach, lung, and prostate. Elevated levels of SRC kinase activity have also been noted in colon cancer specimens, implying that overexpression may also be important in this disease.

5.2.6 DEVELOPMENT OF INHIBITORS OF PROTEIN AND RECEPTOR KINASES

Presently there is significant research activity in the design of small molecule inhibitors of the various kinases thought to be important in the initiation and growth of tumors. Because protein kinases use adenosine triphosphate (ATP) as a source of phosphate, one of the main approaches to date has been to design inhibitors that interact with the ATP-binding site of the protein. This makes it impossible for the kinases to phosphorylate proteins, thus blocking the signaling process. The development of imatinib (Gleevec™) is the best known example of this approach. For receptor kinases, one of the main approaches has been the development of antibodies that bind to the extracellular portion of the kinases, thus blocking their function. The best known example of this approach is the development of trastuzumab (Herceptin™). These and many other kinase inhibitors either in use or in development are described below.

5.2.6.1 BCR-ABL Inhibition: Imatinib

Imatinib (Gleevec™), also known as STI-571, is widely acknowledged to be the classic example of kinase-targeted drug development and the drug that validated the strategy of signal transduction inhibition (see Structure 5.1). Imatinib is targeted against the signaling kinase molecule, BCR-ABL, a result of a genetic abnormality known for a long time to be a chromosomal translocation that creates an abnormal fusion protein (i.e., the kinase BCR-ABL) that signals aberrantly and leads to uncontrolled proliferation of the leukemia cells in CML (see Scheme 5.4).

It is noteworthy that the discovery and development of imatinib represents one of the rare examples of rational drug design in cancer drug discovery. Brian Druker, working at Oregon Health Science University, had extensively researched the abnormal kinase enzyme in CML and had reasoned that inhibiting this kinase with a drug

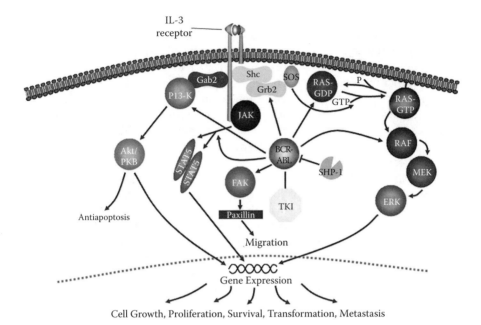

SCHEME 5.4 The key role of BCR-ABL in leukemic cells (ALL and CML).

STRUCTURE 5.1 Structure of imatinib mesylate (Gleevec™).

should control the disease while having little effect on normal cells. Druker collaborated with a chemist at Novartis, Nick Lydon, who developed several candidate inhibitors that could block the ATP-binding site of BCR-ABL and similar kinases. From a relatively small library of compounds, imatinib was found to have the most promise in laboratory experiments and was moved into development.

Druker's group, and then others worldwide, demonstrated that when imatinib is used to treat patients with chronic-phase CML, a remarkable (approximately) 90%

achieve complete hematological remission. Clinical trials were initially carried out with patients in the chronic phase of Philadelphia-chromosome-positive CML refractory to interferon alfa. All patients in these trials treated with 140 mg/day or more of imatinib had a hematological response, but 98% of patients treated at a dose of 300 mg/day or more had a complete hematological response, which was maintained in 96% of these patients for a median follow-up of 265 days. Interestingly, a total cytogenetic response was observed in 54% of patients treated at doses of 300 mg/day or more, with 13% having complete cytogenetic remissions. These results are still considered remarkable and a landmark advance in drug development and, in particular, the treatment of CML.

Two other advantages of imatinib are oral administration, which is convenient for patients, and the absence of some of the more serious side effects commonly observed with cytotoxic agents. Imatinib is now licensed in the U.K. for treating the chronic phase of Philadelphia-chromosome-positive CML after interferon alfa has failed and for treating disease that is in the accelerated or blast crisis phases. Through its use in the clinic, it was discovered that, in addition to inhibiting BCR-ABL, imatinib is also an effective inhibitor of ABL, PDGFR, and c-Kit kinases. In particular, the activity against c-Kit manifested as responses remarkable clinical activity in c-Kit-positive unresectable or metastatic malignant GI stromal tumors (GISTs), for which it is now also licensed.

Although imatinib is reasonably free from serious side effects, the most common adverse events are nausea, vomiting, diarrhea, edema, muscle pain (myalgia), and headache, all of which are mild to moderate in severity. Anemia, thrombocytopenia, neutropenia, and elevation of liver enzymes have also been reported in some patients but do not show any correlation with dose.

5.2.6.2 HER2/neu Inhibition: Trastuzumab

Trastuzumab (Herceptin™) is a recombinant, humanized anti-P185 monoclonal antibody targeted against HER2/neu receptors. It is discussed in this chapter rather than in Chapter 8 (Biological Agents) due to its receptor kinase target. Trastuzumab is a potent inhibitor of the HER2/neu signaling pathway and was developed for use in breast cancer. It is thought to induce internalization of the receptor and to inhibit progression through the cell cycle by mechanisms including up-regulation of p27, an intracellular inhibitor of cyclin-dependent kinase 2 (CDK2).

Several clinical trials have shown that trastuzumab provides a significant antitumor response in HER2-positive patients. For example, one trial has shown that standard chemotherapy plus trastuzumab compared to standard chemotherapy alone can provide a longer median time to disease progression (7.4 versus 4.6 months), a significantly greater overall tumor response (50% versus 32%), a longer median duration of response (9.1 versus 6.1 months), a longer median time to treatment failure (6.9 versus 4.5 months), and a significantly lower death rate after 1 year (22% versus 33%).

Trastuzumab is presently licensed in the U.K. for use with paclitaxel in metastatic breast cancer patients with tumors overexpressing HER2 who have not received drug treatment for metastatic disease and in whom anthracyclines would not be appropriate.

It is also licensed as a monotherapy in the case of metastatic breast cancer patients with tumors overexpressing HER2 who have previously received at least two drug treatments including, where appropriate, an anthracycline and a taxane. Women with estrogen-receptor-positive breast cancer should also have received hormonal therapy before being treated with trastuzumab. However, in 2005, striking clinical results reported in the *New England Journal of Medicine* showed that trastuzumab had eliminated the peak in reappearance of breast cancer in the first 2 to 3 years after surgery. An accompanying editorial suggested that this represented the best evidence of a treatment effect ever seen in oncology, perhaps permanently changing the natural history of the disease. This caused excitement in the media in many countries, with the drug being hailed as a possible cure for breast cancer and, on this basis, one National Health Service (NHS) health authority in the U.K. became the first to state that it would make the drug available to all suitable breast cancer patients even before it was cleared by the regulatory authorities. This has caused a significant political problem for the U.K. government within the context of a fixed NHS budget, as the cost for 1 year's treatment was approximately £21,800 per patient at the time. At present, trastuzumab is still only licensed for advanced breast cancer in the EU, although clinicians can prescribe it for patients with early stage disease at their discretion. However, after an expedited review by the regulatory authorities, the manufacturer (Roche Holding AG) expects to receive European Union approval for use of the drug in early stage HER-positive disease within the second half of 2006.

Despite the proven efficacy of trastuzumab, only a proportion of women with breast cancer benefit from its use because it acts against a protein found in the tumor cells of only 20% to 30% of patients with particularly aggressive breast tumors. Therefore, it is noteworthy that trastuzumab is the first example of an anticancer drug to be used in a so-called "personalized medicine" regimen. A HER2 test kit has been developed and women must be screened for HER2 overexpression before trastuzumab can be administered. It is likely that many more new agents of this type will be codeveloped with a test kit as the move toward personalized medicine regimens gathers momentum (see Chapter 10).

The most common side effects associated with trastuzumab, which is given by intravenous infusion, include fever, chills, pain, asthenia, nausea, vomiting, headache, and possible hypersensitivity reactions, such as anaphylaxis, urticaria, and angioedema. Cardiotoxicity, pulmonary events, and GI disturbances can also occur. Concomitant use of trastuzumab with anthracyclines is associated with cardiotoxicity and should be avoided. Cardiac function should be carefully monitored if coadministration is necessary.

5.2.6.3 EGFR Inhibition: Gefitinib, Erlotinib, and Cetuximab

EGFR is one of a subfamily of closely related receptors that includes itself (ErbB-1), HER2/neu (ErbB-2), HER3 (ErbB-3), and HER4 (ErbB-4). Up-regulated EGFR-mediated signaling can help move cells into a continuous and uncontrolled state of division, thus leading to greater numbers of malignant cells and an increase in tumor size (see Scheme 5.5). Overexpression of the EGFR receptor is common in many types of solid tumors, such as head and neck, lung, and colorectal tumors. Furthermore,

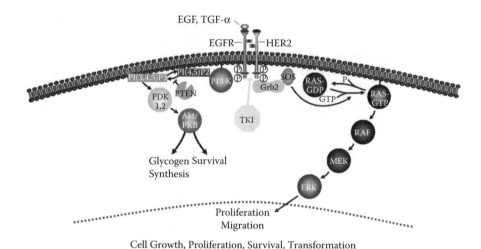

SCHEME 5.5   Growth factor signaling through EGFR/HER2.

EGFR overexpression has been shown to correlate with a poor prognosis, in partic-ular, with an increased risk of metastasis and decreased survival. It can also protect cancer cells from cytotoxic agents and radiotherapy, thus reducing the effectiveness of these treatments. Therefore, it was reasoned that inhibitors of EGFR may have a selective effect on tumor cells compared to healthy ones.

Activation of the EGFR pathway is initiated when an appropriate ligand (e.g., TGF-α or EGF in the case of EGFR) binds to the inactive single units of the receptor. This causes the receptors to pair together to produce a dimer that may be formed either from two identical receptors (e.g., two EGFR-1 receptors pairing to form a *homodimer*) or from nonidentical receptors (e.g., EGFR and HER2/neu receptors pairing to give an *asymmetrical heterodimer*) (e.g., see Scheme 5.5). This pairing process activates the tyrosine kinase enzyme located in the intracellular domain, which then leads to the transphosphorylation of both intracellular domains. This, in turn, initiates a cascade of phosphorylation events that eventually results in a signal arriving at the nucleus.

Gefitinib (Iressa™)   Erlotinib (Tarceva™)

STRUCTURE 5.2   Structures of gefitinib (Iressa™) and erlotinib (Tarceva™).

5.2.6.3.1 Gefitinib

Gefitinib (Iressa™), an 4,6,7-trisubstituted-4-quinazolinamine synthesized and patented by AstraZeneca Pharmaceuticals in the mid- to late-1990s, was the first of a new family of agents that inhibit EGFR by binding to and blocking the ATP binding site (see Structure 5.2). By preventing the activation of EGFR, gefitinib shuts down signaling via this protein, thus leading to cell death. Taken orally, it was initially evaluated (in Phase II) for use in advanced non-small cell lung cancer (NSCLC) for which there is presently no standard treatment once first-line therapy has failed. This disease accounts for approximately 80% of all lung tumors. The overall 5-year survival rate is 10% or less for this highly aggressive disease, and so new forms of treatment are urgently being sought.

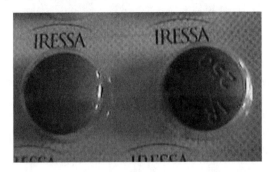

FIGURE 5.1 Gefitinib (Iressa™) tablets, 250 mg.

The results of these early clinical trials appeared promising, and the FDA approved gefitinib in mid-2003 as a third-line treatment for cancer patients who had failed on previous chemotherapy regimens. The critical data for approval came from 142 NSCLC patients in a Phase II clinical trial who had not responded to traditional chemotherapy but had showed an Iressa response rate of approximately 13%, with 10% of participants having a substantial tumor shrinkage response. However, in later Phase III randomized trials (the Iressa Survival Evaluation in Lung Cancer trial) reported in late 2004, the drug appeared to give no survival improvement when added to conventional chemotherapy. This study, which involved 1,692 patients, investigated the survival benefit of 250 mg gefitinib daily (Figure 5.1) as a monotherapy in patients with NSCLC who had not responded to chemotherapy. No increase in survival in the overall population was seen for gefitinib compared with placebo (median 5.6 months versus 5.1 months. This forced the manufacturer into a label revision so that they could keep the drug available to users, of which there were approximately 4,000 in the U.S. alone. The revised label indicated that gefitinib is only to be used in patients who have previously taken Iressa and are benefiting or have benefited from the drug. In addition, AstraZeneca recently announced that it is withdrawing its European marketing authorization application for gefitinib for the treatment of NSCLC.

In 2004, researchers from the Dana-Farber Cancer Institute in the U.S. provided a possible explanation for the discrepancy between the results of the Phase II and Phase III clinical trials. They reported a strong correlation between the responses of

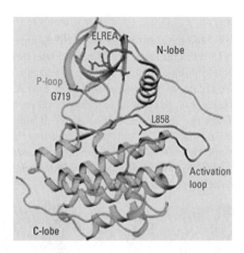

STRUCTURE 5.3   Mutations in various positions of EGFR that have been associated with clinical responses to gefitinib. The L858R mis-sense mutation was also detected in a cell line that was particularly sensitive to the drug. (Adapted from Paez, J.G. et al. *Science,* 304:1497-1500, 2004).

patients and gene mutations discovered in the EGFR kinase domain (Structure 5.3). All of the patients with responses were shown to have mutations, whereas those whose disease had progressed did not. The study also identified a cell line possessing one of the EGFR gene mutations found in patients that showed a dramatic *in vitro* sensitivity toward gefitinib compared with cell lines lacking the mutation. This led to concerns over the structure of the clinical trials carried out to date. For example, experience with related agents underscores the value of stratifying patients in clinical trials. Herceptin™ and Gleevec™ are rationally designed anticancer drugs but, unlike gefitinib, the best potential responders were identified before the trial started through demonstrating expression of the appropriate protein or genetic markers, respectively. Without that benefit, it is possible that the gefitinib trials carried out to date have been underpowered to reveal a significant Phase III outcome.

The ongoing developments with gefitinib exemplify what many health professionals believe is the future of medicine, a new approach to treatment based on *pharmacogenomics*, or *personalized medicine* (see Chapter 10). By understanding how genes influence a drug's pharmacological properties, doctors will be able to prescribe specialized drugs appropriate for an individual's genetic makeup. Meanwhile, several other large pharmaceutical companies are devising the next generation of gefitinib-like compounds for use in cancer that will rely on pharmacogenomic approaches to identify responders, thus optimizing the outcome and maximize the efficiency of future trials.

There are also concerns over the safety of gefitinib as it is known that almost all cells except blood cells have EGFR, and so gefitinib and similar agents may inhibit the replacement of normal cells as well as slowing the growth of cancer cells. In this respect, it is noteworthy that more EGF is required in the cells of tissues that are in the process of repair following injury, and these cells also express more EGFR than normal ones. Interestingly, it has been shown that strains of mice deficient in or lacking EGFR have impaired development of the epithelium in a number of organs, including those in the GI tract, the lungs, and the skin. Also, it can be shown that the healing of a wounded rat cornea can be delayed in a dose-dependent manner after treatment with gefitinib. This suggests that EGFR inhibition may affect the stratification and proliferation of epithelial cells during healing of the cornea, a process which is known to be important for the maintenance of normal corneal epithelial depth.

Finally, only a few months after gefitinib gained approval in Japan (in mid-2002, following an unusually rapid 5-month review) for use in NSCLC, reports began to surface of a number of unusual deaths, mostly involving interstitial pneumonia. To date, more than 25,000 Japanese patients have received the drug and more than 200 have died from pneumonia-like diseases. While investigations into the cause of the fatal lung disease continue, a Japanese health ministry panel has declined to endorse gefitinib. In spite of numerous studies that have deemed the drug effective, the panel concluded that there were insufficient data to recommend it.

5.2.6.3.2 Erlotinib (Tarceva™)

Erlotinib (also known as Tarceva or OSI-774) has a similar molecular structure to gefitinib (i.e., an 4,6,7-trisubstituted-4-quinazolinamine), and was synthesized and patented by Pfizer in the mid- to late-1990s as an oral tablet for daily administration (see Structure 5.2). The drug is now being developed by Roche in partnership with OSI and Genentech. In mid-2002, the U.S. FDA awarded "fast track" status for the treatment of advanced-stage NSCLC patients who had not previously received chemotherapy. This was extended at a later time for second- or third-line single agent use in patients with NSCLC who no longer responded to standard drug regimens.

Erlotinib received formal FDA approval for the treatment of NSCLC in late 2004 and gained the distinction of being the first EGFR inhibitor to provide evidence of a survival benefit in lung cancer patients. It has now been approved for treatment of NSCLC in Canada, Switzerland, and the European Community (EC), where it has been licensed for locally advanced or metastatic NSCLC after a patient has failed on at least one prior chemotherapy treatment. Erlotinib is also the first EGFR inhibitor to receive orphan drug classification for brain cancer treatment.

It is noteworthy that obvious survival advantage associated with erlotinib (as a single agent) was only observed in Phase II studies in chemorefractory NSCLC patients who had been prescreened for EGFR-positive status. Out of 57 patients, approximately 50% achieved stabilization of their disease and 40% met the 12-month survival criterion. The survival benefit has since been confirmed in a double-blind randomized Phase III clinical trial in which erlotinib was compared with a placebo in approximately 730 EGFR-positive patients whose NSCLC had failed to respond to previous drug treatments. The results were highly encouraging, with a 42% improvement in median survival for patients taking the drug (with 1-year survival being enhanced by 45%). Furthermore, improvements occurred in all secondary endpoints (i.e., progression-free survival, and time to symptom deterioration). Interestingly, the earlier so-called

TALENT and TRIBUTE Phase III studies using erlotinib as first-line treatment in combination with cytotoxic drugs had not provided any evidence of survival advantage.

In late 2004, promising data were also obtained from a 450-patient Phase III clinical trial involving pancreatic cancer, a difficult-to-treat disease with a very poor prognosis. In combination with gemcitabine, erlotinib gave better survival figures than with gemcitabine alone (the primary endpoint), thus demonstrating that its activity extends beyond NSCLC. Objective evidence of antitumor activity has also been observed in ovarian and head and neck cancers, and the agent has been granted orphan drug status by the FDA for patients with malignant gliomas.

Erlotinib appears to be generally well tolerated by patients, with rash and diarrhea proving to be the main side effects. These adverse events occurred in 75% of drug-treated patients (compared with placebo) in the erlotinib arm of the main trial providing the registration data. It is now known that skin rashes are the most common adverse reactions associated with EGFR-inhibitors, and many clinicians believe that it may be a useful biomarker of drug activity. Other common side effects include loss of appetite and fatigue; less common adverse events are reported to be mild headache, sore mouth, shortness of breath, cough, nausea and vomiting, and abnormal liver function tests. Rare side effects reported include inflammation or puncture of the cornea and GI bleeding, all of which could be associated with a general down-regulation of EGFR.

Finally, it is worth noting that clinically-relevant signs of the effects of erlotinib treatment (including prolonged survival) have never been demonstrated in EGFR-negative lung tumor patients, and the FDA approvals have been based on data from the crucial Phase III study (Trial BR.21), in which patients were prescreened and selected on the basis of their EGFR-positive status. As noted previously, patients were not prescreened for EGFR status in the gefitinib trials, and so it is possible that this agent may be as efficacious as erlotinib if evaluated under similar conditions. Also, the fact that erlotinib has only been approved for EGFR-positive patients means that a personalized medicine approach must be taken, with all potential patients being prescreened for EGFR status. This has both time and cost implications and, for some patients, a tumor biopsy may not be feasible.

5.2.6.3.3   Cetuximab (Erbitux™)

Unlike gefitinib and erlotinib, cetuximab (Erbitux™) is a chimerized (part human and part mouse) recombinant monoclonal antibody originally developed for use in colorectal cancer. It is made up of the Fv regions of an anti-EGFR mouse antibody with kappa light chain and human IgG1 heavy chain constant regions. Cetuximab binds to the HER1, c-ErbB-1, and EGFR receptors found on the surfaces of both tumor and normal cells. In doing so, it blocks the binding of EGF and related ligands, such as TGF-α, in a competitive manner. The latter has been linked to tumor cell growth in many EGFR-positive tumor types (more than 35% of all solid tumors). Interestingly, preclinical data have indicated that part of the mechanism of action of cetuximab may involve inhibition of angiogenesis in human transitional cell cancers, suggesting that combining cetuximab with radiation may improve the benefit of local irradiation of a tumor.

Cetuximab is used in combination with irinotecan to treat metastatic EGFR-positive colorectal cancer patients who have not responded to traditional irinotecan-based regimens. It is also licensed for single-agent use in colorectal cancer patients who cannot tolerate irinotecan-based regimens. However, the tumors still have to be EGFR-expressing and metastatic. The FDA approval of cetuximab in 2004 was based on three separate clinical trials using patients with metastatic colorectal cancer whose tumors had continued to grow after irinotecan-based therapy. The first trial involved 329 patients given either cetuximab alone (400 mg/m^2 initially, then 250 mg/m^2 weekly until progression or toxicity was observed) or cetuximab with irinotecan. The results demonstrated a response duration of 4.2 and 5.7 months (median) in the single-agent and combination arms, respectively, and analysis showed that subjects randomized to the combination had a significantly longer median time to progression than those who received cetuximab alone. The second trial involved 138 patients treated with a combination of cetuximab and irinotecan and gave a 15% response rate for the overall population versus 12% for those who had previously failed on irinotecan alone (median durations were 6.5 and 6.7 months, respectively). In the third trial, (a single agent, open-label, single-arm study enrolling 57 patients), a response rate of 9% (overall) was obtained for the all-treated group and 14% for those patients who had previously failed on irinotecan alone, with median times to progression of 1.4 and 1.3 months, respectively, and a median duration of response of 4.2 months for both groups. On the basis of these data, the recommended dosage of cetuximab is now an initial loading dose of 400 mg/m^2 administered as a 120-minute intravenous infusion, followed by a weekly maintenance dose of 250 mg/m^2 infused over 60 minutes.

The main adverse event associated with cetuximab is skin rash. However, nausea, vomiting, diarrhea, abdominal pain, asthenia, malaise, fever, and infusion reaction are also common. The manufacturer (ImClone Systems, Inc.) has stated its intentions to apply to the FDA for a supplemental biologics approval for cetuximab to be used either alone or together with radiation to treat squamous cell head and neck cancers.

5.2.6.4 VEGFR Inhibition

Vascular endothelial growth factor (VEGF), and its corresponding receptors such as VEGFR-1, -2, and -3, is the growth factor most often associated with angiogenesis (see Scheme 4.6). In support of this, inhibition of VEGFR-2 in endothelial cells in experimental models can prevent tumor growth. Oxygen-deficient cells use VEGF in the key signaling pathway to promote the growth of new blood vessels. It interacts with specific receptors on endothelial cell surfaces, directing them to build new blood supplies. The cells respond to this message by producing specialized protease enzymes to penetrate the basal lamina, so that they can migrate to oxygen-deficient regions. Once there, the cells replicate and form into tubes, thus creating new capillary pathways.

Both antibody (e.g., Genentech's bevacizumab [Avastin™]) and small-molecule (e.g., Sugen's semaxanib [SU-5416]) (see Structure 5.4) approaches are being taken to inhibit VEGF. Many pharmaceutical companies have small-molecule VEGF inhibitors in development, and nearly all of these agents target the ATP-binding pocket

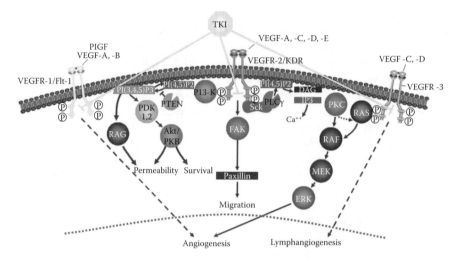

SCHEME 5.6 VEGF receptors in angiogenesis and lymphangiogenesis.

STRUCTURE 5.5 Structure of the VEGF inhibitor semaxanib (SU-5416).

of the kinase enzyme. The most successful antibody agent produced to date has been Genentech's bevacizumab (Avastin™), which is a recombinant humanized antibody. It was designed to bind to and inhibit VEGF and block the formation of new blood vessels at the tumor site (Structure 5.5). Inhibitors of VEGF are discussed in more detail in Chapter 7.

5.2.6.5 PDGFR Inhibition

A large proportion of tumor cells express both PDGF and PDGFR (see Scheme 5.2), and a number of experimental PDGFR inhibitors are presently under evaluation. One example is Sugen's small-molecule inhibitor SU-6688 (see Structure 5.6), which is an inhibitor of PDGFR-β, an angiogenic RTK. However, it is not completely selective, inhibiting Flk-1/KDR and FGFR1 as well, and is thus considered by many to be a broad spectrum kinase inhibitor. In mice, treatment with SU-6668 results in

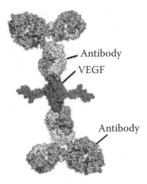

STRUCTURE 5.5 Image of antibodies bound to VEGF which is in the middle of the image (as labeled) and has two antibodies bound at the top and bottom. VEGF itself is made up of two identical subunits; the core at the center forms a dimer that assists in binding to the receptor. From the core of VEGF extends further small domains that bind to heparin and carbohydrates on the surface of cells, thus modifying the activity of VEGF. (Coordinates for this image were obtained from the Protein Data Bank.)

STRUCTURE 5.6 Structure of Sugen's PDGF inhibitor SU-6668.

growth arrest or regression of large human tumor xenografts. Induction of apoptosis in tumor microvessels can be observed within 6 hours of treatment initiation, and dose-dependent reductions in tumor microvessel density can be seen within 3 days of treatment. Furthermore, decreased tumor cell proliferation and enhanced apoptosis accompany these changes. It will be interesting to establish whether these *in vitro* results translate into clinical practice. At present, this is a highly competitive area of research, and many other pharmaceutical companies are searching for selective PDGF inhibitors.

5.2.6.6 Multiple Target Inhibitors

Although initial research programs in the signal inhibition area were focused on selectively modulating one particular pathway, recent experience has suggested that "dirty" inhibitors capable of affecting more than one pathway simultaneously may produce greater clinical benefit. One view is that lessons have been learned from targeted drugs, such as imatinib (Gleevec™), trastuzumab (Herceptin™), gefitinib

(Iressa™), erlotinib (Tarceva™), cetuximab (Erbitux™), and bevacizumab (Avastin™), and that blocking a single signaling pathway may not be sufficient for many patients. In the future, more effective treatments may involve the use of these targeted drugs in combinations in order to simultaneously disrupt the multiple molecular pathways that are aberrant in most tumors. It has been suggested that targeted treatments could eventually be combined in a single dosage form, so that a SRC kinase inhibitor, for example, could be combined with a drug targeting FAK to specifically attack colon cancer cells.

An alternative approach is to develop single agents capable of inhibiting more than one signaling pathway simultaneously in an attempt to obtain greater efficacy and to reduce the possibility of resistance developing through mutations in kinases. It is interesting that, due to the large number of kinase inhibitors that have been synthesized during the last decade in both academic and industrial laboratories, many compounds have already been discovered that are active against two or more kinases. This is perhaps not surprising given the similarity of the ATP-binding sites across a number of different kinase families (some researchers are now targeting non-ATP sites on kinases). One example of this multitarget approach is the investigational drug ARRY-334543, for which an investigational new drug application was submitted by Array BioPharma in 2005. This agent has dual inhibitory activity and can disrupt both ErbB-2 and EGFR. In experimental models of human epidermal, lung, and breast tumors, ARRY-334543 is orally active and has significant antitumor activity. Planned Phase I trials will determine whether this translates into clinical practice. Another example is a recently initiated Phase I clinical trial by the company Exelixis to evaluate XL820, a novel small-molecule anticancer drug that is orally available and capable of simultaneously inhibiting the PDGF, KIT, and VEGF RTKs, all of which are viable clinical targets in their own right.

5.2.6.7 Other Potential Targets

In addition to the kinases discussed above, many other types of kinases might be targeted with potential clinical benefit, and the list is growing ever longer due to the rapidly progressing work of molecular biologists in the cancer field. For example, attempts are being made to develop inhibitors for the insulin-like growth factor (IGF)–1R receptor pathway, which is triggered by the IGF-1 and IGF-2 growth factors and is important for many cellular functions, including transformation and proliferation (see Scheme 5.7). Mutant Flt-3 is also of interest and is known to be important in leukemic (AML and ALL) cells (see Scheme 5.8). Other areas of research include the MET and SRC kinases, and the *Ras*/ERK1-2, AKT and STAT pathways. The attraction of MET kinase is that it is important in the process of metastasis which, if controlled, could allow more focus on the treatment of primary tumors.

Interest in protein kinase B (PKB), additionally known as *Akt,* is also gathering momentum, and AstraZeneca and Astex Therapeutics, Ltd., recently announced a new alliance to discover, develop, and commercialize novel small-molecule inhibitors in this area. Furthermore, Array BioPharma, in collaboration with AstraZeneca, is developing a potent and specific inhibitor of MEK called ARRY-142886 (also known as AZD6244); Phase 1 trials began in 2004.

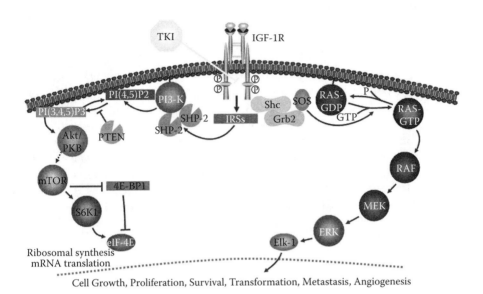

SCHEME 5.7   The IGF-1R signaling pathway.

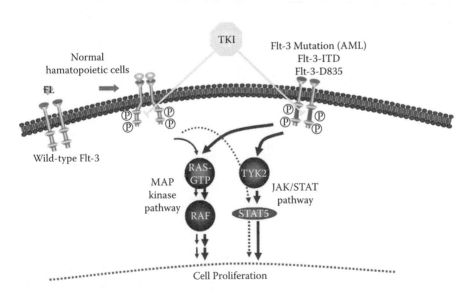

SCHEME 5.8   Mutant and wild-type Flt-3 activation in leukemic (AML, ALL) cells.

5.3 INHIBITION OF *RAS* PATHWAY SIGNALING

5.3.1 INTRODUCTION

Approximately 30% of all human tumors are thought to start from a mutated *Ras* oncogene. For example, approximately 50% of colon cancers and 90% of human pancreatic carcinomas have point mutations in their corresponding *Ras* genes. *Ras* is also proving to be one of the most prevalent oncogenic mutations in myeloma, in which mutations in 39% of newly diagnosed patients have been observed along with a correlation between *Ras* mutation and shorter survival, a characteristic that can be used diagnostically. It is noteworthy that patients with *Ras* mutations have been found to have a 2.1 years median survival, compared with 4 years for those with wild-type *Ras*. Furthermore, a similar incidence of *Ras* mutations at diagnosis can be shown to increase to up to approximately 80% by the time of disease relapse. In addition, subpopulations of tumor cells from all newly diagnosed myeloma patients have been shown to contain N-*ras* 61 mutation-positive cells.

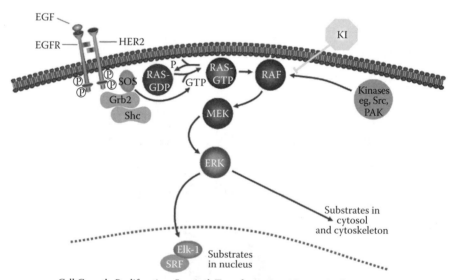

Cell Growth, Proliferation, Survival, Transformation, Metastasis, Angiogenesis

SCHEME 5.9 Activation of *Ras-Raf* oncogenes by mutation or growth factor stimulation.

These mutations of the *Ras* protein cause it to lose its intrinsic GTPase activity (which facilitates the hydrolysis of GTP to GDP), and so the protein stays in the active GTP-bound state, continuously sending signals to the nucleus (Scheme 5.9 and Scheme 5.10). This leads to the uncontrolled cell division characteristic of all cancers. Thus, the forensyl transferase (FTase) protein has been identified as a promising target for cancer chemotherapy, and forensyl transferase inhibitors (FTIs) are designed to decrease the activity of the *Ras* protein by interfering with its anchorage to the plasma membrane. FTIs block the action of the enzyme, preventing

attachment of a lipophilic farnesyl fragment to the *Ras* C-terminus, and thus imped-
ing its binding to the cytoplasmic surface of the cellular membrane.

A number of FTIs are currently undergoing clinical evaluation, including tipifarnib (Zarnestra™), lonafarnib (Sarasar™), and BMS-214662 (see Structure 5.7).
Overall, these agents are proving to be efficacious either as single agents or in
combination, and anticancer activity has been observed in CML, myelodysplastic
syndrome, some acute leukemias, and breast cancer. The compounds are relatively
free from serious toxicities, but adverse effects observed include transaminase elevation, neutropenia, thrombocytopenia, neuropathy, rash, and diarrhea.

Although designed to affect *Ras*, experience from preclinical studies has shown
that these compounds may have other targets. This became obvious when the activities of a number of FTIs in various cell lines, including myeloma, failed to correlate
with their mutated *Ras* status. Also, the discovery that both N-*ras* and K-*ras* can be
prenylated by geranylgeranyl transferase suggested that *Ras* may not be the only
target, and preclinical models containing both these mutations have been found to
be sensitive to FTIs. As a consequence, a number of other proteins have been
proposed as potential targets. However, regardless of the uncertainty surrounding
the real targets of FTIs, they have significant anticancer activity in humans when
administered either as single agents or when given in combination with certain
cytotoxic drugs. Finally, FTIs are known to synergize with gamma radiation and
may possess some chemopreventive properties.

5.3.2 Mechanism of *Ras* Signaling Pathway

Ras oncoproteins are members of the *Ras* (Rat-Adeno-Sarcoma) super-family, and
many are important components of the signal transduction cascade that eventually
leads to gene expression and subsequent cell differentiation and proliferation. The
membrane-localized monomeric *Ras* protein is 21 kD in size and is a G-protein
signal transducer. Because it does not have a transmembrane domain as part of its
own structure, it can only function properly when bound at the internal surface of
the cell membrane near to a growth factor receptor. For this reason, farnesylation
of the protein is required to enhance its hydrophobicity and ability to bind to the
plasma membrane (see Scheme 5.10).

Both normal and mutated *Ras* proteins need to anchor to the cell membrane in
order for signal transduction to occur. Attachment to the membrane occurs through
a number of posttranslational modifications. First, the enzyme FTase transfers the
farnesyl group of farnesyl pyrophosphate (FPP) to the mercapto function of a
cysteine. This cysteine is one amino acid constituent of the "CaaX-box" found at
the *Ras* C-terminus, where "C" = cysteine, "a" = an aliphatic amino acid, and "X"
= serine, methionine, alanine, or glutamine. The binding mode at the active site of
FTase has been established by X-ray studies of crystallized complexes of a CaaX
mimetic, FPP, and FTase. For example, the crystal structures of R115777 and BMS-
214662 complexed with mammalian FTase have both been reported. The deproto-
nation of the cysteine to its thiolate is brought about by a central zinc ion. The
thiolate then displaces the FPP's pyrophosphate group to form a C-S bond that
connects the farnesyl residue to the *Ras* protein. Next, a specific endoprotease

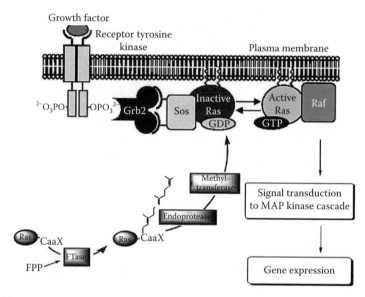

SCHEME 5.10   Details of the pathway of *Ras* activation leading to gene expression.

removes the "aaX" component, and a methyltransferase then methylates the C-terminus of free S-farnesylcysteine, thus allowing the *Ras* protein to attach to the membrane. Overall, these modifications involve the covalent addition of either far-nesyl (i.e., 15-carbon) or geranylgeranyl (i.e., 20-carbon) chains to the *Ras* protein's conserved carboxy-terminal cysteine moiety by the FTase and geranylgeranyl trans-ferase enzymes, respectively.

The membrane-bound *Ras* protein thus represents a "molecular switch" that allows transport of a signal (e.g., a growth factor) from the external environment of a cell to its nucleus. The first stage of this process involves an extracellular ligand stimulating a monomeric RTK that then dimerizes. Next, Grb2, an initial adaptor protein, identifies and interacts with a binding site, which in turn allows recruitment of "Son of sevenless," or Sos, a second adaptor protein. The latter causes the inactive GDP-carrying *Ras* to become active by substituting GDP for GTP. After this, the signal can be transmitted downstream by the activated *Ras* to other effectors, such as *Raf*. In the MAP-kinase signaling pathway, the *Raf* protein is the first kinase that eventually results in the expression of different genes.

5.3.3 CLASSES OF INHIBITORS OF FARNESYLTRANSFERASE

A number of different ways to inhibit FTase have been suggested. For example, one approach is to make use of the "CaaX" motif to simulate the C-terminal tetrapeptide and compete with *Ras* for farnesylation. Examples were produced by simply varying certain amino acids, such as Cys-Val-Phe-Met, which gave inhibitors with *in vitro* IC$_{50}$ values in the 25 nM region. Although peptides of this type are not very efficient at crossing cell membranes and can be rapidly degraded by hydrolysis *in vivo*, combinatorial chemistry approaches coupled with high throughput screens eventually afforded some highly effective inhibitors. For example, BMS-214662 has *in vitro* IC_{50} values in the 0.7 nM region. Farnesyl pyrophosphate analogs have also been considered but have significantly lower activities and are, in general, less attractive because they are competitive in alternative enzymatic pathways (e.g., those relating to squalene synthase) as substrates with FPP. On the basis that a complex must initially form between both substrates (CaaX-box and FPP) and the enzyme before catalysis can occur, bisubstrate inhibitors combine both of the strategies discussed above. Other types of inhibitors have been designed and synthesized that incorporate a farnesyl and a CaaX mimetic, and have enhanced selectivity and sub-mM activity. Other inhibitors have been identified from natural sources — for example, pepticin-namin E, which is produced by a *Streptomyces* species and is active in the low to mid mM region. More recent studies have attempted to make metal-chelating FTIs constructed from a metal-chelating core (e.g., zinc) and a specialized chemical moiety designed to recognize aromatic groups adjacent to the zinc ion in the active site of the FTase. However, such compounds are so for only active in the low mM region.

Tipifarnib (Zarnestra™, R115777) Lonafarnib (Sarasar™, SCH-66336) BMS-214662

STRUCTURE 5.7 Structures of the *Ras* inhibitors tipifarnib (Zarnestra™), lonafarnib (Sarasar™), and BMS-214662.

5.3.4 TIPIFARNIB

Tipifarnib (Zarnestra, R115777) is a potent imidazole-containing heterocyclic non-peptidomimetic inhibitor of FTase that was first synthesized by Janssen in 1997 (Structure 5.7). *In vitro*, it is inhibitory in a number of human tumor cell lines (including those with mutant or wild-type *Ras*) in the 1.7 to 50 nmol range. It is also active *in vivo*, where it produces growth inhibition of pancreatic and colon tumors in human xenografts in a dose-dependent manner with an observable anti-angiogenesis effect, coupled with decreased proliferation and enhanced apoptosis. Doses of up to 1300 mg twice a day have been given to patients in Phase I clinical trials without any significant toxicities other than a dose-limiting reversible myelo-suppression. Most importantly, clear signs of antitumor activity have been observed in Phase I and Phase II investigations of AML, myelodysplastic syndrome, and metastatic breast cancer. Interestingly, pharmacodynamic (PD) studies demonstrated that although protein farnesylation was efficiently blocked in all patients during the trials, it did not in fact correlate with antitumor activity. However, a correlation was shown between disease stabilization in the small number of patients examined and reduced levels of phosphorylated STAT3 and Akt in the bone marrow of patients in whom these tumor survival pathways were constitutively active. Interestingly, in ovarian cancer cell lines, the PI 3-kinase/AKT2 pathway is also known to be a target for FTI-induced apoptosis.

Tipifarnib has also been investigated at the Phase II level for activity in advanced myeloma, a disease in which *Ras* mutations are highly prevalent and represent the most common gene mutation. It was tolerated well, found to be relatively free from adverse effects, and provided disease stabilization in more than 60% of patients. However, again no correlation of disease stabilization with inhibition of FTase was observed, suggesting that alternative non-*Ras* pathways may be involved. A clinical trial evaluating the combination of tipifarnib and imatinib in myeloid and chronic leukemia is also underway.

As a result of the lack of PD end-point data in the clinical trials, in mid-2005 a recommendation against accelerated approval status for tipifarnib as a new treatment for newly diagnosed elderly patients with AML was made by the Oncologic Drugs Advisory Committee of the FDA. This was being sought by the sponsor Johnson & Johnson ahead of completing Phase III clinical trials. Instead, the committee decided that complete Phase III data would be required in order to consider approval.

5.3.5 LONAFARNIB

Lonafarnib (Sarasar™, SCH-66336) has a different structure to tipifarnib, being comprised of a multisubstituted tricyclic benzpyrridinoheptane nucleus (see Structure 5.7). Together with paclitaxel in metastatic NSCLC patients who had failed previous taxane regimens, lonafarnib was found to have significant antitumor activity in 11 of the 29 patients studied, providing examples of stable disease and with 3 patients achieving partial responses. Overall, a 16-week (average) disease-free progression survival time was obtained, with an average overall survival of 39 weeks. Minimal side effects were observed with this combination of Taxol™ and lonafarnib, with fatigue, diarrhea, and dyspnea being the most frequently reported side effects. Only one patient had a significant reduction in white blood cell count, and reports of more serious side effects, including respiratory insufficiency and acute respiratory failure, were infrequent.

5.3.6 BMS-214662

This drug, like lonafarnib, also contains a seven-membered ring (see Structure 5.7). Low concentrations of BMS-214662 in human myeloma cell lines efficiently inhibit protein farnesylation but do not affect the activation of Akt. Other effects include the development of apoptotic morphology; a reduction of Mcl-1 levels; caspase activation; induced proapoptotic conformational changes of Bax and Bak; $\Delta\Psi_m$ loss; increased levels of p53; up-regulated modulator of apoptosis (PUMA); the BH3-only protein; cytochrome C release; apoptosis inducing factor nuclear translocation; and phosphatidylserine exposure. These results suggest that the apoptosis triggered by BMS-214662 might occur through a mechanism dependent on PUMA-Bax-Bak-(Mcl-1). The drug is presently being studied in clinical trials.

5.3.7 OTHER NOVEL *RAS* PATHWAY INHIBITORS

Further down the *Ras* pathway are the key regulators of cell signaling, MEK1 and MEK2, which are at the hub of several pathways (see Scheme 5.9). MEK1 and MEK2 are able to activate further proteins called ERK1 and ERK2 that can, among other actions, initiate cell division and activate some genes associated with tumors. Overactivation of MEK occurs in several forms of cancer, including breast tumors. Array BioPharma, in collaboration with AstraZeneca, is developing a potent and specific inhibitor of MEK known as ARRY-142886 (also called AZD6244); Phase I trials began in 2004.

5.4 CELL CYCLE INHIBITORS

The D-type cyclins and corresponding partner kinases, CDK4 and CDK6, act as central integrators of extracellular signals and operate during the G1 phase of the cell cycle by phosphorylating the tumor suppressor protein pRb, thus contributing to its inactivation. Mutations that can influence the operation of cyclins CDK4 and CDK6, their regulating proteins, or pRB can be found in most human tumors. Furthermore, cyclin D1 expression can be up-regulated by the *Ras* signaling pathway, which is itself up-regulated in many cancer cells.

STRUCTURE 5.8 Structures of flavopiridol, olomoucine, and (*R*)-roscovitine (CYC-202).

Flavopiridol is the prototype inhibitor of CDKs, and is capable of inducing apoptosis in some tumor cells (see Structure 5.8). It is a flavone, synthetically derived from rohitukine, a plant alkaloid isolated from the leaves and stems of *Dysoxylum binectariferum* and *Amoora rohituka*. Both of these plants are native to India and are widely used in traditional medicine. Although flavopiridol was developed as far as clinical evaluation through a collaboration between Aventis and the NCI, all studies were terminated in early 2004 following poor results from intermediate-stage Phase II clinical trials. Several other CDK inhibitors have been described such as olomoucine and roscovitine, the chiral (*R*) version (CYC-202) of which was developed by Cyclocel Ltd. and is presently completing Phase II clinical trials (Structure 5.8). One trial is exploring the potential activity of CYC-202 in stage IIIB/IV NSCLC when coadministered with cisplatin and gemcitabine, and another is evaluating its use in combination with capecitabine (an oral prodrug of 5-fluorouracil) in advanced breast cancer. A further trial is studying the effect of CYC-202 in glomerulonephritis, an inflammatory disease of the kidney.

Another class of CDK inhibitors are the paullones, which inhibit various CDKs, including CDK1 (cyclin B), 2 (cyclin A), 2 (cyclin E), and 5 (p25), with different IC_{50} values (0.4, 0.68, 7, and 0.85 μM, respectively). Examination of these compounds in the NCI's 60 cell line screen has shown that there is generally a lack of correlation between CDK inhibitory potency and cytotoxicity. Although one derivative, alsterpaullone (Structure 5.9) has shown a much higher activity against CDK1 and is also associated with greater cytotoxicity and antitumor activity, the molecule has now been shown to potently inhibit glycogen synthase kinase-3 and CDK5/p25, suggesting that it may work through multiple targets.

STRUCTURE 5.9   Structure of alsterpaullone.

In general, one problem with all the classes of CDK inhibitors studied to date is that there is often a discrepancy between *in vivo* activity and *in vitro* and cellular data. For example, *in vitro*, IC_{50} values for inhibiting CDK2 in the low nM range typically translate into cellular IC_{50} values in the micromolar range in order to inhibit progression of cells from G1-phase to S-phase. Similarly, disproportionately higher concentrations are often required to observe effects *in vivo*. Although in some cases poor transport across cell membranes or poor bioavailability are possible explanations, it is also likely that some molecules do not inhibit CDK2 in the cell as effectively as they do *in vitro*. In this context it is worth noting that most of the lead molecules under development bind at the ATP-docking site of the kinase enzymes. The ATP concentration within a cell is between 5 and 15 mM, whereas the ATP concentration in *in vitro* CDK2 assays is usually between 10 and 400 μM. Therefore, it is likely that the much higher concentration of ATP in cells significantly reduces the activity of CDK inhibitors.

5.5 PROTEASOME INHIBITORS

The principal route for the degradation of proteins (including signaling proteins) within cells is the *ubiquitin-proteasome* pathway. Proteins destined for degradation are initially "tagged" with a polyubiquitin chain. The tagged protein is then recognized by the *proteasome,* a large multimeric protein found within all eukaryotic cells, which degrades it to the constituent peptides and free ubiquitin. This disposal pathway can be crucial for tumorigenesis, tumor growth, and metastasis because the sequenced and temporal degradation of many key control proteins, such as tumor suppressors and cyclins, is critical for cell cycle progression and mitosis. Hence, in some cases, proteasome inhibitors should arrest or retard cancer progression by interfering with degradation of these regulatory molecules. One example is that a reduction in expression of the NF-κB-dependent cell adhesion molecule should make dividing cancer cells more sensitive to apoptosis. Therefore, as the proteasome is involved in activating NF-κB by degrading its inhibitory protein IκB, inhibition of proteasome-mediated IκB degradation may increase levels of NF-κB and promote apoptosis in tumor cells. Another example is that NF-κB plays an important role in the response of cells to environmental stress or exposure to cytotoxic agents by promoting the production of inhibitors of apoptosis to maintain cell viability. Therefore, stabilization of the IκB protein, by blocking its degradation with proteasome inhibitors, should increase NF-κB levels, thus making cells less susceptible to apoptosis. In summary, proteasome inhibitors can act through many different mechanisms, perhaps simultaneously, to reduce or block tumor growth, angiogenesis, and metastasis. However, dosing regimens may need to be carefully optimized in order to limit the potentially adverse effects of inhibiting the proteasome in healthy cells.

A number of potent boron-containing proteasome inhibitors have been discovered that work by inhibiting the chymotryptic activity of the proteasome in a selective and reversible manner. Importantly, for these protease analogs, the inhibitory potency data (Ki) correlate with both *in vitro* cytotoxicity profiles and *in vivo* antitumor activity, thus supporting their proposed mechanism of action.

STRUCTURE 5.10 Structure of bortezomib (Velcade™).

Bortezomib (Velcade), developed by Millennium Pharmaceuticals, Inc., represents a new class of proteasome inhibitor and is used to treat multiple myeloma, a cancer of the plasma cell, which is an important component of the immune system that produces disease-fighting antibodies (see Structure 5.10). Early results in 2005 from an international Phase III trial involving relapsed multiple myeloma patients demonstrated that bortezomib is more effective at delaying disease progression than a standard treatment of high-dose dexamethasone. Although certain serious side effects were worse for patients taking bortezomib, more patients taking the drug were alive after 1 year than those on the standard treatment. On this basis, the FDA recently approved bortezomib injection as a treatment for multiple myeloma, having taken less than 4 months to review the license application under the accelerated approval program. Due to the success of bortezomib, which has validated the proteasome as a drug target, it is likely that many more types of proteasome inhibitors will be discovered and developed in the near future.

5.6 mTOR INHIBITORS

mTOR is a cellular enzyme that plays a key role in cell growth and proliferation as part of the mTOR signaling pathway. It is found at the center of a complex regulatory network that senses the nutrient and growth factor status of the intracellular and extracellular environments, respectively, and transduces the acquired information into growth and proliferative decisions (see Scheme 5.11). Inhibition of mTOR delivers the false signal that the cell is starved of nutrients and lacks growth factor stimulation. This initiates the cellular starvation response, which includes dramatic metabolic reprogramming, prevention of cell growth, and arrest of cell division.

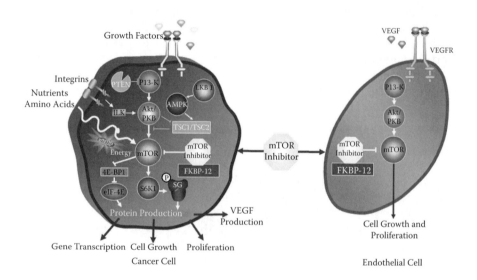

SCHEME 5.11 mTOR function: A central controller of tumor cell growth and angiogenesis.

In particular, mTOR is a crucial component of the transmission of signals mediated by the phosphatidylinositol 3-kinase (PI3K) pathway, a signaling cascade that is aberrant in more than 70% of tumors. Activation of PI3K by growth factors signals the cell to grow and proliferate. This signal is transmitted to mTOR through the cell by AKT, also known as *protein kinase B* (see Scheme 5.12). mTOR also initiates growth in response to the presence of nutrients, including glucose. It does this by activating the ribosomal S6 kinase (S6K) and the eukaryotic initiation factor 4E binding protein 1 (4E-BP1) via phosphorylation. These translation factors are

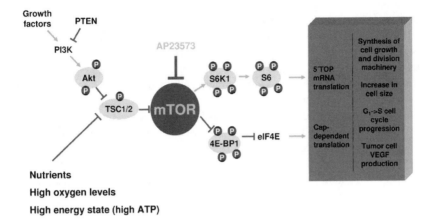

SCHEME 5.12 Details of the mTOR pathway.

crucial for the production of proteins involved in cell cycle progression; thus, mTOR serves as a master switch by interpreting extracellular and intracellular signals, and transducing these to the cellular protein translation machinery to stimulate cells to grow and divide.

Activated S6K or its substrate, the ribosomal protein (S6), as well as 4E-BP1 (that is, phospho-S6K, phospho-S6, and phospho-4E-BP1, respectively), are currently being evaluated for use as pharmacodynamic biomarkers in tissues because their presence or absence should theoretically reflect whether the mTOR signaling pathway is active. Pre-clinical studies suggest that sensitivity to mTOR inhibition may correlate with aberrant activation of the phosphatidylinositol 3'-kinase pathway or with aberrant expression of cell-cycle regulatory or anti-apoptotic proteins.

STRUCTURE 5.11 Structures of rapamycin (Rapamune™), AP23573, everolimus (Certican™), and temsirolimus.

The prototype mTOR inhibitor rapamycin (also called *sirolimus*) was isolated from a bacterium, *Streptomyces hygroscopicus,* discovered in soil taken from Rapa Nui (Easter Island) (see Structure 5.11). It is a macrolide antibiotic that has been known since the mid-1970s as an antifungal agent. In the 1990s, after the discovery of its immunosuppressive effects, it was developed by Wyeth and approved as an antirejection drug for use in kidney transplant patients under the brand name Rapamune™. Its antitumor properties were observed subsequently, although a clinical development program for rapamycin itself in the cancer area was not pursued due to its unfavorable pharmacological profile. However, a number of rapamycin analogs have now been developed for use as anticancer agents, including AP23573, everolimus (Certican™) temsirolimus, and RAD001 (see Structure 5.11).

Rapamycin and its analogs bind with high affinity to the cellular protein FK506-binding protein 12 (FKBP12), and the resultant complex directly inhibits mTOR. In addition, rapamycin appears to inhibit angiogenesis, which is consistent with the observation that PI3K signaling is required for expression and secretion of the proangiogenic VEGF from endothelial cells. Also, many of the genetic translocations that underlie sarcomas seem to result in aberrant activation of the mTOR pathway, and many sarcomas produce large amounts of VEGF. Interestingly, certain sarcoma cell types are more sensitive than other human cancer cells to the growth inhibitory effects of rapamycin-based drugs. Rapamycin can also induce apoptosis in sarcoma cell lines under some conditions, and genetic experiments using mutant mTORs have confirmed that mTOR is the critical target for rapamycin-induced cell death. Thus, mTOR inhibitors based on rapamycin may inhibit tumor cell growth through multiple mechanisms, and the previously noted observations suggest that they could be particularly useful in the treatment of sarcomas.

AP23573 (Ariad Pharmaceuticals) was designed to retain the antitumor activity of rapamycin but to have a more favorable pharmacology (see Structure 5.11). It has shown promise in a broad range of tumors, especially whereas temsirolimus and everolimus (2-Hydroxyethyl)rapamycin) have demonstrated clinical activity in other cancers as well as sarcomas. In preclinical studies, as well as being cytostatic, AP23573 reduced tumor size in mouse models and reduced cell counts in human tumor cell lines. In cell culture, the drug has shown positive additive effects when combined with doxorubicin, docetaxel, cisplatin, and topotecan. In addition, additive effects are obtained in combination with breast cancer–targeted therapies, including trastuzumab and EGFR inhibitors, as well as imatinib when studied in CML and GIST. Thus, AP23573 could potentially contribute to the efficacy of combination chemotherapy regimens which are currently the mainstay for treating several differ-ent cancer types. It can also inhibit the secretion of VEGF in some cell lines, suggesting that it exerts part of its antitumor effects through the inhibition of angiogenesis.

Clinical trial results to date show that mTOR inhibitors are generally well tolerated and may induce prolonged disease stability and even tumor regressions in a subset of patients, although questions remain regarding optimal dose, schedule, patient selection, and combination strategies. So far, promising efficacy has been observed for AP23573 in sarcoma patients in Phase I and Phase II clinical trials. It is compatible with either intravenous or oral delivery and is known to cross the blood–brain barrier in experimental animals, suggesting that it may be useful for treating brain cancers. In 2005, AP23573 was granted fast-track approval status by the FDA for development in treating soft tissue and bone sarcomas, a decision that was based on positive clinical trial results in refractory sarcoma patients for whom few useful treatments currently exist. AP23573 is now being evaluated alone in a number of Phase Ib and Phase II clinical trials involving hematological malignancies (i.e., lymphomas and leukemias) and in certain solid tumors (i.e., glioblastoma multiforme [GBM], prostate tumors and sarcomas).

Minor responses and examples of stable disease in sarcoma patients have also been reported for the other mTOR inhibitors in Phase I and Phase II clinical trials. Everolimus (Certican™), obtained by chemical modification of sirolimus, exhibits improved relative bioavailability and has a shorter half-life than sirolimus. Temsirolimus (CCI-779 [Wyeth]) has a slightly more complex side chain and has provided promising results in a Phase II trial in locally advanced metastatic breast cancer in heavily pretreated patients. Furthermore, Phase III studies of temsirolimus in combination with the aromatase inhibitor letrozole are underway. The drug is also being evaluated in patients with soft tissue sarcoma or GISTs, and in combination with imatinib in patients with GISTs, with promising early results. Clinical studies involving AML and other types of leukemia, prostate and endometrial cancer, and GBM are also underway, as are combination studies with paclitaxel and capecitabine. Novartis is also developing an mTOR inhibitor known as RAD001, which has good activity in tumor cell lines and in animal models, and displays antiangiogenic activity. It is now in Phase II clinical studies involving several tumor types, including breast cancer.

Clinical studies carried out to date have shown that AP23573 has a favorable safety profile coupled with broad anticancer activity. In the clinic, its dose-limiting toxicity is mucositis, or inflammation of the mucosal lining of the oral cavity and GI tract. Mild to moderate adverse effects include reversible anorexia, diarrhea, fatigue, rash, anemia, leukopenia, neutropenia, and thrombocytopenia.

Research is also underway to develop predictive biomarkers to identify which tumors and patients are most likely to respond to AP23573. One challenge here is that surgical removal of tumor tissue for analysis is not a practical option for most sarcoma patients. Therefore, there is an emphasis on establishing blood-based or other types of biomarkers.

FURTHER READING

Adjei, A.A. "Blocking Oncogenic *Ras* Signaling for Cancer Therapy," *J. Nat. Cancer Inst.*, 93:1062-74, 2001.

Bernd, N. "Farnesyltransferase-Inhibitors: New Drugs for Cancer Treatment," *Asian Stud. Med. J.*, Institut für Organische Chemie, Universität Regensburg, Universitätsstrasse 31, D-93051 Regensburg, Germany, 2003. http://asmj.netfirms.com/article1203.html.

Dancey, J.E. "Agents Targeting *Ras* Signaling Pathway," *Curr. Pharm. Des.*, 8:2259-67, 2002.

Dancey, J.E. and Chen, H.X. "Strategies for Optimizing Combinations of Molecularly Targeted Anticancer Agents." *Nature Reviews Drug Discovery*, 5(8):649–659, 2006.

Dobrusin, E.M., and Fry, D.W. "Protein Tyrosine Kinases and Cancer," *Ann. Rep. Med. Chem.*, 27:169-78, 1992.

Druker, B. "Perspectives on the Development of a Molecularly Targeted Agent," *Cancer Cell,* 1:31-36, 2002.

Faivre, S. et al. "Current Development of mTor Inhibitors as Anticancer Agents." *Nature Reviews Drug Discovery,* 5(8):671–688, 2006.

Herbst, R.S., and Hong, W.K. "IMC-C225, an Anti-Epidermal Growth Factor Receptor Monoclonal Antibody for Treatment of Head and Neck Cancer," *Sem. Oncol.,* 5 (Suppl 14): 18-30, 2002.

Lynch, T.J., et al. "Activating Mutations in the Epidermal Growth Factor Receptor Underlying Responsiveness of Non-Small-Cell Lung Cancer to Gefitinib," *NEJM,* 350(21):2129-39, 2004.

Nagar, B., et al. "Crystal Structures of the Kinase Domain of *C-Abl* in Complex with the Small Molecule Inhibitors PD173955 and Imatinib (StI-571)," *Cancer Res.,* 62:4236-43, 2002.

Paez, J.G., et al. "EGFR Mutations in Lung Cancer: Correlation with Clinical Response to Gefitinib Therapy," *Science,* 304(5676):1497-1500, 2004.

Parker Hughes Cancer Center. "Tyrosine Kinase Inhibitors: Molecules with an Important Mission," 2000. http://www.ih.org/pages/tyrosine_kinase_inhibitors.html.

Schlessinger, J. "Cell Signaling by Receptor Tyrosine Kinases," *Cell,* 103:211-25, 2002.

Sclabas, G.M., et al. "Restoring Apoptosis in Pancreatic Cancer Cells by Targeting the Nuclear Factor-Kappa B Signaling Pathway with the Anti-Epidermal Growth Factor Antibody IMC-C225," *J. Gastrointestinal Surg.,* 1:37-43, 2003.

Shawver L., et al. "Smart Drugs: Tyrosine Kinase Inhibitors in Cancer Therapy," *Cancer Cell,* 1:117-23, 2002.

Vignot, S., et al. "mTOR-Targeted Therapy of Cancer with Rapamycin Derivatives," *Ann. Oncol.,* 16(4):525-37, 2005.

6 Hormonal Therapies

6.1 INTRODUCTION

It has long been recognized that tumors derived from hormone-dependent tissues, such as the breast, endometrium, and testes, are themselves dependent on the same hormones. This is demonstrated by the remissions observed in premenopausal breast cancer following ovariectomy (surgical removal of the ovaries) and in prostatic cancer following orchidectomy (surgical removal of the testes). Hormonal treatments thus play a large role in the treatment of these tumors. While not curative, they may provide excellent palliation of symptoms in selected patients, sometimes for many years. As with all treatments, tumor response and toxic side effects should be carefully monitored and treatment changed if progression occurs or side effects exceed the benefits.

6.2 BREAST CANCER

The management of patients with breast cancer involves surgery, radiotherapy, drug therapy, or a combination of these treatments. Originally, the most common cytotoxic chemotherapy regimen for both adjuvant use and metastatic disease was a combination of cyclophosphamide, methotrexate, and fluorouracil. However, anthracycline-containing regimens are now increasingly used and are regarded as standard therapy unless contraindicated (e.g., in cardiac disease). In metastatic disease, the chemotherapy regimen chosen reflects whether the patient has previously received adjuvant treatment and also the presence of any comorbidity. For patients who have not previously received chemotherapy, either cyclophosphamide, methotrexate, and fluorouracil or an anthracycline-containing regimen is the standard initial therapy for metastatic breast disease. However, patients with anthracycline-refractory or resistant disease should now be considered for treatment with a taxane either alone or in combination with trastuzumab (Herceptin™) if their tumors overexpress human epidermal growth factor 2 (HER2). Other drugs licensed for breast cancer include capecitabine, gemcitabine, raltitrexed, mitoxantrone, mitomycin, and vinorelbine. In cancers that overexpress HER2, trastuzumab alone is an option for cytotoxic chemotherapy-resistant disease.

6.2.1 EARLY BREAST CANCER

All women with early breast cancer should be considered for adjuvant therapy following surgical removal of the tumor because adjuvant therapy can help eradicate the micrometastases that cause relapse. The choice of adjuvant treatment is determined by the risk of recurrence, the estrogen-receptor status of the primary tumor, and menopausal status.

Tamoxifen, an estrogen-receptor antagonist, is presently the preferred choice of adjuvant hormonal treatment for all women with estrogen-receptor-positive breast cancer. It is supplemented in selected cases by cytotoxic chemotherapy. Premenopausal women may also benefit from treatment with a gonadorelin analog or ovarian ablation. Treatment with tamoxifen delays the growth of metastases and increases survival. If tolerated, treatment should be continued for 5 years. Tamoxifen also lowers the risk of tumor formation in the other breast.

Anastrozole is also licensed for the adjuvant treatment of estrogen-receptor-positive early breast cancer in postmenopausal women who are unable to take tamoxifen (due to a high risk of thromboembolism or endometrial abnormalities). For both premenopausal and postmenopausal women with estrogen-receptor-negative breast cancer, treatment with cytotoxic agents is preferred.

6.2.2 Advanced Breast Cancer

Tamoxifen is used in postmenopausal women with estrogen-receptor-positive tumors, patients with long disease-free intervals following treatment for early breast cancer, and those with disease limited to bone or soft tissues. However, aromatase inhibitors, such as letrozole or anastrozole, may be more efficacious and are regarded as the preferred treatment in postmenopausal women. Ovarian ablation or a gonadorelin analog should be considered in premenopausal women. Progestogens such as medroxyprogesterone acetate continue to be used in advanced breast cancer in postmenopausal women. They are as effective as tamoxifen but are not as well tolerated. However, they are less effective than the aromatase inhibitors. Cytotoxic chemotherapy is preferred for advanced estrogen-receptor-negative tumors and for aggressive disease, particularly where metastases involve visceral sites (e.g., the liver) or where the disease-free interval following treatment for early breast cancer is short.

6.2.3 Role of Estrogen in Tumor Growth

Estrogen has effects on the growth and function of reproductive tissues and also preserves bone mineral density, thus reducing the risk of osteoporosis and protecting the cardiovascular system by reducing cholesterol levels. Estrogens act as promoters, rather than initiators, of breast tumor development and can also facilitate tumor invasiveness by stimulating the production of proteases, which can degrade the extracellular matrix.

Estrogen action is conducted through estrogen receptors (ERs). There are two types, ERα and ERβ, although the role of the latter in breast cancer is unclear. ERs can be detected in 60% to 80% of human breast cancers, and their presence (i.e., ER-positive or ER[+]) is related to the sensitivity of the tumor to anti-endocrine treatment, with 60% of ER(+) breast cancer patients responding.

Estradiol (E2), the principal estrogen, is synthesized in the ovaries from androstenedione by aromatase. However, in postmenopausal women for whom ovarian synthesis has ceased, local synthesis occurs in adipose and other tissues, especially the breast, where E2 levels are 20-fold higher than in plasma. After entering the cell, E2 binds to the estrogen receptor, which leads to a conformational

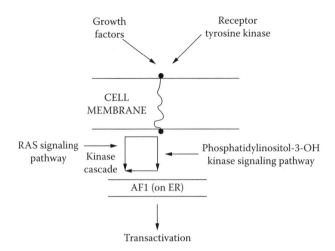

SCHEME 6.1 Simplified view of growth factor activation of estrogen receptor (ER) by phosphorylation of AF1. (Ali, S., and Coombs, R.C. "Endocrine-Responsive Breast Cancer and Strategies for Combating Resistance," *Nature Rev. Cancer,* 2:101-12, 2002.)

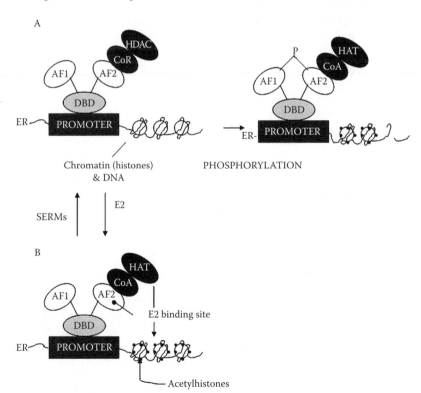

SCHEME 6.2 ER-gene binding showing (A) resting state, and (B) estrogen-activated state. (Ali, S., and Coombs, R.C. "Endocrine-Responsive Breast Cancer and Strategies for Combating Resistance," *Nature Rev. Cancer,* 2:101-12, 2002.)

change that allows the ER-dimer, through its deoxyribonucleic acid (DNA)–binding domain, to bind to the promoter (activator) region of the gene. Gene transcription is controlled by two areas on the ER: AF1, where activity is regulated by phosphorylation, and AF2, where ligand-binding of E2 (activating) and anti-estrogens (e.g., tamoxifen-deactivating) occurs. Also, cofactors complex to ER that have either activating (CoA) or deactivating (corepressor, CoR) effects on transcription (see Scheme 6.1 and Scheme 6.2).

The dormant CoR-ER complex recruits histone deacetylases, which maintain the histones that make up the chromatin in a nonacetylated state such that transcription from the associated DNA cannot occur. The activated CoA-E2-ER complex formed upon E2 binding recruits histone acetyltransferases that acetylate the histones, leading to chromatin decondensation and transcription. E2 binding at the ligand-binding pocket produces a conformational change that allows the CoA binding surface on ER to appear. Estrogen binding facilitates interaction of CoA at the AF2 site, whereas anti-estrogen (e.g., tamoxifen) binding inhibits CoA interaction thus, in part, mediating its anticancer effects.

In the absence of estrogen, a number of growth factors can still produce ER activity, thus providing a possible pathway for endocrine-resistant breast tumors, in which case, upon prolonged treatment with endocrine agents, a patient can relapse with disease progression. This alternative ER stimulation is considered to be due to phosphorylation of the AF1 region of the ER by a cascade of events in at least two pathways, some involving kinases, initiated at the receptor tyrosine kinase external to the cell. One of these kinases, HER2 (an epidermal growth factor receptor), when overexpressed in tumors can lead to resistance to tamoxifen. Trastuzumab (Herceptin™), a recombinant humanized mouse monoclonal antibody against HER2, acts as an inhibitor and is used in the treatment of metastatic breast cancer in combination with paclitaxel in situations where HER2 is overexpressed (see Chapter 5).

Some other hormonal agents are also used in the treatment of breast cancer. For example, trilostane is licensed for postmenopausal breast cancer. It is well tolerated but diarrhea and abdominal discomfort can be a problem. Trilostane causes adrenal hypofunction, and so corticosteroid replacement therapy is required. Also, the use of bisphosphonates in patients with metastatic breast cancer may prevent skeletal complications of bone metastases.

6.2.4 Anti-Estrogens

After surgery and associated radiation therapy to remove a tumor mass, adjuvant therapy with anti-estrogens is usually initiated to prevent the growth of metastases (some of which may be ER-(+)). The best known agents in this group are tamoxifen (Nolvadex™) and toremifene (Fareston™), although the more selective SERMs (selective ER modulators) are also discussed below.

6.2.4.1 Tamoxifen

Tamoxifen (Nolvadex™) is a nonsteroidal estrogen antagonist that is used as a first-line anti-estrogen (Structure 6.1). It is licensed in the U.K. for the treatment of breast cancer as well as anovulatory infertility. The structure was first reported by (the then) ICI in 1964, and the separated geometrical isomers 2 years later. The first stereospecific synthesis was reported in 1985. Tamoxifen competes with estrogen for the ERs at AF2, thus preventing estrogen activation and subsequent tumor growth. Approximately one third of nonselected postmenopausal patients respond and the rate is higher (50% to 60%) for patients with ER-(+) tumors.

STRUCTURE 6.1   Structure of tamoxifen (Nolvadex™).

Whereas tamoxifen has a beneficial effect due to its agonist actions on preserving bone mineral density and protection of the cardiovascular system by reducing plasma lipids, its agonist effect on the gynecological tract presents a small but significant risk of generating endometrial cancer, possibly due to nonprevention of activation of AF1 since this is more significant in the uterus. Pure anti-estrogens (e.g., ICI 164384 and Faslodex™) prevent activation of both AF1 and AF2 and have only antagonist effects, in addition to reducing the half-life of ER.

The potential side effects of tamoxifen include hot flashes, vaginal discharge and bleeding, suppression of menstruation in some premenopausal women, pruritic vulvae, gastrointestinal (GI) disturbances, headache, light-headedness, tumor flare, and decreased blood platelet count. Occasionally, cystic ovarian swellings in premenopausal women can occur. Most importantly, tamoxifen can increase the risk of thromboembolism, particularly during times of immobility or during and immediately after major surgery. Patients should be alerted to the symptoms of thromboembolism and advised to report sudden breathlessness and any pain in the calf of one leg. Increased endometrial changes, including hyperplasia, polyps, cancer, and uterine sarcoma, can also occur.

Despite the small risk of endometrial cancer, clinical trials have shown that 20 mg of tamoxifen daily substantially increases the survival rate in early breast cancer, and so the benefits generally outweigh the risks. Furthermore, recent evidence suggests that the risk of breast cancer developing in women at high risk for the disease can be reduced by tamoxifen prophylaxis. However, the adverse effects preclude its routine use in most healthy women. Therefore, research is underway to find new agents with the beneficial chemopreventive effects of tamoxifen but without the accompanying side effects.

6.2.4.2 Toremifene

Toremifene (Fareston™) is a nonsteroidal anti-estrogen structurally similar to tamox-ifen. It was first reported in 1983 by Farmos. It is licensed in the U.K. to treat hormone-dependent metastatic breast cancer in postmenopausal women, although it is not often used due to its side effects. When given orally, toremifene can cause hot flashes, vaginal bleeding or discharge, dizziness, edema, sweating, nausea, vom-iting, chest or back pain, fatigue, headache, skin discoloration, weight increase, insomnia, constipation, dyspnea, vertigo, asthenia, paresis, tremor, reversible corneal opacity, pruritus, and anorexia. Thromboembolic events have also been reported. Rarely, dermatitis, alopecia, emotional lability, depression, jaundice, and stiffness can occur. Hypercalcemia may also result, especially if bone metastases are present. Finally, patients on toremifene are at increased risk for such endometrial changes as hyperplasia, polyps, and cancer; therefore, any abdominal pain, vaginal bleeding, or similar symptoms should be immediately investigated. Toremifene should not be used in women with a history of endometrial hyperplasia, severe hepatic impairment, or thromboembolic events.

STRUCTURE 6.2 Structure of toremifene (Fareston™).

6.2.4.3 Novel Selective ER Modulators

Much research has gone into the discovery and research of novel selective ER modulators (known as "SERMs") as replacements for tamoxifen. Ideally, SERMs should possess increased potency at AF2 and have agonist effects on bone and plasma lipids but should lack any unwanted agonistic effects on the uterus. They fall into three general categories: (1) nonsteroidal compounds with a triphenylethylene struc-ture, thereby resembling tamoxifen, and toremifene, (2) nonsteroidal compounds based on a benzothiophene structure, and (3) steroidal pure anti-estrogens.

The first group, those with a triphenylethylene structure, includes compounds such as toremifene, droloxifene, and idoxifene (see Structure 6.3). Some of these drugs have shown limited individual advantages over tamoxifen in preclinical eval-uations. For example, some have shown a greater potency, a shorter half-life, or a reduced estrogenic effect in the rat uterus. Others do not cause the hepatocarcinogenicity associated with tamoxifen. However, in clinical trials, these compounds have proved to be no more efficacious or safe than tamoxifen.

STRUCTURE 6.3 First-generation SERMs: Tamoxifen-like triphenylethylene structures.

Drugs in the second (benzothiophene) group, include raloxifene. Initially developed for the treatment of osteoporosis, raloxifene was found to diminish the incidence of breast cancer by a significant amount and so was subsequently studied in breast cancer trials. Arzoxifene (LY353,381), a more potent drug than raloxifene with an improved side effect profile, is also in development.

STRUCTURE 6.4 Second-generation SERMs: Benzothiophene structures.

Third generation SERMs are represented by the steroidal anti-estrogen fulvestrant (ICI 182,780, Faslodex™), which is essentially an estradiol structure with a long 7β-hydrophobic chain that not only impairs the necessary ER dimerization for agonist action but also down-regulates the ER, thereby acting as a potent pure anti-estrogen in all tissues (breast, uterus, and probably bone) (see Structure 6.5). A second example is EM-800, an orally active prodrug of the benzopyrene EM-652 (SCH 57068), which had the required profile in preclinical studies, with no agonist activity on the uterus and, like fulvestrant, down-regulated ER levels. These

X = $\overset{O}{\underset{\|}{\wedge\wedge\wedge\wedge\wedge}} S \wedge\wedge CF_2CF_3$ EM-652 (SCH 57068)

Fulvestrant (Faslodex™)

STRUCTURE 6.5 Third-generation SERMs: Steroidal pure anti-estrogens.

molecules are undergoing further studies as first-line adjuvant treatments for breast cancer.

Other analogs being studied include the orally active SR-16234 and ZK-191703. However, it is presently unclear how important the latest generation of SERMs will be for the treatment of breast cancer.

6.2.5 AROMATASE INHIBITORS

Aromatase inhibitors work predominantly by blocking the formation of estrogens from androgens in the peripheral tissues. They do not inhibit ovarian estrogen synthesis and thus should not be used in premenopausal women. Androstenedione and testosterone are converted by the cytochrome P450 aromatase enzyme ($P450_{AROM}$) to estrone and estradiol, respectively, as the final step in the steroidogenesis pathway from cholesterol (see Scheme 6.3). Selective inhibition of aromatase leads to reduced estrogen plasma levels without affecting other hormones produced by the steroidogenesis pathway. In postmenopausal women for whom estrogen synthesis has ceased in the ovaries and is carried out mainly in adipose and other tissues, the breast can have up to a 20-fold higher concentration of estrogen than does plasma due to local synthesis by the action of the sulfatase enzyme on estrone sulfate. The various generations of aromatase inhibitors are described below.

First-Generation Inhibitors

The first-generation aromatase inhibitor, aminoglutethimide (see Structure 6.6), was discovered serendipitously when it was being developed originally as an anticonvulsant agent. During clinical trials, it was found to cause adrenal insufficiency that was serious enough to prevent further development. Subsequent studies showed, however, that the drug inhibited a number of cytochrome P450 enzymes in the adrenal steroidogenesis pathway, and so it was redeveloped for use in advanced breast cancer as a form of medical adrenalectomy. Unfortunately, use of aminoglutethimide for this purpose was limited due to a number of adverse reactions, including rashes and drowsiness. However, the discovery that its activity was essentially due to inhibition of aromatase stimulated the development of many new types of inhibitors in the late 1980s and early 1990s. These are known as second- or third-generation inhibitors based on the chronological order of their appearance, with

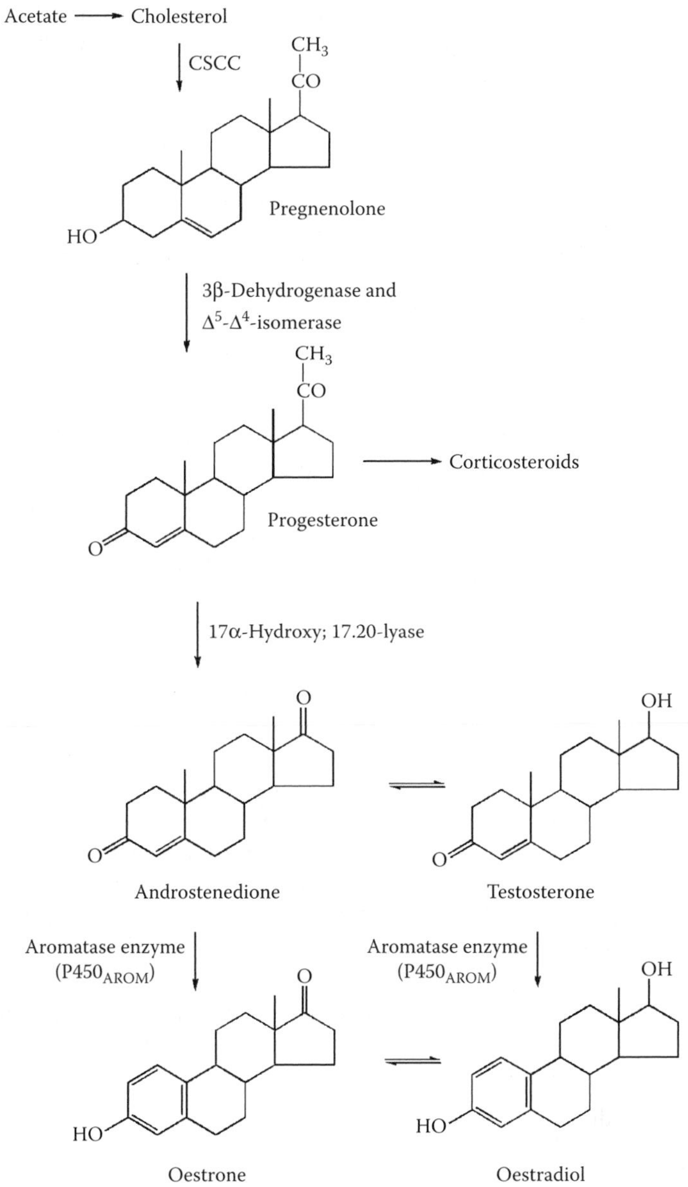

SCHEME 6.3 The steroidogenesis pathway.

further classification as Type 1 or Type 2, depending on their mechanism of action. Type 1 inhibitors are known as *enzyme inactivators* and are steroidal derivatives of androstenedione. They interact with the aromatase enzyme at the same site as andros- tenedione but, unlike the latter, they are converted by the aromatase to reactive intermediates that then bind irreversibly. Type 2 inhibitors have nonsteroidal structures and bind in a reversible manner to the heme group of the aromatase enzyme via a basic nitrogen. Anastrozole and letrozole (see Structure 6.7 and Structure 6.8) are examples of Type 2 (third-generation) inhibitors that bind to the heme of the enzyme via the nitrogens of their triazole rings.

Second-Generation Inhibitors

Formestane (4-hydroxyandrostenedione) and fadrozole are examples of second-generation Type 1 and Type 2 aromatase inhibitors, respectively. Both are effective in the clinic, but formestane has the disadvantage of requiring administration by intramuscular injection. Fadrozole has a different disadvantage of causing the suppression of aldosterone, which limits its administration to doses that provide only approximately 90% inhibition of aromatase activity. A number of other second-generation inhibitors have been studied in clinical trials, but none have been licensed.

Third-Generation Inhibitors

Examples of orally administered third-generation inhibitors include exemestane (steroidal Type I, irreversible) (Structure 6.10) and the nonsteroidal triazole-containing agents anastrozole, letrozole ((Structure 6.7 and Structure 6.8), and vorozole (Type II, reversible) (Structure 6.9). In contrast to aminoglutethimide and fadrozole, these third generation inhibitors discovered in the early 1990s have not presented any selectivity issues at clinical doses and are 10^3- to 10^4-fold more potent than aminoglutethimide, with negligible effects on basal levels of aldosterone or cortisol. Anastrozole and letrozole are at least as effective as tamoxifen when used in postmenopausal women for the first-line treatment of metastatic breast cancer.

The nonsteroidal 1,2,4-triazoles, such as anastrozole and letrozole, are usually stable to metabolism *in vivo*. They are very potent, *in vivo,* with letrozole (2.5 mg daily) and anastrozole (1 mg daily) reducing serum estrogen levels by 99% and 97%, respectively (beyond the limit of detection in many patients). This performance is superior to that of aminoglutethimide (1000 mg daily, 90% reduction) but comparable with exemestane (25 mg daily, 97% reduction). Advantages of these third-generation inhibitors include improved tolerance with fewer side effects, rather than improved response rates (11% to 24%) and durations (18 to 23 months) compared to aminoglutethimide (12% to 30%, 13 to 24 months) or the progestin megestrol (8% to 17%, 12 to 18 months). However, it is not yet known whether the benefits of aromatase inhibitors persist over the long term.

6.2.5.1 Aminoglutethimide

$P450_{AROM}$ inhibition was discovered serendipitously in the late 1950s, when the experimental anticonvulsant agent aminoglutethimide (Orimeten™) was observed to block adrenal steroidogenesis. Follow-up clinical trials in advanced breast cancer showed that a 20% to 40% response rate and a duration of remission of 6 to 12 months could be achieved. However, the agent caused a number of serious side effects, including adrenal hypofunction (particularly under conditions of stress such as during acute illness, trauma, or surgery), which required corticosteroid replacement therapy. Aminoglutethimide has now been largely replaced by newer, more-specific aromatase inhibitors, such as anastrozole, which are much better tolerated.

STRUCTURE 6.6 Structure of aminoglutethimide (Orimeten™).

Aminoglutethimide lacks selectivity for $P450_{AROM}$, and side effects include central nervous system depression (e.g., drowsiness, lethargy) and rash, which is sometimes accompanied by fever. Occasionally, dizziness, nausea, and other minor side effects occur. Aminoglutethimide inhibits several other P450 enzymes on the steroidogenesis pathway, thus blocking production of various adrenal steroids and requiring coadministration of glucocorticoid. Given orally, it is used to treat advanced breast or prostate cancer. When used for prostate cancer, it is often coadministered with a glucocorticoid and sometimes with a mineralocorticoid. It is also used to treat Cushing's syndrome arising from malignant disease.

6.2.5.2 Anastrozole

Anastrozole (Arimidex™) was first reported in 1989 by (the then) ICI (Structure 6.7). Given orally, it is used in postmenopausal women who are unable to take tamoxifen because of a high risk of thromboembolism or endometrial abnormalities. It is used as an adjuvant treatment for early estrogen-receptor-positive breast cancer and is recommended for a 5-year duration of treatment only. If menopausal status is unknown, a laboratory test is recommended. The drug should be used cautiously in women who are susceptible to osteoporosis, and bone mineral density should be assessed before and after treatment when necessary.

Possible side effects of anastrozole include bleeding, vaginal dryness, hot flashes, rash (including Stevens-Johnson syndrome), hair thinning, nausea, anorexia, vomiting, diarrhea, asthenia, bone fractures, arthralgia, and drowsiness. A slight increase in total cholesterol has also been reported during treatment.

6.2.5.3 Letrozole

Letrozole (Femara™), a nonsteroidal aromatase inhibitor structurally related to fadrozole, was first reported in 1987 by (the then) Ciba-Geigy and is used mainly for the treatment of advanced breast cancer in postmenopausal women (Structure 6.8). It is also used in postmenopausal women with localized hormone-receptor-positive breast

STRUCTURE 6.7   Structure of anastrozole (Arimidex™).

cancer as a preoperative measure prior to breast-conserving surgery. Letrozole is not recommended for use in premenopausal women or for those with severe hepatic impairment.

STRUCTURE 6.8   Structure of letrozole (Femara™).

Given orally, the drug can cause a range of side effects, including hot flashes, nausea, vomiting, dyspepsia, constipation, diarrhea, abdominal pain, anorexia (and sometimes weight gain), dyspnea, chest pain, coughing, dizziness, fatigue, headache, infection, muscular skeletal pain, peripheral edema, rash, and pruritus.

6.2.5.4 Vorozole

Vorozole (Rivizor™) is a nonsteroidal aromatase inhibitor closely related to anastrozole and letrozole that was first reported in 1988 by Janssen (see Structure 6.9). Resolution of the enantiomers of vorozole was reported in 1994.

Vorozole is not currently licensed for use in the U.K. It did not demonstrate a survival advantage over megestrol acetate and aminoglutethimide in two recent Phase III trials investigating these agents as second-line therapy after tumor progression in patients on tamoxifen. Interestingly, because of its ability to stimulate ovulation, vorozole is also being developed as a new treatment for ovulation induction and *in vitro* fertilization.

STRUCTURE 6.9 Structure of vorozole (Rivizor™).

6.2.5.5 Exemestane

Exemestane (Aromasin™), the first example of an irreversible aromatase inhibitor, was reported in 1987 by Farmitalia (Structure 6.10). It is used for the treatment of advanced breast cancer in postmenopausal women in whom anti-estrogenic therapy has failed.

STRUCTURE 6.10 Structure of exemestane (Aromasin™).

Given orally, the main adverse effects of this agent include sweating, hot flashes, dizziness, nausea, and fatigue. Other less frequently observed side effects include rash, alopecia, insomnia, depression, vomiting, dyspepsia, headache, peripheral edema, anorexia, constipation, and abdominal pain. More rarely, thrombocytopenia and leukopenia can occur. The drug should be used with caution in women with hepatic and renal impairment, and it is not indicated for use in premenopausal women.

6.3 PROSTATIC CANCER

Prostatic cancer is mainly hormone dependent, being promoted by the androgen dihydrotestosterone (DHT), which is derived from testosterone by the action of

5α-reductase. Thus, one obvious form of treatment involves removal of the DHT stimulus. Prostatic cancer is usually well developed on presentation, so survival rates are low and treatment is aimed at increasing the survival time and quality of life. Removal of the DHT stimulus to tumor growth can be achieved either by blocking its synthesis or the action at its receptor. Surgical removal of the prostate or testes (orchidectomy) is less prevalent, having been largely replaced by endocrine therapy (estrogens) and, in more recent years, by treatment with anti-androgens and luteinizing hormone releasing hormone (LH-RH) analogs. One problem with surgical removal of the prostate is that it can lead to incontinence and sexual dysfunction.

Hormonal therapy does not provide a cure for prostate cancer because the tumor usually becomes resistant (refractory) to treatment by a process that is not completely understood. A prostate cancer cell population usually contains both androgen-sensitive (i.e., dependent on DHT for growth and viability) and androgen-insensitive cells. It is thought that, over time, hormonal treatment suppresses the sensitive cells, leaving the insensitive cells to flourish. Resisting programmed cell death, they gradually become the majority in the population.

Androgen receptors are proteins with a molecular weight of 120 kDa, and their synthesis is guided by genes on the X chromosome. In cells, the androgen-binding monomer (4.4S) is bound to a receptor-associated protein and to other small molecules to produce a 9S inactive oligomer. DHT binds to the androgen receptor (AR), as does testosterone but with lower affinity. The AR-DHT complex then binds to the nuclear androgen response element sequence of the gene relevant, thereby promoting transcription of messenger ribonucleic acid (mRNA) and leading to protein synthesis and maintenance of cellular growth of the prostate (Scheme 6.4).

The zinc fingers of the AR, each consisting of a zinc atom and four cysteine amino acids, are separately responsible for recognition of the response element and the AR dimer formation area (see Scheme 6.5). The zinc fingers insert into DNA at

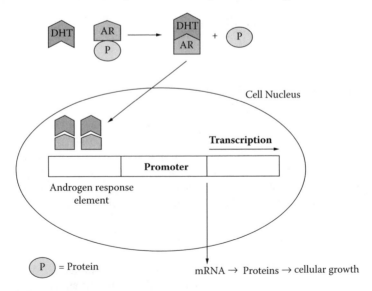

SCHEME 6.4 DHT activation of the androgen response.

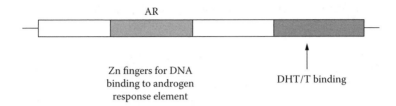

Zn fingers for DNA
binding to androgen
response element

DHT/T binding

SCHEME 6.5   Schematic view of the androgen receptor (AR).

two six base-pair runs of particular sequences separated by a three-base spacer. Two
have been identified as binding to ATAGCAtctTGTTCT or AGTACTccaAGAACC
sequences (Scheme 6.6).

The binding of testosterone/DHT to the hormone-binding site near the terminal
carboxyl group activates the receptor complex so that it can bind to the response
element. This is achieved by disaggregation of the macromolecular complex and
release of an accessory protein, consequently exposing the DNA binding site of the
4.4S receptor. Anti-androgens competitively prevent DHT from binding to the AR,
and the AR-anti-androgen complex formed cannot become correctly aligned to bind
to the response element, thus preventing dimerization of the monomer AR and
subsequent promotion of transcription.

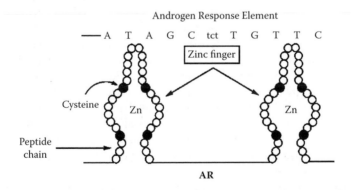

Androgen Response Element

— A   T   A   G   C   tct   T   G   T   T   C

Zinc finger

Cysteine

Zn   Zn

Peptide
chain

AR

SCHEME 6.6   Zinc fingers of AR binding to nucleotide sequences of the androgen response
element.

Metastatic prostate cancer normally responds to androgen-depleting hormonal
strategies. Standard treatments include bilateral subcapsular orchidectomy or use of
gonadorelin analogs, such as buserelin, goserelin, leuprorelin, or triptorelin. In these
cases, responses can last for 12 to 18 months. No entirely satisfactory therapy exists
for disease progression (i.e., hormone-refractory prostate cancer) but, occasionally,
patients respond to other methods of hormone manipulation, such as the use of an
anti-androgen. Bone metastases can often be palliated with radiation or, if widespread, with strontium, aminoglutethimide, or prednisolone.

6.3.1 GONADORELIN ANALOGS (LH-RH ANALOGS; AGONISTS AND ANTAGONISTS)

Luteinizing hormone-releasing factor (LH-RF), also known as luteinizing hormone-releasing hormone (LH-RF), gonadorelin, gonadotropin-releasing factor, gonado-liberin, or luliberin, is a neurohumoral hormone of molecular weight 1182.29 kDa produced in the hypothalamus (see Structure 6.11). It stimulates production of the pituitary hormones, luteinizing hormone (LH), and follicle-stimulating hormone (FSH), which in turn produce changes that result in females, in the induction of ovulation. In males, it leads to the production of testosterone. LH-RF was first isolated from porcine hypothalamic extracts and then structurally elucidated and synthesized in the early 1970s. It has been previously commercialized under the names Fertagyl™ (Intervet), Fertiral™ (Hoechst Marion Roussell), Kryptocur™ (Hoechst Marion Roussell), Relefact LH-RH™ (Hoechst Marion Roussell), Cys-torelin™ (Abbott), Hypocrine™ (Tanabe Seiyaku), Lutrelef™ (Ferring), and Lutrepulse™ (Ortho).

5–oxoPro–His–Trp–Ser–Tyr–Gly–Leu–Arg–Pro–GlyNH$_2$

STRUCTURE 6.11   Structure of luteinizing hormone-releasing factor (LH-RH).

Analogs of gonadorelin are of similar effectiveness to diethylstilbestrol (DES) treatment or even orchidectomy but are expensive and require parenteral administration, at least initially. These agents work by initially stimulating but then depressing release of LH by the pituitary gland. In the initial stages of treatment (1 to 2 weeks), increased production of testosterone may stimulate progression of the disease and, in patients affected, this tumor flare can lead to increased bone pain, ureteric obstruction, or spinal cord compression. When such problems are anticipated, alternative treatments (e.g., orchidectomy) or concomitant use of anti-androgens, such as flutamide or cyproterone acetate, is recommended. In this case, anti-androgen treatment should be started 3 days before the gonadorelin analog and continued for 3 weeks. For this reason, gonadorelin analogs should be used with caution in men at risk for tumor flare, and these patients should be observed carefully in the first month after treatment.

The introduction of LH-RH analogs (see Structure 6.12 through Structure 6.15) has provided an alternative, without significant side effects, to the use of estrogens and/or orchidectomy in the treatment of advanced prostate cancer. Naturally occurring LH-RH (S-oxoPro-His-Trp-Ser-Tyr-Gly-Leu-Arg-Pro-Gly-NH$_2$) has a short half-life (and a pulsatile action on the receptor) but, by substituting the amino acid at the 6th position, deleting the amino acid at the 10th position, and adding an ethylamide group to the proline residue at the 9th position, the first synthetic analog, leuprolide (Lupron™), was produced (see Structure 6.14). This early agent had a greatly increased potency together with prolonged activity and a nonpulsatile action on the receptor.

The initial effect of LH-RH agonists is to stimulate the secretion of LH and FSH, leading to increased testosterone synthesis in males. This action can provide elevation of serum testosterone to between 140% and 170% of basal levels within several days, and an accompanying resurgence of symptoms (i.e., tumor flare), with 3–17% of patients affected. Continuous administration, however, leads to dramatic inhibitory effects through a process of down regulation of LH-RH pituitary membrane receptors and a reduction in gonadal receptors for LH and FSH, thus resulting in suppression of testosterone secretion comparable to surgical castration. Thus, chronic administration causes the pituitary gland to become refractory to additional stimulation by endogenous LH-RH, and so testicular androgen production is ablated.

The most commonly used LH-RH agonists are buserelin (Suprefact™), goserelin (Zoladex™), leuprorelin acetate (Prostap SR™), and triptorelin (Decapeptyl SR™) (see Structure 6.12 through Structure 6.15). Apart from the initial tumor flare (treated by coadministration of anti-androgens), other side effects include atrophy of the reproductive organs, loss of libido, and impotence. Because they are only able to cause a decrease in testicular androgens, while leaving adrenal androgen production unaffected (i.e., incomplete androgen clearance), the use of anti-androgens in combination with LH-RH agonists has been studied and found to give greater survival rates than the use of LH-RH agonists alone.

Gonadorelin analogs are also used to treat breast cancer in women and other related indications. They should be used cautiously in women with metabolic bone disease because decreases in bone mineral density may occur. Gonadorelin analogs cause side effects similar to those caused by menopause in women and orchidectomy in men, including hot flashes, sweating, sexual dysfunction, vaginal dryness or bleeding, and gynecomastia or other changes in breast size. Other side effects include hypersensitivity events (rashes, pruritus, asthma and, rarely, anaphylaxis), injection site reactions, headache (rarely migraine), mood changes, sleep disorders, weight changes, GI disturbances, visual disturbances, peripheral edema, dizziness, hair loss, myalgia, and arthralgia.

6.3.1.1 Buserelin

Buserelin (Suprefact™) is a synthetic nonapeptide agonist analog of LH-RH, first reported by Hoechst in 1976 (Structure 6.12). It is used for the treatment of advanced prostate cancer. Side effects of buserelin include musculoskeletal pain, hearing disorders, increased thirst, worsening hypertension, anxiety, memory and concentration disturbances, fatigue, nervousness, palpitations, leukopenia, thrombocytopenia, altered blood lipids, and glucose intolerance. The drug is normally given by subcutaneous injection for the first 7 days, followed by administration by nasal spray, although the latter can cause nasal irritation, nosebleeds, and an altered sense of taste and smell. Buserelin should not be given to patients with depression.

5–oxoPro–His–Trp–Ser–Tyr–D–Ser(t-Bu)–Leu–Arg–ProNHCH$_2$CH$_3$

STRUCTURE 6.12 Structure of buserelin (Suprefact™).

6.3.1.2 Goserelin

Goserelin (Zoladex™) is a synthetic peptide gonadorelin analog agonist of LH-RH first synthesized by (the then) ICI in 1977 (Structure 6.13). It is licensed in the U.K. for the treatment of prostate cancer, estrogen-receptor-positive early breast cancer, and advanced breast cancer in premenopausal women.

5–oxoPro–His–Trp–Ser–Tyr–D–Ser(t-Bu)–Leu–Arg–Pro–NHNHCONH$_2$

STRUCTURE 6.13 Structure of goserelin (Zoladex™).

For breast or prostate cancer, a 3.6-mg acetate implant (Zoladex™) is administered every 28 days into the anterior abdominal goserelin wall by subcutaneous injection. A longer-acting version (Zoladex LA™) containing 10.8 mg of the acetate is licensed for prostate cancer and can be administered every 12 weeks. Side effects include the usual ones experienced with agonist analogs of LH-RH in addition to transient changes in blood pressure, paraesthesia and, rarely, hypercalcemia in patients with metastatic breast cancer. Goserelin is contraindicated in women who are pregnant or have undiagnosed vaginal bleeding.

6.3.1.3 Leuprorelin Acetate

Leuprorelin (Prostap SR™), also known as leuprolide (and originally marketed as Lupron™), is a synthetic nonapeptide agonist analog of LH-RH first reported by Takeda in 1975 (Structure 6.14). Licensed for the treatment of advanced prostate cancer, leuprorelin is administered every 4 weeks by subcutaneous or intramuscular injection as a 3.75-mg microsphere formulation (Prostap SR™). Alternatively, a larger subcutaneous formulation (11.25 mg) can be given every 3 months. In addition

5–oxoPro–His–Trp–Ser–Tyr–D–Leu–Leu–Arg–ProNHC$_2$H$_5$

STRUCTURE 6.14 Structure of leuprorelin acetate (Prostap SR™).

to the usual side effects common with agonist analogs of LH-RH, other adverse reactions include fatigue, muscle weakness, paraesthesia, hypertension, palpitations, and alteration of glucose tolerance and blood lipids. Thrombocytopenia and leukopenia have also been reported.

6.3.1.4 Triptorelin

Triptorelin (Decapeptyl SR™) is a synthetic peptide agonist analog of LH-RH, first reported in 1976 (Structure 6.15). Used to treat advanced prostate cancer and endometriosis, this drug is administered in the form of a 3-mg aqueous copolymer microsphere suspension (Decapeptyl SR™) by intramuscular injection every 4 weeks. In addition to the usual side effects of agonist analogs of LH-RH, other reported adverse effects include transient hypertension, dry mouth, excessive salivation, paraesthesia, and increased dysuria.

5–oxoPro–His–Trp–Ser–Tyr–D–Trp–Leu–Arg–Pro–GlyNH$_2$

STRUCTURE 6.15   Structure of triptorelin (Decapeptyl SR™).

6.3.2 LH-RH Receptor Antagonists

LH-RH receptor antagonists can suppress LH release from the pituitary at the outset, thus circumventing the tumor flare effect observed with gonadorelin analogs. However, the relatively low potency of antagonists and their adverse effects due to histamine release has delayed their introduction into clinical practice.

Ser–Tyr–D–Cit–Leu–Arg–Pro–D–AlaNl–D–Ala–NH$_2$

Cetrorelix

STRUCTURE 6.16   Structure of cetrorelix.

STRUCTURE 6.17   Examples of potent orally active nonpeptide LH-RH antagonists (from Takeda). The fluorinated analog (right) has 50 times the affinity of LH-RH for its receptor.

Several peptide antagonists with low adverse effects have been evaluated in clinical trials (e.g., Cetrorelix [SB 75300] (see Structure 6.16) and Abarelix). However, many have the drawback that, due to insufficient bioavailability, they need to be administered either by daily injection, intranasal spray, or a sustained-release delivery system.

Potent orally active nonpeptide antagonists described by Takeda have been designed to mimic the shape of the type II β-turn involving residues 5 to 8 (Tyr-Gly-Leu-Arg) of LH-RH (p-Glu-His-Trp-Ser-Tyr-Gly-Leu-Arg-Pro-Gly-NH₂) (Structure 6.17). These agents were discovered by screening a library of compounds active at G-protein receptors. Interestingly, the fluorinated analog has 50 times the affinity of LH-RH for its receptor and is equiactive with the agonist leuprolide.

Merck has also described several series of nonpeptide antagonists (e.g., Structure 6.18). Chemical library screening identified a lead quinoline molecule that was optimized by changes including conversion of the pyridine ring to a piperidine and the addition of nitro and methyl groups.

STRUCTURE 6.18   A lead quinolone LH-RH receptor antagonist identified by Merck (left), and a resulting optimized lead (right).

STRUCTURE 6.19 Indole- (left) and pyrrolopyridine-based (right) LH-RH receptor antagonists.

Other types of structures with LH-RH antagonist properties have also been described. For example, indole- and pyrrolopyridine-based analogs have been described that have significant antagonist properties (Structure 6.19).

6.3.3 ANTI-ANDROGENS

Anti-androgens inhibit the binding of DHT and other androgens to the androgen receptors in target tissues. Target cells are located in all areas of the body that depend on androgens, such as the male genital skin, the seminal vesicles, the prostate, fatty tissues and breast tissue as well as the hypothalamus and pituitary gland.

Anti-androgens bind to the androgen receptor, creating a receptor–anti-androgen complex that is unable to elicit a response at the response element, and so androgen-dependent gene transcription and protein synthesis are not stimulated. Therefore, anti-androgens can block the tropic effect of all androgens, not only peripherally (e.g., in the prostate) but also centrally (e.g., in the hypothalamus and pituitary gland). They are used in conjunction with LH-RH agonists to reduce the effects of androgens produced by the adrenals that are not under LH control.

Anti-androgen agents are divided into two families based on their structures (steroidal or nonsteroidal). The most extensively studied steroidal agent is cyproterone acetate (Cyprostat® or Androcur®) (Structure 6.20). In prostatic cells, this drug acts as a competitive inhibitor of the binding of DHT to androgen receptors, and also acts centrally, suppressing the corticotropic axis and thereby reducing effective plasma testosterone levels due to its progestin-like activity as well as its own low agonist activity, which is approximately 1/10 that of testosterone.

Nonsteroidal anti-androgens have, such as flutamide (Drogenil™), bicalutamide (Casodex™), and nilutamide (Anandron™) (see Structure 6.21 through Structure 6.23), more limited central activity (depending on their individual structures) but all have significant and potent peripheral effects. They displace testosterone and DHT from the androgen receptor in the prostate and can do the same in the hypothalamus. If such agents are used alone, this central blocking effect can potentially lead to an increase in the release of LH-RH by feedback control and subsequent LH production, thus causing a slow but gradual rise in serum testosterone levels to overcome the blockade (see Scheme 6.7). This is undesirable because it can stimulate prostatic tumor growth.

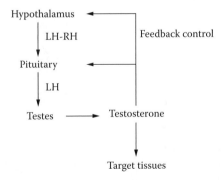

SCHEME 6.7   Control of androgen levels by the hypothalamus-pituitary axis.

Cyproterone acetate, flutamide, and bicalutamide can be used to inhibit the tumor flare that may occur after commencing treatment with gonadorelin analogs. Cyproterone acetate and flutamide are also licensed in the U.K. for use alone in patients with metastatic prostate cancer refractory to gonadorelin analog therapy. Bicalutamide is used for prostate cancer either alone or as an adjunct to other therapies, according to the clinical circumstances.

The nonsteroidal anti-androgens are interesting from a medicinal chemistry standpoint in that bicalutamide and the active metabolite of flutamide (hydroxyflutamide) both contain a pseudo five-membered ring stabilized by a cyclic NH- - -O hydrogen bond. However, nilutamide contains a fully formed bioisosterically equivalent imidazolidinedione ring of similar three-dimensional shape.

6.3.3.1 Cyproterone Acetate

Cyproterone acetate (Cyprostat™ or Androcur™) is the main steroidal anti-androgen in clinical use today. The free alcohol form was first reported in 1965 by Schering AG. Interestingly, although the alcohol form is an anti-androgen in which the acetate (i.e., Cyprostat™) is both an anti-androgen and a progestogen, thus affecting both hormone secretion and spermatogenesis. In target prostatic cells, cyproterone acetate acts as a competitive inhibitor of the binding of DHT to androgen receptors. It also acts centrally, suppressing the corticotropic axis and thereby reducing plasma testosterone levels through its progestin-like activity as well as its agonist activity (approximately 1/10 testosterone).

Cyproterone acetate is licensed in the U.K. for the treatment of prostate cancer and is given orally for the tumor flare that occurs initially with gonadorelin therapy. Also, where oral therapy is preferred or where orchidectomy or gonadorelin analogs are contraindicated or are not likely to be well tolerated, it can be used for long-term palliative therapy. Cyproterone can also be used to control the hot flashes that sometimes occur with gonadorelin therapy or after orchidectomy. Clinical trials have shown that, with cyproterone acetate, objective responses can be obtained in 33% of patients with advanced prostate cancer, and stabilization of the disease can be achieved in 40% of patients.

STRUCTURE 6.20   Structure of cyproterone acetate (Cyprostat™ or Androcur™).

One of the most serious side effects of this agent is direct hepatotoxicity, including jaundice, hepatitis, and hepatic failure, which can occur after several months of therapy in patients on large doses of 200 to 300 mg daily. For these patients, careful monitoring is necessary with liver function tests before and after treatment. Furthermore, patients on lower doses who show symptoms of hepatotoxicity should be monitored, and the drug should not be given to patients with any type of liver problem unless the benefits are judged to outweigh the risks. Lethal cardiovascular events can also occur with cyproterone acetate, and at a slightly higher rate than with diethylstilbestrol (DES). Blood counts should be monitored initially and throughout the treatment. In addition, skilled or performance-based tasks such as driving may be affected by fatigue and lassitude. There is also a risk of recurrence of severe depression, sickle-cell anemia, diabetes mellitus, and thromboembolic disease. In addition, adrenocortical function should be monitored regularly throughout the treatment period.

6.3.3.2 Flutamide

Flutamide (Drogenil™ or Eulexin™), the first nonsteroidal anti-androgen to be developed, was reported by Sherico in 1967 (Structure 6.21). It is a prodrug, with the active metabolite hydroxyflutamide acting by inhibiting the nuclear binding of testosterone and DHT to the androgen receptor. Hydroxyflutamide has both peripheral and central activity on all androgen target cells.

Flutamide   Hydroxyflutamide (Active metabolite)

STRUCTURE 6.21 Structure of flutamide (Drogenil™ or Eulexin™) and its active metabolite hydroxyflutamide.

Molecular modeling studies of hydroxyflutamide have attributed its binding affinity, which is greater than that of flutamide, to the dominant conformation in which the NH-proton is hydrogen bonded to the tertiary hydroxyl function. The related agent bicalutamide is able to form a comparable internal hydrogen bond and thus take up a similar three-dimensional shape. In the case of nilutamide, the structure contains a fully formed imidazolidinedione ring of similar three-dimensional shape.

Given orally, flutamide is licensed in the U.K. for use in advanced prostate cancer, although it should not be given to patients with cardiac disease or hepatic impairment. Due to the potential for hepatotoxicity (hepatic injury with occasional resulting deaths have been reported), periodic liver function tests are important when the first indications of liver disorders appear or for those on long-term therapy. One of the most common side effects is gynecomastia, with approximately 60% of patients affected, 10% of whom suffer severely. Nausea, vomiting, diarrhea, increased appetite, insomnia, and tiredness can also be significant. Other reported side effects include lymphedema, systemic lupus erythematosus–like syndrome, decreased libido, hemolytic anemia, rash, pruritus, gastric and chest pain, blurred vision, dizziness, inhibition of spermatogenesis, edema, and headache.

6.3.3.3 Bicalutamide

Bicalutamide (Casodex™) is a nonsteroidal peripherally active anti-androgen first reported by (the then) ICI in 1984 (Structure 6.22). Interestingly, the structure allows a hydrogen-bonded pseudo five-membered ring to form in a similar manner to the active metabolite of flutamide (hydroxyflutamide), and this is thought to contribute to its activity.

STRUCTURE 6.22   Structure of bicalutomide (Casodex™).

Licensed in the U.K. for the treatment of prostate cancer, bicalutamide is well absorbed orally with a half-life of 5 to 7 days. It selectively blocks peripheral androgen receptors, which is a potential advantage over centrally acting agents in that DHT synthesis is not stimulated by LH-RH release from the hypothalamus. Side effects include jaundice, cholestasis, hematuria, dry skin, pruritus, dyspepsia, hirsutism, nausea, vomiting, diarrhea, hot flashes, alopecia, decreased libido, breast tenderness, gynecomastia, depression, asthenia, and abdominal pain. Other less-common side effects include cardiovascular events and GI disturbances. Bicaluta-mide should not be given to patients with hepatic impairment.

6.3.3.4 Nilutamide

Nilutamide (Anandron™) was first reported in 1977 by Roussel-UCLAF (Structure 6.23). It is a nonsteroidal anti-androgen with structural similarities to the bicalut-amide and active metabolite of flutamide (hydroxyflutamide). However, unlike the pseudo five-membered rings of hydroxyflutamide and bicalutamide, which are held in place by hydrogen bonds, nilutamide (which does not require metabolic activation) contains a fully formed five-membered imidazolidinedione ring of similar three-dimensional shape.

STRUCTURE 6.23   Structure of nilutamide (Anandron™).

Nilutamide is well absorbed and has a much longer half-life than flutamide (i.e., 45 hours compared to 5 to 6 hours for flutamide), permitting once-daily dosing. It is both centrally and peripherally active, and is therefore associated with a rise in plasma LH and subsequently plasma testosterone. Nilutamide is particularly useful in patients who are intolerant to flutamide.

6.4 NEUROENDOCRINE TUMORS: SOMATOSTATIN ANALOGS

Neuroendocrine (particularly carcinoid) tumors (and also acromegaly) are usually stimulated by various hormone-releasing factors. Therefore, agonist analogs of the hypothalamic release-inhibiting hormone somatostatin (e.g., octreotide and lanreotide) can slow down their growth.

$$Ala-Gly-Cys-Lys-Asn-Phe-Phe-Trp$$
$$|\qquad\qquad\qquad |$$
$$Cys-Ser-Thr-Phe-Thr-Lys$$

STRUCTURE 6.24 Structure of somatostatin (Hypothalamic release-inhibiting hormone).

Somatostatin (also known as somatotropin release-inhibiting factor [SRIF or SRIF-14], growth hormone-release inhibiting factor, or hypothalamic release-inhibiting hormone) is a widely occurring cyclic tetradecapeptide with a molecular weight of 1637.91 kDa (Structure 6.24). Together with other peptides (e.g., somatoliberin), it mediates the neuroregulation of somatotropin secretion. In particular, it inhibits release of growth hormone, insulin, and the glucagons, and is also a potent inhibitor of a number of systems including central and peripheral neural, GI, and vascular smooth muscle. It was first isolated from ovine hypothalamic extracts and then structurally elucidated and synthesized in the early 1970s.

Because somatostatin is a naturally occurring neuropeptide that degrades rapidly in the bloodstream, attempts have been made to produce agonist analogs with drug-like properties. This approach has been successful, and octreotide (Sandostatin™) and lanreotide (Somatuline LA™) are licensed for the relief of symptoms associated with neuroendocrine (particularly carcinoid) tumors and acromegaly. Octreotide is also licensed for the prevention of complications following pancreatic surgery, and can also be useful in reducing vomiting in palliative care and in stopping variceal bleeding.

It should be noted that growth hormone–secreting pituitary tumors can expand during treatment with somatostatin analogs, leading to serious complications; therefore, patients should be carefully monitored for tumor expansion. The gallbladder can also be affected, and the duration and depth of hypoglycemia produced should be monitored carefully. Adverse effects with these agents mainly involve GI disturbances, including anorexia, nausea, vomiting, abdominal pain and bloating, flatulence, diarrhea, and steatorrhea. In addition, postprandial glucose tolerance may be impaired, gallstones have been reported after long-term treatment and, more rarely, pancreatitis has been observed. Lastly, pain and irritation may occur at the injection site.

6.4.1 OCTREOTIDE

Administered subcutaneously, octreotide (Sandostatin™) is a long-acting octapeptide analog of somatostatin and was first reported in 1981 (Structure 6.25). It is licensed in the U.K. for the treatment of carcinoid tumors with features of carcinoid syndrome, VIPomas, and glucagonomas.

Octreotide (Sandostatin™)

STRUCTURE 6.25 Structure of octreotide (Sandostatin™).

A microsphere powder for reconstitution to an aqueous suspension of a depot preparation known as Sandostatin LAR™ is also available, and is licensed as a longer-acting preparation for neuroendocrine tumors and acromegaly. Sandostatin LAR™ is designed for deep intramuscular injection into the gluteal muscle.

6.4.2 LANREOTIDE

Also known as angiopeptin, lanreotide (Somatuline LA™) is an octapeptide disulfide analog of somatostatin that was first reported in 1987 (Structure 6.26). In the form of copolymer microparticles for reconstitution as an aqueous suspension for intramuscular injection (i.e., Somatuline LA™), the drug is licensed in the U.K. for the treatment of neuroendocrine tumors. It is also available as an acetate in a gel form suitable for direct injection (Somatuline Autogel™).

STRUCTURE 6.26 Structure of lanreotide (Somatuline LA™).

Side effects of lanreotide include asthenia, fatigue, and increased bilirubin levels. Other less common symptoms include skin nodules, hot flashes, leg pains, malaise, headache, tenesmus, decreased libido, drowsiness, pruritus, increased sweating and, rarely, hypothyroidism.

6.5 ESTROGEN THERAPY

Estrogen is rarely used to treat prostate cancer because of its adverse reactions but is occasionally used to treat breast cancer in postmenopausal women if no other options are available. In prostate cancer, an estrogenic agent such as DES works by inhibiting the hypothalamic-pituitary system through a negative feedback (shut off) mechanism. A shortage of testosterone/DHT reaching the cells signals of an increased production of testosterone by the testes (some of which will reach cells and occupy receptors) which results in a fall of output of LH from the pituitary of LH and a subsequent decrease in testosterone synthesis by the testes (an outcome known as "chemical castration").

However, toxicities with this treatment are common, with dose-related adverse reactions including arterial and venous thrombosis, nausea, and fluid retention. Furthermore, impotence and gynecomastia always occur in men, and withdrawal bleeding can be a problem in women. Estrogen may also cause bone pain and hypercalcemia when used to treat breast cancer. Ethinylestradiol, also used to treat breast cancer, is the most potent estrogen available for this purpose.

6.5.1 DIETHYLSTILBESTROL

Diethylstilbestrol (DES), also known as *stilbestrol* or *DHS*, is a synthetic nonsteroidal estrogen first reported in 1938. Its structure is interesting from the medicinal chemistry perspective, as the two ethyl and two phenolic substituents arrange themselves around the central double bond to mimic a steroidal four-ring system in a similar manner to the estrogen inhibitor tamoxifen.

STRUCTURE 6.27 Structure of diethylstilbestrol (DES).

The use of DES in prostate cancer has now lost favor due to the cardiovascular complications and the feminizing side effects associated with estrogen. Other adverse effects include sodium retention with edema, thromboembolism, and jaundice. DES is well absorbed orally and is licensed in the U.K. for use in both breast and prostate cancer.

6.5.2 ETHINYLESTRADIOL

Ethinylestradiol is a synthetic steroid with the higher known estrogenic potency that is well absorbed orally (Structure 6.28). Its synthesis from estrone was first reported in 1938.

STRUCTURE 6.28 Structure of ethinylestradiol.

Unlike other estrogenic agents, it is metabolized very slowly in the liver. In the U.K. it is licensed for the treatment of breast cancer; however, it is also a component of numerous contraceptive products that are in use worldwide.

6.6 PROGESTOGEN THERAPY

Progestogens are used in the treatment of breast carcinoma and other hormone-dependent cancers, including endometrial tumors and renal and prostatic adenocarcinomas. Four agents — gestonorone caproate, medroxyprogesterone acetate, megestrol acetate, and norethisterone — are licensed in the U.K. for these uses. Medroxyprogesterone and megestrol are the agents of choice and can both be given orally. In general, side effects of progestogen therapy are mild but may include nausea, fluid retention, and weight gain.

6.6.1 GESTONORONE CAPROATE

Gestonorone caproate (Depostat™), also known as *gestronol hexanoate,* is a steroidal progestogen first reported in 1969 (Structure 6.29). It is used in the treatment of prostatic hypertrophy.

STRUCTURE 6.29 Structure of gestonorone caproate (Depostat™).

Administered by intramuscular injection in an oily formulation, it is licensed in the U.K. for the treatment of benign prostatic hypertrophy and endometrial cancer.

6.6.2 MEDROXYPROGESTERONE ACETATE

Medroxyprogesterone acetate (Provera™ or Farlutal™) is an orally active progestogen formerly used in combination with other steroids as an oral contraceptive (Structure 6.30). It was first reported by the then Upjohn Company in 1958.

STRUCTURE 6.30 Structure of medroxyprogesterone acetate (Provera™, Farlutal™).

In addition to the general adverse effects caused by progestogens, glucocorticoid effects at higher dose levels may lead to cushingoid syndrome. Given orally or by deep intramuscular injection into the gluteal muscle, medroxyprogesterone acetate is licensed in the U.K. for the treatment of breast carcinoma and other hormone-dependent cancers, including endometrial cancer and renal and prostatic adenocarcinomas.

6.6.3 MEGESTROL ACETATE

Megestrol acetate (Megace™) is an orally active progestin, formerly used in combination with other steroids as an oral contraceptive, and first reported by the then Searle in 1959 (Structure 6.31). It is mainly used for the palliative treatment of breast and endometrial cancers. Megestrol acetate causes the usual side effects associated with progestogens.

STRUCTURE 6.31   Structure of megestrol acetate (Megace™).

6.6.4 NORETHISTERONE

Norethisterone, also known as *norethindrone,* was first reported by Syntex in the mid-1950s (Structure 6.32). Given orally, norethisterone is licensed in the U.K. for use in breast cancer.

STRUCTURE 6.32   Structure of norethisterone (norethindrone).

FURTHER READING

Ali, S., and Coombs, R.C. "Endocrine-Responsive Breast Cancer and Strategies for Combating Resistance," *Nature Rev. Cancer,* 2:101-12, 2002.

Cuzick, J., et al. "Overview of the Main Outcomes in Breast-Cancer Prevention Trials," *Lancet,* 361:296-300, 2003.

Jones, K.L., and Buzdar, A.U. "A Review of Adjuvant Hormonal Therapy in Breast Cancer," *Endocrine-Related Cancer,* 11:391-406, 2004.

7 Tumor-Targeting Strategies

7.1 INTRODUCTION

One of the greatest challenges in the discovery and development of new therapeutic agents and strategies for the treatment of cancer is the achievement of selectivity between tumor cells and healthy tissues. With few exceptions, all agents in clinical use today are associated with varying degrees of side effects (i.e., toxicities) due to their collateral action on healthy cells and tissues. Therefore, many ingenious methods have been devised to target drugs to tumor cells or tumor masses.

One obvious way to achieve targeting is through antibodies directed towards markers on the surface of tumor cells. This method has given rise to four strategies. First, an antibody can be used as a single entity; this approach has been validated through the development of such agents as trastuzumab (Herceptin™) and bevacizumab (Avastin™). Second, a drug can be directly attached to a tumor-specific antibody to guide it to tumor cells, an example of which is gemtuzumab ozogamicin (Mylotarg™). A third antibody strategy involves the use of antibodies to guide an attached radioactive element or ligand to the tumor. This approach was exemplified by pemtumomab (Theragyn™), an antibody labeled with yttrium-90 (no longer in development). A fourth strategy involves the use of antibodies to guide a prodrug/drug-releasing enzyme to tumor cells, such as in antibody-directed enzyme prodrug therapy (ADEPT).

A different approach to achieving selectivity involves targeting tumor vasculature. Anti-angiogenic agents work by blocking the ability of tumors to develop new vasculature vital for their growth and survival; healthy tissues are relatively unaffected by these drugs because their vasculature is already in place. This approach has been recently validated by the success of bevacizumab (Avastin™) in the clinic. An alternative strategy involves targeting the existing blood supply of tumors, which can be functionally distinct from the vasculature of healthy tissue. Drugs of this class are known as vascular disrupting agents (VDAs). In particular, because tumor vasculature is usually less mature than that of other tissues and organs, the cells are more prone to drugs such as combretastatin due to their lack of actin and relatively exposed tubulin.

Other triggers, such as oxygen concentration, are also being used to selectively release drugs at tumor sites. For example, it is well established that the centers of most tumor masses are hypoxic compared to healthy tissue due to an abnormal vasculature. Therefore, prodrugs that are activated under these hypoxic (i.e., bioreductive) conditions have been designed. This approach is exemplified by AQ4N, a prodrug that produces a DNA–interactive agent (AQ4) upon bioreduction. In addition, it has been observed that polymers such as polyethylene glycol (PEG) tend to accumulate in tumor masses due to their abnormal vasculature, and this phenomenon has been exploited in the form of experimental polymer-drug conjugates that accumulate at the tumor site and then release active drug. More elaborate two component systems based on this phenomenon involving both polymer-bound drugs and polymer-bound releasing enzymes (i.e., polymer-directed enzyme prodrug therapy [PDEPT]) have also been investigated but not yet fully developed.

Enzymes have also been utilized in various tumor-targeting strategies. For example, the experimental biphasic X-DEPT therapies involve a two-component system whereby an enzyme is localized at the tumor site via an antibody, viral, or polymer vector. Once there, the enzyme can selectively activate a prodrug administered separately. The X-DEPT approaches are still being evaluated in the clinic, and none has been successfully commercialized to date. An alternative enzyme-based strategy, which is also still being evaluated in the clinic, relies on the use of an enzyme constitutively expressed by tumor cells (but not healthy cells) to activate a prodrug either without further intervention or after systemic administration of an enzyme cofactor. Yet another approach involves the use of the enzyme asparaginase (Erwinase™) to break down asparagine in the bloodstream and tissues. The targeting is based on the fact that asparagine is required for survival by leukemic cells which, unlike healthy cells, are unable to synthesize their own supply. This treatment has been successfully used to manage certain types of leukemia for many years.

The concept of activating systemically administered prodrugs by a light source selectively at the tumor site has been established for some time and is based on a similar approach used for the treatment of psoriasis (i.e., psoralen ultraviolet A [PUVA] therapy). Although originally restricted to phototreatment of the skin (e.g., melanomas), this area has developed rapidly in recent years due to advances in laser technology and the introduction of "key hole" surgery techniques that allow flexible laser light sources to be introduced into body cavities such as the gastrointestinal (GI) tract and trachea. A related strategy, boron neutron capture therapy (BNCT), uses a neutron beam to activate a prodrug specifically at the tumor site (usually in the brain).

Finally, in the last decade, significant advances have been made in drug delivery techniques, novel approaches, such as those involving gene therapy, nanotechnology, intracranial wafers, and ultrasound, are being used experimentally to release established anticancer drugs specifically at the site of tumors without harming healthy tissues. All of these approaches are described below.

7.2 ANTIBODY-BASED APPROACHES

Monoclonal antibodies (MAbs) have been developed for both the diagnosis and treatment of cancer, and several MAbs are already commercially available as cancer therapies. MAbs can be used as single agents, paired with powerful cytotoxics or radiopharmaceuticals to create tumor-specific agents, or they can be used in an X-DEPT approach such as ADEPT. Therapies based on MAbs have been slow to emerge, partly due to the technical challenges involved but also because early studies with vinblastine-antibody conjugates were disappointing. However, the clinical potential of MAbs has been recently validated by the success of such agents as trastuzumab (Herceptin™) and bevacizumab (Avastin™). Even so, in practice, these MAbs tend to be most efficacious when used in combination with traditional cytotoxic agents. Also, for solid tumors, debulking may be initially required.

Rather than use MAbs alone, an alternative approach to enhance efficacy is to attach a cytotoxic agent through a chemical linker. The linker can be designed to cleave specifically at the tumor site, thus releasing the cytotoxic agent. For example, conjugates have been reported that contain linkers designed to cleave on exposure to the enzyme cathepsin, which is overexpressed in some tumor cell types. With this type of construct, in which exposure to the cytoplasm within the cell is important for drug release, it is necessary to demonstrate that, once bound to the tumor cell, the drug-antibody conjugate is internalized. If internalization does not occur, then the full cytotoxic effect may not be achieved, although some drug may still be released in the vicinity of the tumor. If some release occurs externally to the cancer cell, then the drug is free to diffuse into neighboring cells, a phenomenon known as the *bystander effect*. A further consideration is that, for maximum efficacy, as many linker-drug units as possible should be attached to a single antibody without affecting its antigen-recognition or pharmaceutical properties, a technically challenging feat. For example, in the case of gemtuzumab ozogamicin (Mylotarg™), a loading of calicheamicin of 4 to 6 moles per mole of antibody over 50% of the antibody's surface is achieved. Another approach is to ligate the MAb to a radiopharmaceutical to create an agent for use in radioimmunotherapy. This combines the advantages of targeted radiation therapy and specific immunotherapy.

7.2.1 Antibodies as Single Agents

Antibodies used as single agents to target markers on tumor cells are discussed in detail in other sections of this book. For example, trastuzumab (Herceptin™) is a humanized recombinant anti-P185 monoclonal antibody targeted to human epidermal growth factor receptor 2 (HER2)/neu receptors (see Chapter 5). It inhibits the HER2/neu signaling pathway and was developed for use in breast cancer. Cetuximab (Erbitux™), a chimerized monoclonal antibody that is part mouse and part human, targets and inhibits epidermal growth factor receptor (EGFR), and has proven useful in EGFR-positive colorectal cancer (see Chapter 5). Similarly, bevacizumab (Avastin™), which targets the vascular endothelial growth factor receptor (VEGFR), was the first angiogenesis inhibitor to reach the market (in 2004) for first-line use together with 5-fluorouracil (5-FU) for patients with metastatic colorectal cancer (see Section 7.3.1.1).

Many other types of antibody-based single-agent therapies are under development. For example, IGN311 (Aphton) is a humanized monoclonal antibody currently in clinical evaluation that targets the tumor-associated antigen of Lewis Y, a marker expressed in as many as 90% of all epithelial cancers, including pancreatic, gastric, colon, and breast tumors. Data from Phase I studies suggest that this antibody significantly decreases the number of tumor cells circulating in peripheral blood.

Furthermore, IGN311 has favorable safety and tolerability profiles and a serum half-life of more than 20 days. Similarly, a CD4-targeted antibody called HuMax-CD4 (Genmab AS/Serono SA) is being evaluated for the treatment of T-cell lymphomas. This antibody has been granted fast-track status by the U.S. FDA and is currently being evaluated in a Phase III clinical trial.

7.2.2 Antibody-Drug Conjugates

Historically, many attempts have been made to conjugate MAbs to cytotoxic "warheads" in order to achieve greater tumor selectivity. The first studies in this area involved conjugation of vinca cytotoxic agents to MAbs; however, the results were disappointing, with no significant improvement in therapeutic index. Only recently have more encouraging results been obtained with the experimental agent gemtuzumab ozogamicin (Mylotarg™), which was approved by the FDA in 2000.

7.2.2.1 Gemtuzumab Ozogamicin

Gemtuzumab ozogamicin (Mylotarg™), originally developed by American Home Products, was the first licensed chemotherapeutic agent to use MAb technology targeted directly at tumor cells (Structure 7.1). Gemtuzumab was given accelerated approval by the U.S. FDA for use in CD33-positive acute myeloid leukemia (AML) in patients age 60 years or older in first relapse who are not considered suitable for cytotoxic chemotherapy. This agent, which is comprised of the cytotoxic antitumor antibiotic calicheamycin conjugated to a humanized recombinant IgG₄ kappa antibody, has a molecular weight of 151 to 153 kDa. Calicheamycin is a natural product isolated from the bacterium *Micromonospora echinospora* sp. *calichensis* discovered in caliche clay (a type of soil found in Texas) by Wyeth-Ayerst researchers. The

STRUCTURE 7.1 Structure of gemtuzumab ozogamicin (Mylotarg™).The antibody is depicted as a circle (top left) labeled as "hP67.6."

CD33 antigen to which gemtuzumab binds is a sialic acid–dependent adhesion protein. It is an attractive target for antibody therapy because it appears on the surface of immature normal cells of myelomonocytic lineage and the surface of leukemic myeloblasts but not on normal hematopoietic stem cells.

Approximately 98.3% of the amino acid sequence of the anti–CD33 antibody fragment of gemtuzumab is human in origin, including the framework and constant regions. However, the CD33-recognizing complementarity-determining regions come from a murine p67.6 antibody. The antibody component is made in a mammalian myeloma NS0 cell suspension culture and then purified under conditions that either remove or inactivate viruses. The early development work on the antibody component was carried out by researchers at the Fred Hutchinson Cancer Research Center (Seattle, WA), who licensed it to Wyeth-Ayerst. It was subsequently humanized by the Celltech Group, and a bifunctional linker was used to join it to N-acetyl-gamma calicheamicin, with a loading of 4 to 6 moles of calicheamicin per mole of antibody over 50% of its surface.

Gemtuzumab works by targeting and binding to the CD33 antigen, which is expressed on the surface of hematopoietic cells but not on the surfaces of pluripotent hematopoietic stem cells or nonhematopoietic cells. Thus, CD33 is expressed on normal and leukemic myeloid colony-forming cells, including leukemic clonogenic precursors. For example, more than 80% of patients with AML express CD33 on their leukemic blasts. When gemtuzumab binds to the CD33 antigen, the complex formed is subsequently internalized, which is followed by release of calicheamicin inside the lysosomes of the cells. After reaching the nucleus, calicheamicin binds in the minor groove of DNA, causing DNA double-strand breaks that lead to cell death. (See Chapter 3 for more details.)

The most serious side effect of gemtuzumab is substantial myelosuppression through a cytotoxic effect on normal myeloid precursors. However, because pluripotent hematopoietic stem cells are spared, this toxicity is reversible. Other side effects reported include allergic reactions, nausea, vomiting, flu-like symptoms, high or low blood pressure, sore mouth, taste changes, lowered resistance to infection, bruising or bleeding, anemia, and liver toxicity.

7.2.3 ANTIBODY-RADIONUCLIDE CONJUGATES

Rather than conjugating an antibody to a cytotoxic agent, an alternative strategy is to attach a radionuclide as the "warhead." One of the best-known examples of this approach is pemtumomab (Theragyn™), which gave encouraging results in early clinical trials but was not developed any further due to disappointing results in Phase III evaluations.

7.2.3.1 Pemtumomab

Pemtumomab (Theragyn™) is an yttrium-90 labeled antibody developed by Anti-Soma Ltd., targeted to the aberrantly glycosylated polymorphic epithelial mucin (PEM) in ovarian cancer. In this tumor type, a number of antigenic targets have been exploited, including the folate receptor CA-125 and PEM. MAbs to PEM were developed in the early 1980s; one of the best known is HMFG-1, the antibody that forms the basis of pemtumomab.

Research into the tumor-associated mucin molecule led to the discovery of a gene on chromosome 1 of the human genome, coding for the core protein of PEM. This gene, called the MUC-1 gene, codes for the MUC-1 antigen (a 20 amino-acid antigen) repeated in tandem, which forms the core protein of PEM. In normal epithelial tissue, the MUC-1 gene product is fully glycosylated (covered in carbohydrates) and is therefore "hidden." However, in many tumors of epithelial origin, including ovarian cancer, the glycosylation is incomplete, which results in exposure of amino acid sequences of the MUC-1 gene product, which was exploited by pemtumomab.

During development, the fact that ovarian cancer is normally confined to the peritoneal cavity was also exploited, and so the radiolabeled antibody was administered intraperitoneally. This strategy, known as *regional immunotherapy,* results in increased localization of the antibody compared to that achievable when given systemically, thus increasing the radiation dose to the tumor. Initially, iodine-131 was used as the isotope because it had been successfully used itself in the treatment of other tumor types (e.g., thyroid cancer). However, the major disadvantage of iodine-131 was that, being a γ-emitter, patients had to be isolated and kept in the hospital for at least 5 days. Exposure of hospital staff to the radiation was also a potential hazard. To overcome these disadvantages, the pure β-emitting isotope yttrium-90 was used instead. This isotope had already been used on its own to treat liver cancer (hepatoma) and raised none of the same concerns regarding exposure, isolation, or long hospitalization. However, in order to use high amounts of this radioactive metal, new chemical chelates had to be developed that could bind yttrium to the antibody in an almost nonreversible manner.

The initial clinical trials of the yttrium-90 labeled MAb (pemtumomab) gave promising results. Patients underwent surgery followed by chemotherapy and then waited for some time before receiving pemtumomab. Those who received the agent 4 to 6 weeks after their chemotherapy, and at a time when they had no evidence of disease, survived longer than patients who only received surgery and chemotherapy. Five-year survival data on patients in a Phase II trial surpassed any previously reported therapy for the treatment of ovarian cancer, which led to a multinational, multicentered Phase III clinical trial to evaluate pemtumomab in a much wider setting using a greater number of patients. Unfortunately, the results of this trial, published in 2004, suggested that the benefits to patients were not significant enough to warrant continued development.

7.3 VASCULAR-TARGETING STRATEGIES

The concept that solid tumors require a functioning network of blood vessels to sustain growth through the supply of oxygen and nutrients and to avoid a buildup of toxic by-products of cellular metabolism has been under consideration for more than 50 years. The impetus to target tumor angiogenesis (the formation of new blood vessels from the endothelium of existing vasculature) arose from key observations by Folkman in the early 1970s. It is now recognized that for any tumor to grow beyond a volume of 1 to 2 mm³, a so-called "angiogenic switch" must be present, prompting the formation of new vasculature (i.e., neovascularization) (Figure 7.1 and Scheme 7.1). Since Folkman's original observations, key molecules in the angiogenesis process have gradually been identified, such as VEGF and its receptors, culminating in the recent clinical proof of the concept of targeting VEGF in colorectal cancer with the humanized MAb bevacizumab (Avastin™). Many small-molecule inhibitors of VEGF receptors are also now in clinical development (e.g., SU11248 and PTK787/ZK22854).

An alternative but complementary strategy to anti-angiogenic agents is that of targeting established tumor vasculature (Scheme 7.1). This approach was first suggested as long ago as the 1920s, but research in this area was given impetus by Denekamp in the early 1980s and has led to a class of drugs now known as Vascular Disruptive Agents (VDAs), whose role is to cause a rapid and selective shutdown of existing tumor blood vessels. The first known agent of this type was combretastatin (and its derivatives), which was followed by related prodrug forms, such as CA4P and Oxi4503. 5,6-Dimethylxanthenone-4-acetic acid (DMXAA), an agent with a similar effect on existing tumor vasculature, is presently in late-stage clinical trials. One attraction for clinicians is that agents such as combretastatin and DMXAA cause a rapid and selective vascular shutdown in tumors in a matter of minutes to hours, and so can be given in intermittent doses to potentiate the activity of conventional chemotherapeutic agents (see below). Many other potential drugs of this type are at the discovery and development stage; examples include ZD6126 and AVE8062A.

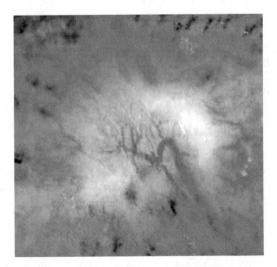

FIGURE 7.1 Photomicrograph showing the process of angiogenesis. The image shows the formation of numerous new blood vessels in the area of a metastasis on the surface of a human lung.

Anti-Angiogenic Agents   **Vascular Disrupting Agents (VDAs)**

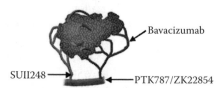

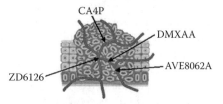

Small solid tumors with new blood
vessels. Anti-angiogenic agents
have major effect on tumor
periphery inhibiting endothelial
proliferation and migration.

Larger solid tumors with
established blood vessels.
VDAs have a major effect on the
central part of the tumor causing
vessel occlusion and necrosis.

SCHEME 7.1 Mechanism of action of anti-angiogenic agents and VDAs. (Adapted with permission from Kelland, L.R., *Curr. Cancer Ther. Rev.*, 1:1-9, 2005).

7.3.1 ANTI-ANGIOGENIC AGENTS

The concept of blocking the growth of new tumor vasculature was first described in the early 1970s, although a practical application has only just emerged, more than 30 years later. Early enthusiasm for angiogenesis inhibitors was lost after a number of promising agents were unsuccessful in increasing survival in pivotal Phase III clinical trials. Only more recently have the colorectal cancer trials of bevacizumab established a survival benefit through this mechanism of action. This has not only demonstrated proof of concept, but has also provided more treatment options for colorectal cancer patients with metastatic disease. As a result, many more anti-angiogenic agents, both antibody-based and small molecules, are now in development.

7.3.1.1 Bevacizumab

Bevacizumab (Avastin™) was hailed as a major advance in cancer chemotherapy when it became the first angiogenesis inhibitor to gain U.S. FDA approval in early 2004 for first-line use, in combination with 5-FU, for the treatment of metastatic colorectal cancer. Developed jointly by Roche and Genentech, it is a humanized MAb directed against VEGF, a growth factor important in both tumor angiogenesis and in the maintenance of existing tumor blood vessels. By binding to VEGF, bevacizumab blocks its interaction with receptors such as Flt-1 and KDR on the surface of endothelial cells, a process that normally triggers the proliferation of endothelial cells and the formation of new blood vessels. This interferes with the blood supply to tumors, hence reducing growth and potentially inhibiting the process of metastasis. Human tumor xenograft studies in a colon model initially confirmed that bevacizumab was able to reduce microvascular growth at the tumor site and also inhibit metastasis. A large, randomized, placebo-controlled Phase III clinical trial involving colorectal cancer patients then demonstrated a clear extension of median survival of approximately 5 months when bevacizumab was given together with an IFL regimen (5-FU/CPT-11/leucovorin) compared to the IFL alone (20.3 versus 15.6 months). This was one of the greatest improvements in survival ever reported for colorectal cancer patients in a randomized Phase III trial, and led to FDA approval in early 2004 for first-line use of an intravenous bevacizumab/5-FU–based combination for patients with metastatic colon or rectal cancer. European approval was obtained in early 2005 for a similar indication.

Bevacizumab is normally administered as an intravenous infusion once every 14 days, at a dose level of 5 mg/kg. Administration is halted when disease progression occurs. As anticipated for an agent that inhibits the actions of VEGF, wound healing complications and GI perforations are the most serious side effects and, in some cases, can prove fatal. Other potentially serious adverse effects may include congestive heart failure, nephrotic syndrome, hypertensive crises, and hemorrhage. Hemoptysis has also been observed in non-small-cell lung cancer (NSCLC) patients treated with both bevacizumab and other types of chemotherapy. Less serious adverse effects include proteinuria; exfoliative dermatitis; anorexia; asthenia; pain, including abdominal pain; stomatitis; headache; hypertension; nausea; vomiting; diarrhea; constipation; upper respiratory infection; leukopenia; epistaxis; and dyspnea.

Bevacizumab is also being evaluated for use in other types of solid tumors, including renal cell, breast, non-small-cell lung, prostate, and ovarian cancers; melanoma; and some hematologic malignancies. Based on the proof-of-concept that bevacizumab has provided for the anti-VEGF approach to cancer therapy, approximately 50 novel anti-angiogenesis inhibitors are presently in various stages of development worldwide, demonstrating the high level of interest in this new area.

7.3.2 VASCULAR DISRUPTIVE AGENTS (VDAs)

Another group of drugs that is currently of significant interest is the VDAs, not only because of their potential as single agents but also because they can be used for so-called adjunct or "complementary" therapy. This involves using a VDA, such as DMXAA, to target the central regions of a tumor via its blood supply in combination with cytotoxic agents or radiotherapy which preferentially affect the well-oxygenated outer rim of the tumor. In addition, there is the possibility of pharmacokinetically "trapping" cytotoxic drugs in the tumor by scheduling a VDA to close the tumor blood vessels after cytotoxic agents have been administered. Bioreductive agents (see Section 7.3.3) should be particularly useful in this context because they themselves target the centers of larger tumors, and so a synergistic effect may be obtained. Combretastatin and its derivatives were the first VDAs to be identified, followed by the prodrugs Oxi4503 and CA4P. Flavone-8-acetic acid (FAA) and the related DMXAA were discovered later, and DMXAA is presently in late-stage clinical development.

7.3.2.1 Combretastatins

The combretastatins are a group of *cis*-stilbene compounds (Structure 7.2) isolated from the bark and stem wood of the African bush willow tree (*Combretum caffrum*) (Figure 7.2) that inhibit tubulin polymerization. They were identified in the early

1980s by George R. Pettit at Arizona State University (U.S.A.) based on the observation that extracts of the tree had been used in folk medicine and shown to possess anti-leukemic properties. The highly attractive African bush willow only grows on the banks of rivers in the Eastern Cape Province of South Africa, and historians believe that Arabians traded for the bark with the San people (Bushmen) for more than 2000 years. At the time, the bark was probably used as a general tonic because, apart from its anticancer properties, it allegedly produces a feeling of improved general well-being. However, it is also known that Zulu warriors used this shrub to prepare a substance that they used both as a poison for their arrow tips and also as a "charm" to protect themselves against enemies. Structurally, the combretastatin molecules are biaryls connected in the *cis*(z) configuration by an ethylene bridge, and structure activity relationship (SAR) studies have shown that the Z-configuration is essential for their antitumor activity. Furthermore, restricted

FIGURE 7.2 (a) African bush willow tree; (b) mohair tapestry produced in the Zulu village of Howick in Kwa-Zulu (Natal, South Africa) showing the gently sweeping elegance of the bush willow along the Mooi River in Northern Natal.

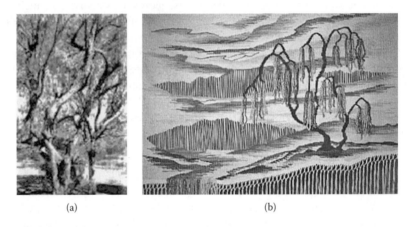

Combretastatin A-1 (Oxi4500): R = OH   CA4P   Oxi4503
Combretastatin A-4: R = H

STRUCTURE 7.2 Structures of the combretastatins A-1 and A-4, and related prodrugs CA4P and Oxi4503.

rotation about the olefinic bridge is crucial for biological activity, as is the distance between the two rings.

The interesting feature of the combretastatins is their selectivity for tumor vasculature rather than healthy tissue. Combretastatin A-1(CA-1) is known to bind to tubulin with a higher affinity for β-tubulin than colchicine at or near colchicine-binding sites, thus destabilizing the tubulin cytoskeleton. When this tubulin structure is disrupted, the endothelial cells change from a flat streamlined profile to a round swollen shape because the internal tubulin skeleton is no longer able to mechanically support their elongated shape. Due to this shape change, the cells effectively plug the capillaries, thus restricting tumor blood flow. In addition, a process of apoptosis is initiated in the cells of tumor blood vessels. This appears to first kill cancer cells in the core of the tumor but with the apoptotic effect then radiating out, thus starving the tumor of nutrients which, in turn, leads to exposure of the basement membrane, hemorrhage, and coagulation (i.e., tumor necrosis). The selectivity arises because, compared to mature endothelial cells that line blood vessels in normal tissues, tumor blood vessels are usually newly formed and immature, and only vessels of this type can be affected by combretastatin. Actin, a protein not present in immature endothelial cells, protects the tubulin in mature endothelial cells from the effects of combretastatin, and hence their shape is maintained. Interestingly, actin does not appear for a number of days after the formation of new endothelial cells, and so this provides a window of opportunity to achieve selectivity toxicity toward the tumor microvasculature.

In addition to these fundamental differences between immature and mature endothelial cells, characteristics of the tumor microcirculation, such as high interstitial fluid pressure, procoagulant status, vessel tortuosity, and heterogeneous blood flow distribution, assist the selectivity process. Reduced blood flow results in significant hypoxia within the tumor, leading to cell death and regression. The study of these agents in the clinic has been facilitated by the development of techniques using positron emission tomography (PET) and magnetic resonance imaging (MRI) that allow tumor blood flow to be measured in patients during treatment.

It is noteworthy that neither of the two side effects commonly associated with cytotoxic agents (i.e., leukopenia and alopecia) occur during or after treatment with combretastatin. However, one of the most serious side effects is tumor pain and, at high doses (more than 90 mg/m^2), some patients show signs of lung and heart toxicity. Other more minor adverse effects observed in clinical trials include faint flush, transient blood pressure changes, abdominal pain, nausea, vomiting, fever, light-headedness, headache, diarrhea, bradycardia, and tachycardia. Finally, on an electrocardiogram, nonspecific ST-T wave changes have been observed that peak between 3 and 5 hours after infusion.

Despite extensive clinical evaluation of the original combretastatin compounds, efficacy proved disappointing. The poor aqueous solubility of these molecules was thought to be partly responsible for their poor performance in the clinic, and so a number of phosphate prodrugs have prepared, such as Oxi4503 and CA4P (see Section 7.3.2.2). These prodrugs are presently being evaluated in the clinic. Other types of tubulin depolymerizing agents are also in clinical development (e.g., ZD6126 and AVE8062), along with DMXAA, which has a different mechanism of action.

7.3.2.2 Combretastatin Prodrugs: CA4P and Oxi4503

CA4P and Oxi4503 are monophosphate and diphosphate prodrugs of combretastatin A-4 and A-1, respectively. Oxi4503 is 10 times more potent than the monophosphate CA4P and has superior *in vivo* activity to A-4. Also, it is active in various cell lines *in vitro*, including those expressing the multidrug resistant (MDR) phenotype. Oxi4503, which is presently in clinical trials, was designed to improve the pharmaceutical properties of the parent molecule, particularly by enhancing its water solubility. After injection, it hydrolyzes to the parent A-1 by enzymes in the bloodstream, which then rapidly penetrates the endothelial cells that line the tumor blood vessels to produce an antitumor effect.

7.3.2.3 Flavone-8-Acetic Acid (FAA) and DMXAA

Interest in this nontubulin class of VDAs began in the mid-1980s, with the serendipitous discovery of the antitumor properties of flavone-8-acetic acid (FAA), a molecule originally synthesized as a nonsteroidal anti-inflammatory agent by Lyonnaise Industrielle Pharmaceutique (Structure 7.3). FAA was found to be active (even curative) against a wide variety of preclinical solid tumor models, and its mechanism of action appeared to be related to the production of tumor necrosis factor (TNF). Initial Phase I studies using an ester prodrug (LM985) of FAA showed that the dose-limiting toxicity was acute, reversible hypotension occurring during drug infusion at doses of 1500 mg/m^2. Later Phase I studies using FAA itself (LM975) showed that hypotension was again dose-limiting but at the significantly higher dose of 10 g/m^2. However, no evidence of efficacy was apparent from any of these trials, including a Phase II study in NSCLC. A Phase I trial of FAA in combination with recombinant interleukin-2 gave one complete and two partial responses, but severe hypotension (grade 3 or 4) was observed after the third dose of FAA, with no sign of increased tumor necrosis in several tumor biopsies. This led to the clinical abandonment of FAA on the grounds that it did not elicit similar antivascular effects in humans as those seen in many murine tumor models.

A synthetic program to identify a more potent inducer of TNF-α was initiated by Denny, Baguley, and colleagues at the University of Auckland, New Zealand,

Flavone-8-acetic acid (FAA)

DMXAA

STRUCTURE 7.3 Structures of flavone-8-acetic acid (FAA) and DMXAA.

which resulted in DMXAA (AS1404, Antisoma) (see Structure 7.3). Encouragingly, DMXAA was shown to induce TNF-α (mRNA) in both murine and human cells (whereas FAA only induced TNF in murine cells). However, the mechanism of action of DMXAA appears to be more complex than simply induction of TNF-α, and data support an early direct effect on endothelial cells (possibly through involvement of nuclear factor κB, although the exact target remains to be identified), leading to rapid apoptosis. This is followed by indirect effects involving the release of vasoactive agents within tumor tissue, such as serotonin (from platelets) and TNF-α, possibly other cytokines and, even later, nitric oxide. A comparison of blood perfusion effects in a murine tumor (C3H) for DMXAA and CA4P showed that blood flow reduction with CA4P was more rapid (maximum effect only 1 hour after administration for CA4P versus 6 hours for DMXAA) but more transient, as perfusion had returned to normal 24 hours later with CA4P but not with DMXAA.

DMXAA showed single-agent antitumor activity, including cures in some models such as the syngeneic colon 38, especially when administered as a loading dose of 25 mg/kg followed by two supplementary maintenance doses of 5 mg/kg 4 and 8 hours later. As with CA4P, DMXAA showed at least additive, even synergistic antitumor effects in combination with a variety of treatment modalities, including radiation, radioimmunotherapy, and hyperthermia. In addition, marked potentiation was observed with a variety of chemotherapeutic drugs in a syngeneic mouse mammary tumor model. The therapeutic gain was most striking with paclitaxel but was also apparent with vincristine, etoposide, carboplatin, cyclophosphamide, doxorubicin, and cisplatin. In mice, the therapeutic index for DMXAA is rather narrow; with vascular targeting/tumor necrosis effects typically being observed at doses greater than 15 mg/kg, relatively close to the maximum tolerated dose of approximately 25 mg/kg. However, fortunately, there appears to be a wider therapeutic index in man.

DMXAA (AS1404) entered clinical trials in 1996 both in the U.K. (weekly schedule) and in New Zealand (3-weekly schedule), using a 20-minute intravenous infusion. In the U.K. study, 46 patients received DMXAA over a dose range of 6 to 4900 mg/m² and, unlike with conventional chemotherapeutic agents, dose-limiting toxicities included urinary incontinence, visual disturbance (as was reported for FAA), and anxiety. The New Zealand study reported similar findings, with the rapidly reversible toxicities of confusion, tremor, slurred speech, visual disturbance, anxiety, urinary incontinence, and possibly left ventricular failure (seen at the 4900 mg/m² dose level). A transient prolongation of cardiac QTc interval was also observed in 13 patients administered doses of 2000 mg/m² and greater. Dose-dependent plasma increases in the serotonin metabolite 5-hydroxyindoleacetic acid were seen at doses above 650 mg/m², thus providing a possible pharmacodynamic marker of vascular targeting for this agent. A number of patients in these trials had unconfirmed partial responses.

Based on the findings that DMXAA is well tolerated using both weekly and 3-weekly schedules of administration, and that adverse affects appear manageable, a number of combination trials were planned, particularly involving coadministration with taxanes. These trials are still underway.

7.3.3 BIOREDUCTIVE AGENTS

The importance of oxygen in cancer therapy (particularly radiotherapy) has been understood for many years. Low oxygen levels (hypoxia) make cells resistant to being killed by radiation and chemotherapy, thus reducing the efficacy of treatment. Furthermore, oxygen levels are lower in solid tumors (especially their centers) than in healthy tissues, and tumors over a certain size (i.e., greater than 2 mm^3) suffer from disordered vasculature and contain hypoxic fractions (Scheme 7.2). *In vitro* data indicate that, compared to normally oxygenated (i.e., "oxic") cells, when hypoxic fractions are present in a tumor, a much lower response (approximately one third the normal) to radiotherapy and conventional chemotherapeutic agents is obtained. One explanation for this is that, compared to their oxic counterparts, hypoxic cells are quiescent (i.e., not replicating) and are thus less vulnerable to antiproliferative agents such as DNA-damaging drugs and antimetabolites. Therefore, if drugs could be designed to be optimally effective at low oxygen levels, then they would be useful for killing tumor cells in the centers of larger tumors and those that remain after radiotherapy. They may also be less toxic towards "oxic" normal cells, thus providing a greater therapeutic index (TI).

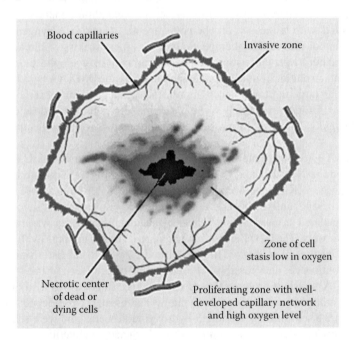

SCHEME 7.2 Section through a solid tumor showing the necrotic center and the development of a network of small blood vessels.

One approach to the design of hypoxia-targeted agents is to incorporate a "bioreductive trigger" as part of a prodrug so that a linker can be cleaved and an active drug is released only under hypoxic conditions (see Scheme 7.3). Such bioreductive prodrugs can be made to work preferentially at low oxygen levels if an intermediate

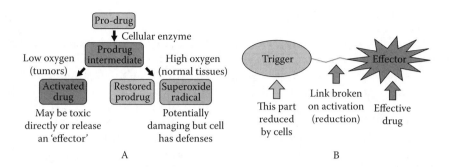

SCHEME 7.3 (a) Schematic diagram showing the bioreductive prodrug activation pathway; (b) design of bioreductive prodrug.

in the activating process is oxygen-reactive. In this case, the relatively higher oxygen level in healthy cells and tissues protects the prodrug from reduction by reacting with an intermediate (usually a short-lived "free radical"), thus restoring the drug to its original (prodrug) form.

A number of examples of bioreductive drugs exist, some of which are either in clinical use or are currently undergoing preclinical or clinical studies. The first group consists of molecules containing quinone moieties; best-known members include mitomycin C (which contains an indolequinone nucleus) and the experimental agent EO9. The mechanism of action of mitomycin is described in detail in Chapter 3. The second group of bioreductive agents includes molecules based on aromatic N-oxides (e.g., tirapazamine) and aliphatic N-oxides (e.g., AQ4N). The latter is presently being evaluated in the clinic, and preliminary pharmacodynamic and pharmacokinetic results convincingly demonstrate that it is working through a bioreductive mechanism (see Section 7.3.3.1). Another group comprises molecules containing nitroheterocycles and includes experimental agents such as CB-1954, SN-23862, NITP, SR4554, and pimonidazole, the latter of which is also being evaluated in the clinic for use as a hypoxia marker.

7.3.3.1 AQ4N

Apart from the natural product mitomycin C, AQ4N is the first example of a bioreductive agent to provide encouraging results in the clinic (see Structure 7.4). It is a water-soluble anthracene derivative that, when administered intravenously, targets the hypoxic compartments of tumors. Although AQ4N enters all cells freely, in theory it is only converted to a charged (at physiological pH) tertiary amine species in hypoxic tumor cells. This amine species becomes trapped in the hypoxic cells, killing them through DNA intercalation and inhibition of topoisomerase II. Clinicians are generally enthusiastic about this mechanism of action because treating the hypoxic fractions of tumors has the potential to significantly enhance selective toxicity and thus improve therapeutic index.

AQ4N itself is a bis(N-oxide) which, because it is uncharged, crosses cell membranes with ease. However, in hypoxic tumor cells, it is selectively and irreversibly reduced to AQ4, the charged quaternary amino form, where it remains

AQ4N
(N-Oxides are not charged overall and allow molecule to cross cell membrane)

AQ4
(After reduction, tertiary amines cause molecule to become "trapped" in cell due to charge)

STRUCTURE 7.4 Structures of AQ4N and its biologically active metabolite (AQ4).

localized. The bioreduction process is enhanced in the absence of oxygen but inhibited strongly in its presence. It is thought that when radiotherapy or chemotherapy kills oxygenated cells on the outside of a tumor, the more-central quiescent AQ4-containing cells become reoxygenated. When these cells resume replication, they are rapidly killed by the antitopoisomerase II activity of AQ4. As demonstrated in a recent Phase I clinical trial, systemic toxicity is thus reduced to a very low level (i.e., the agent has a high maximum tolerated dose) because the prodrug form (AQ4N) is relatively nontoxic and is only transformed into the active agent (i.e., AQ4) when it is inside hypoxic tumor cells.

Phase I clinical trials involving patients with non-Hodgkin's lymphoma and advanced solid tumors are presently underway. AQ4N is also being evaluated in the treatment of esophageal cancer in combination with radiation therapy. During these trials, it has been possible to confirm by studying biopsy material that the drug is converted into its tertiary amine form (AQ4) in hypoxic cells. One curious but nonproblematic side effect observed with AQ4N in these clinical trials was that the mucous membranes of patients acquired a blue coloration, as did body fluids such as urine and saliva. Interestingly, this effect was welcomed by many patients, who felt that the unusually visible sign of the drug in their bodies might be associated with a beneficial effect on their tumors.

7.3.4 POLYMER-DRUG CONJUGATES

Polymer-drug conjugates are based on the principle that a tumor's blood supply and lymphatic drainage capability are usually poor, and so polymer-bound drugs can become trapped and accumulate. The linker between the drug and polymer may then be cleaved enzymatically, releasing the drug specifically at the tumor site and thus avoiding toxicities associated with wider distribution of the active agent. Polyethylene glycol (PEG) is a popular choice of macromolecule for studies of this type because it is highly water soluble, nontoxic, and nonimmunogenic, and it has been previously used in the modification of proteins to increase their biological half-life.

There has been a significant amount of research carried out on the biodistribution of inert macromolecules and polymers in humans in relation to their hydrodynamic radius. For example, early studies demonstrated a greater uptake of macromolecules in tumor cells than in healthy ones, a phenomenon ascribed to a higher rate of constitutive pinocytosis in tumor cells. However, a more general effect is now recognized in which circulating macromolecules accumulate passively in solid tumors. This phenomenon, known as the *enhanced permeation and retention* (EPR) effect, is thought to be the result of three properties of solid tumors: (1) enhanced vascular permeability, (2) limited macromolecular recovery via postcapillary venules, and (3) poor lymphatic drainage. The optimum size range of macromolecules for this effect is thought to be 40 to 70 kDa. Conjugation of cytotoxic drugs, such as doxorubicin, to polymers has been shown to result in an increased concentration of the drug in tumors, and a number of clinical trials of such conjugates have been undertaken during the past decade. Although PEG is an ideal pharmaceutically-compatible polymer with which to form drug conjugates, research is also ongoing to identify polymers with more favorable properties and to attach other types of cytotoxic agents to PEG and related polymers. Important areas of concern with new types of conjugates include stability in plasma, and the extent of selective uptake by tumors compared to normal tissues, and the efficiency of drug release at the tumor site. Although a number of Phase I and Phase II clinical trials have been carried out with various polymer-drug conjugates, none has yet demonstrated any clear benefits in the clinic.

One ingenious variant of this approach, PDEPT, involves the use of a two-component system of polymer-linked drug and polymer-linked enzyme, which are both administered and subsequently accumulate in the tumor. The linker attaching the drug to the polymer is designed to be cleaved by the polymer-bound enzyme, thus providing drug release specifically at the tumor site (see Section 7.4.3).

A related polymer uptake effect can be utilized by entrapping anticancer drugs in liposomes of an appropriate size. For example, Caelyx™ is a liposomal form of doxorubicin that is licensed for advanced AIDS-related Kaposi's sarcoma and for advanced ovarian cancer when platinum-based chemotherapy has failed. A similar product, Myocet™, is licensed for use with cyclophosphamide for metastatic breast cancer. Due to the different pharmacokinetic profiles of these liposomal preparations compared to doxorubicin itself (i.e., accumulation at the tumor site), the incidence of cardiotoxicity is also lowered, as is the potential for local necrosis at the site of administration (see Chapter 3).

7.4 X-DEPT (BIPHASIC) STRATEGIES

The usefulness of traditional cytotoxic chemotherapeutic agents is usually restricted by their low therapeutic indices. One approach to improving this situation has been the development of strategies that allow the conversion of an inactivated form of an anticancer drug (i.e., a prodrug) to an active agent specifically at the tumor site. One such tactic relies on the activation process being carried out by an enzyme that has been targeted to the surface of tumor cells via a suitable antibody. This therapy, known as antibody-directed enzyme prodrug therapy (ADEPT), is presently in Phase I evaluation. An alternative tactic, also being evaluated in Phase I, involves activation of a prodrug by an enzyme caused to be expressed (i.e., not naturally expressed) within the tumor cells. This therapy, known as gene-directed enzyme prodrug therapy (GDEPT) or virus-directed enzyme prodrug therapy (VDEPT), involves the introduction of enzyme-producing genes through viral or nonviral vectors. Finally, the concept of using separate polymers to concentrate both the enzyme activity and prodrug in the tumor mass simultaneously led to the experimental PDEPT approach, although this is no longer being developed.

7.4.1 ADEPT

Bagshawe and co-workers first introduced ADEPT in the early 1990s. In this two- or three-component approach, an antitumor antibody linked to an enzyme is first used to target the enzyme to tumor cells. A second antibody selective for the antibody-enzyme conjugate may then be introduced to clear the conjugate from general circulation, thus reducing systemic toxicity. A prodrug form of a cytotoxic agent is then administered, which is transformed into the active agent by the antibody-bound enzyme selectively at the tumor site (Scheme 7.4 and Scheme 7.5). Collateral damage to healthy tissue is theoretically avoided because the enzyme chosen has no human equivalent (e.g., a bacterial enzyme). Variations of this approach (GDEPT and VDEPT) were later established in which the activating enzyme is introduced through organ-specific gene therapy or other means rather than through an antibody-enzyme conjugate.

The ADEPT system originally proposed by Bagshawe utilizes the bacterial enzyme carboxypeptidase G2 (CPG2), which is a metalloenzyme isolated from a *Pseudomonas* species. It has no mammalian counterpart and catalyzes the conversion of both reduced and nonreduced folates into pteroates and L-glutamic acid. It is also known that CPG2 is capable of cleaving the amidic bond of methotrexate and folic acid at the benzoylglutamate linkage. Thus, a conjugate of CPG2 with an appropriate monoclonal antibody was designed for the selective activation of chemically "masked" nitrogen mustards. The original experimental mustard prodrugs were bifunctional alkylating agents incorporating either chlorine- or sulfonate-leaving groups in which the activating effect of the ionizable carboxyl group had been inhibited by protection with a glutamic acid moiety connected through an amide linkage that formed the substrate for CPG2 (Scheme 7.4).

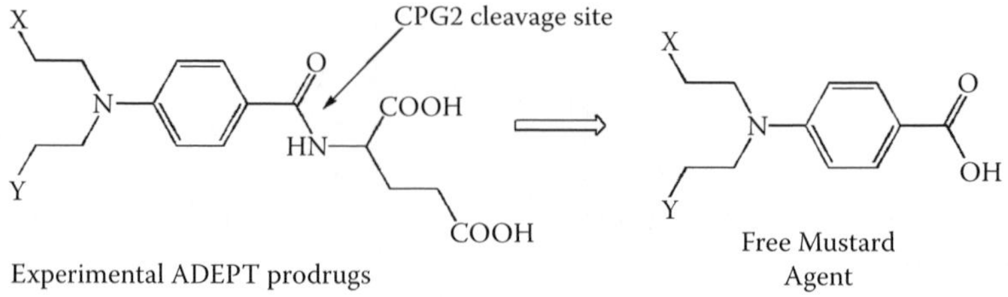

Experimental ADEPT prodrugs

1: X = Y = Cl
2: X = Cl, Y = OSO_2CH_3
3: X = Y = OSO_2CH_3

SCHEME 7.4 Examples of ADEPT prodrugs and their conversion to active agents. The original experimental mustard prodrugs were bifunctional alkylating agents incorporating either chlorine or sulfonate leaving groups (1-3).

To improve the clinical effectiveness of these prodrugs, extensive mechanistic and structure-activity studies were carried out, with three factors being identified as important: (1) the electronic character of the substituent *para-* to the aromatic ring of the nitrogen mustard, (2) the electronic character of other substituents in the aromatic ring (i.e., positions 2- and 3-), and (3) the nature of the leaving groups. As a result, further novel nitrogen mustards were synthesized and evaluated in order to optimize their properties based on the influence of these three factors. This led to the di-iodomustard prodrug, ZD2767P, which entered two Phase I clinical trials that have produced encouraging results (Scheme 7.5).

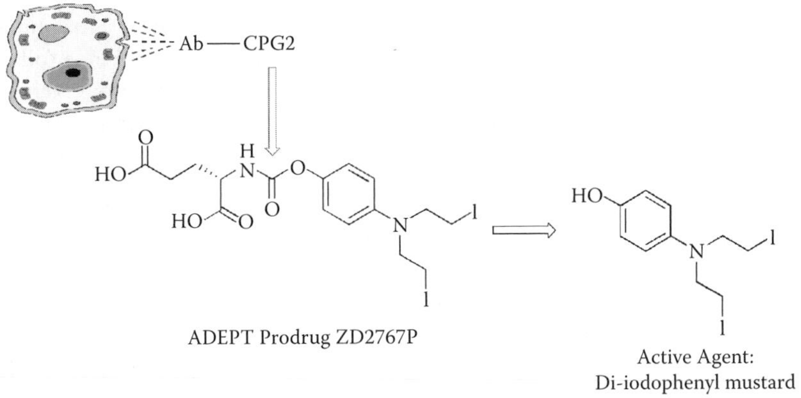

SCHEME 7.5 The principle of ADEPT, showing the conversion of the di-iodomustard prodrug ZD2767P, presently in Phase I clinical trials, into the active di-iodophenyl mustard at the tumor site by the Ab-CPG2 conjugate.

In both trials, patients with colorectal carcinoma expressing carcinoembryonic antigen received an A5B7 F(ab')(2) antibody conjugated to CPG2, followed later by administration of a galactosylated antibody directed against the active site of CPG2 (SB43-gal) to clear and inactivate circulating enzyme. The prodrug (ZD2767P) was then administered after plasma of enzyme conjugate levels had fallen to a predetermined safe level. It proved possible to measure tumor enzyme levels through quantitative gamma camera imaging (using labeled antibody conjugated) and from direct measurements in plasma and tumor biopsies, which demonstrated that the median tumor-to-plasma ratio of enzyme exceeded 10,000:1 prior to prodrug administration. Moreover, enzyme concentrations in the tumor were shown to be sufficient to generate cytotoxic levels of active drug, and so the concentration of prodrug needed for optimal conversion was achieved. Drug release was confirmed by establishing detectable levels in plasma; one patient was observed to have a partial response, and six had stable disease for a median of 4 months after previous tumor progression. Therefore, conditions for effective antitumor therapy were met, and evidence of tumor response in colorectal cancer obtained.

DNA interstrand cross-links, the cytotoxic lesions produced by these types of mustard prodrugs, can be observed in individual cells by a modification of the

COMET assay, and the extent of genomic damage at any time can be defined for a population of tumor or normal cells. Experiments with the ZD2767/ADEPT system in tissue culture showed that cross-links were produced in colon carcinoma cells in a dose-dependent fashion that related to the degree of cell kill. However, the cross-links were repaired within 24 hours in a sufficient proportion of cells, suggesting a high probability of tumor regrowth in the clinic. This finding was supported by similar findings in nude mice bearing xenografts of human colon carcinoma. Therefore, a third clinical trial utilizing ZD2767 focused on studying DNA cross-links. Tumor biopsies and peripheral blood lymphocytes were obtained on a small number of patients for COMET assay. One patient with evidence of tumor response had extensive cross-linking in tumor cells but no increase over controls in lymphocytes. However, a patient with no evidence of response had no significant cross-links in either his or her tumor or lymphocyte cells. The fact that tumor progression followed in the patient with evidence of response is consistent with human tumor xenograft model findings of repair of the cross-links induced by ZD2767. Therefore, research is currently underway to develop novel nonmustard prodrugs suitable for ADEPT that produce DNA adducts resistant to repair.

7.4.2   GDEPT (or VDEPT)

GDEPT, also known as VDEPT, is a two-phase polymer-leased strategy to treat solid tumors. First, a gene coding for a nonhuman enzyme is targeted to the tumor in a vector suitable for expression. In the second phase, a nontoxic prodrug capable of being activated by the tumor-associated enzyme is administered systemically. The important feature of this concept is that the enzyme genes are expressed exclusively, or with a relatively high abundance, in the tumor cells compared to healthy cells.

Unfortunately, current vectors for gene delivery are incapable of providing expression of a foreign enzyme specifically in all tumor cells. Therefore, a *bystander effect* (BE) is helpful whereby active agents, once produced, can diffuse and kill neighboring tumor cells not expressing the foreign enzyme. In addition, as with ADEPT, the prodrugs used need to be tailored to respond to the particular nonhuman enzyme being used and, preferably, the enzyme should have no human homologs so that prodrug conversion occurs only at the tumor site.

A large number of enzyme systems have been developed for GDEPT. Ideally, they should have high catalytic activity toward their respective prodrugs without the need for cofactors, which could become rate-limiting in target tumor cells. Also, these enzymes must achieve concentrations sufficient to activate the prodrugs under physiologic conditions. Enzymes suitable for GDEPT can be characterized into two major classes. The first are those of nonmammalian origin, although they may have human counterparts. Examples include viral TK, CPG2, and bacterial or yeast CD. The second class comprises enzymes of human or other mammalian origins that are found only in low concentrations in tumor cells or that may be completely absent, such as CYP450, β-Glu, or CPA.

The enzyme genes can be manipulated for either intracellular or extracellular expression in the recipient tumor cells. A key point concerning intracellular expression is that not only must the prodrug be able to penetrate a cell to become activated, but the activated agent must be able diffuse back across the cellular membrane in order to elicit a BE. This is not a requirement when the enzyme is expressed on the outer cell surface because it can activate the prodrug extracellularly. Although the latter mechanism can enhance a BE, the disadvantage is that activated drug can potentially leak back into the general circulation and lead to toxic effects, thus lowering the therapeutic index.

The prodrugs used for GDEPT and VDEPT should ideally possess such characteristics as limited cytotoxicity until activated, good chemical stability under physiological conditions, a favorable ADME profile, the ability to kill both proliferative and quiescent cells after activation, and the capacity to induce a BE. Also, for intracellular activation, the prodrugs should have sufficient lipophilicity to penetrate cell membranes or otherwise utilize an active transport mechanism. Many prodrugs in current use, such as the nucleoside analogs (e.g., 5-FC, CP, and CMDA) and the alkylating agent CB-1954, penetrate cells by passive diffusion.

The precise mechanism of activation of prodrugs is also important. The prodrugs used in GDEPT are commonly based on antimetabolites that require cycling cells (e.g., in S phase) for cytotoxicity and are not active in quiescent cells. Therefore, it has been suggested that alkylating prodrugs may have an advantage over, for example, purine nucleosides or 5-FC in that they are cytotoxic to noncycling as well as proliferative cells. In non-self-immolative prodrugs, the active agent is formed directly following a one-step process. Most prodrugs used in clinical trials to date have been based on licensed anticancer drugs whose characteristics are already well known. Examples of activation mechanisms exploited include scission reactions (e.g., CPG2, CE, CPA, PNP, PGA, MDAE, TP, β-GAL, β-L, and β-Glu), reductions (e.g., nitroreductase [NR] and DT-diaphorase), phosphorylation (e.g., HSV-TK, VZV-TK, and dCK), hydroxylation (e.g., CYP4B1), functional group substitution (e.g., NH_2 to OH; CD), deoxyribosylation (e.g., XGPRT), or oxidation (e.g., DAAO).

A challenge related to the choice of prodrug is that the enzymes used in GDEPT impose rigid structural requirements on the prodrug substrates, which limits the choice of anticancer drugs employed One approach to this problem is to design a self-immolative prodrug that, when activated, forms an unstable intermediate that extrudes the active agent through a series of subsequent degradative steps. For self-immolative prodrugs, the initial activation step is generally enzymatic in nature and distinct from the extrusion step, which is usually chemical and relies on spontaneous fragmentation. Examples of self-immolative mechanisms include 1,4- or 1,6-eliminations or cyclizations. A major advantage of this approach is that the structure of the active drug is independent of the substrate requirements of the enzyme being used in the system. Therefore, a broad range of drugs of various structural classes can be converted to self-immolative prodrugs.

It is also an advantage if the expressed enzyme can activate the prodrug directly without multiple catalytic steps because the requisite host endogenous tumor enzymes can become defective or deficient in some tumor cells. For example, alkylating agent prodrugs such as CP and CB-1954 (5-[aziridin-1-yl]-2,4-dinitrobenzamide) have an activation cascade that depends on endogenous enzymes. In the latter case, nitroreductase (NR) activates CB-1954 to the 5-aziridinyl-4-hydroxylamino-2-nitrobenzamide intermediate, which is followed by a further enzymatic

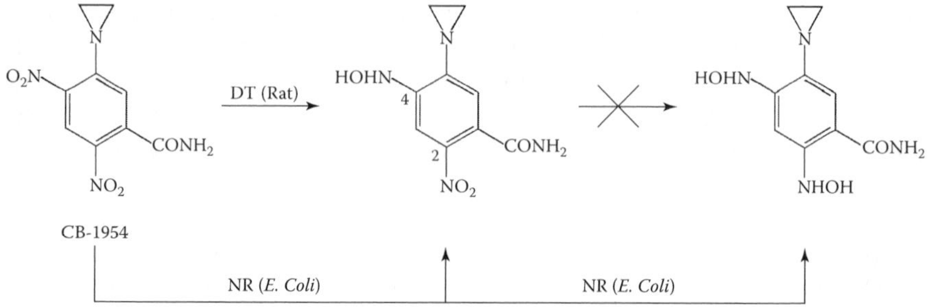

SCHEME 7.6 Activation of prodrug CB-1954 by rat DT diaphorase [DT (rat)] or *E. coli* NR (NR [*E. coli*]).

step that converts it to a powerful electrophile that alkylates DNA. An optimal half-life for the active agent produced is also critical; it should be short enough to prevent drug leakage from the tumor site but long enough to allow a local BE.

A recent GDEPT clinical trial involving NR to activate the prodrug CB-1954 used a replication-deficient adenovirus vector expressing NR from the CMV promoter (CTL102). Patients with operable primary or metastatic liver tumors were treated with increasing doses of CTL102 given as a single intratumor injection prior to radical surgery. This trial demonstrated the tolerability of the vector and the ability of CTL102 to induce a dose-dependent NR expression in the tumor. At the highest dose of 5×10^{11} virus particles, NR expression occurred in up to 50% of the tumor cells and was detectable in more than 50% of the specimen slides.

A detailed knowledge of the relevant enzymology is crucial when developing X-DEPT approaches. For example, it is known that reduction of the prodrug CB-1954 by rat DT-diaphorase results in conversion of the 4-nitro group to a hydroxyl-amine moiety, leading to an agent capable of cross-linking DNA. Human DT-diaphorase can also perform this reduction but at a slower rate, which explains why degrees of activation are reported to be up to 10,000-fold in rodent cell lines, whereas an activation of only 670-fold is observed in human cell lines. This finding probably explains, in part, why CB-1954 is devoid of antitumor activity in humans. Interestingly, *Escherichia coli* NR is able to reduce both the 2- and 4-nitro groups efficiently and thus provides a good choice of enzyme for GDEPT studies (Scheme 7.6).

Another recent example of GDEPT is a clinical trial involving prostate cancer, in which researchers are using CTL102 delivered via multiple intraprostatic injections in combination with intravenous CB-1954. These trials have demonstrated the safety of the treatment and have provided preliminary evidence of biological activity.

7.4.3 PDEPT

PDEPT is an experimental two-phase antitumor strategy that employs a combination of polymer-prodrug and polymer-enzyme conjugates to generate a cytotoxic species selectively within the tumor mass. The treatment is based on the principle that polymeric materials are known to accumulate in tumors due to their poor vasculature (the so-called enhanced permeation and retention (EPR) effect).

A degree of proof-of-concept was demonstrated in preliminary preclinical *in vivo* experiments in which polymeric prodrug conjugate PK1 (N-[2-hydroxy-propyl]methacrylamide [HPMA] copolymer-Gly-Phe-Leu-Gly-doxorubicin) and the HPMA copolymer-cathepsin B were studied in combination. Following polymer conjugation (with a yield of 30% to 35%), HPMA copolymer-cathepsin B retained approximately 20% to 25% of native enzymatic activity *in vitro*. To investigate pharmacokinetics *in vivo*, ^{125}I-labeled HPMA copolymer-cathepsin B was administered intravenously to B16F10 tumor-bearing mice. It was found to exhibit a longer plasma half-life (free cathepsin B $t_{1/2}\alpha$ = 2.8 h; bound cathepsin B $t_{1/2}\alpha$ = 3.2 h) and a 4.2-fold increase in tumor accumulation compared to the free enzyme. Also, when PK1 (10 mg kg^{-1} doxorubicin equivalent) was injected intravenously into C57 mice bearing B16F10 tumors, followed after 5 hours by HPMA copolymer-cathepsin B, a rapid increase in the rate of doxorubicin release occurred within the tumor (3.6-fold increase in the AUC compared to that seen for PK1 alone). Furthermore, when this PDEPT combination was used to treat established B16F10 melanoma tumors (single dose; 10 mg kg^{-1} doxorubicin equivalent), the antitumor activity observed was 168% compared to 152% seen for PK1 alone and 144% for free doxorubicin. The PDEPT combination also showed activity against a COR-L23 xenograft, which PK1 did not.

These preclinical results provide some limited evidence that PDEPT may work in the clinic. This approach also has certain advantages compared to ADEPT and GDEPT. For example, the relatively short plasma residence time of the polymeric prodrug allows subsequent administration of polymer-enzyme without fear of prodrug activation in the circulation. In addition, polymer-enzyme conjugates have reduced immunogenicity compared to enzyme alone. Although the PDEPT strategy is not being further developed at present, advances in polymer chemistry and drug delivery technologies may see this approach revisited in the future.

7.5 ENZYMATIC TARGETING

In addition to the use of enzymes in the X-DEPT therapies previously described, three other distinct strategies involving enzymes are discussed in this section. The first relies on the use of an enzyme constitutively expressed by tumor cells (but not healthy cells) to activate a prodrug without further intervention. The second is a similar approach but requires systemic administration of an enzyme cofactor in order to activate the enzyme. The third approach involves use of the enzyme asparaginase (Erwinase™) to break down asparagine in the blood and tissues. Asparagine is required for survival by leukemic cells.

7.5.1 PRODRUG ACTIVATION BY CONSTITUTIVELY OVEREXPRESSED TUMOR ENZYMES

There is growing evidence that certain tumor cells constitutively overexpress some enzymes relative to healthy cells. For many years, it has been a goal to use these enzymes to activate a systemically administered prodrug selectively at the tumor site. To realize this goal, the first challenge is to identify a suitable enzyme or enzyme family, and the second challenge is to design prodrugs that are not toxic to normal tissues but that can be converted to cytotoxic species by the relevant enzyme.

In terms of the choice of enzyme, recent attention has focused on the cytochrome P450 family which, in addition to endogenous substances, metabolizes many drugs and other xenobiotics (i.e., foreign) substances. The P450s are mixed function oxidases displaying selective aromatic hydroxylase activity that can be exploited by medicinal chemists to activate prodrugs. One particular family member, CYP1B1, is thought to be overexpressed in a wide variety of human tumors but not in normal tissues, and so is presently being studied more than any other family members in the context of drug activation.

It is worth noting that subtle genetic modifications (i.e., base pair changes) in the gene coding for the P450 drug metabolizing enzymes are often used to explain differences between the response of individuals to cancer chemotherapy agents. Thus, it is thought that relatively modest genetic changes can have an impact on the pharmacodynamic and pharmacokinetic characteristics of any anticancer drug and may have an especially significant effect on those agents with narrow therapeutic indices. The detection of variations in the P450 genes is therefore growing in importance as a means to select the most beneficial anticancer drugs for an individual, and the most appropriate doses and dose schedules to use. This approach has led to the identification of many different types of P450 enzymes, such as those in the CYP2C and CYP3A families, as well as CYP1B1 and CYP2D6. In particular, studies on the CYP2C and CYP3A families have shown that polymorphisms (i.e., subtle changes of base-pair sequence) in individual genes can lead to significant changes in the cytotoxic potential of certain drugs. In addition, the etiology of a range of human cancers is now thought to be related to polymorphisms in the CYP1B1 gene.

Prodrugs based on this approach are still at the discovery stage and so are only just beginning to reach the clinical evaluation. A number of biotech companies have been set up to exploit this new therapeutic strategy, and so the structures of novel prodrugs are not always freely available. However, prodrugs based on the duocarmycin DNA minor-groove alkylating agents are in development that require hydroxylation in order to become DNA reactive. Similarly, Phortress, a novel agent that has just reached Phase I trials, is activated by the CYP family of enzymes to give an intermediate that spontaneously fragments into DNA-damaging species.

The major criticism of this therapeutic approach is that, given the heterogeneity of tumors, it is highly unlikely that all tumor cells in a tumor mass will robustly overexpress the chosen enzyme and will therefore succumb to the drug released. However, this problem may be somewhat alleviated by a bystander effect. A second criticism is that it is unlikely that the chosen enzyme would not be expressed elsewhere in the body. Therefore, taken together, these points suggest that the therapeutic index of such a therapy is likely to be narrower in practice than expected.

7.5.2   COSUBSTRATE-MEDIATED PRODRUG THERAPY

This therapeutic strategy is similar to the one previously described but requires a systemically administered cofactor to activate an enzyme constitutively expressed by tumor cells. One such activation system, presently being evaluated in Phase I clinical trials, involves the systemic administration of a small synthetic cosubstrate molecule, dihydronicotinamide ribose (NRH), which can activate the latent enzyme NAD(P)H quinone oxidoreductase 2 (NQO2). The enzyme is a flavin-containing oxidoreductase that catalyzes two- or four-electron reductions of a variety of quinone compounds. Crucially, it has a unique activation mechanism that relies on the synthetic electron donor cosubstrate NRH.

The prodrug used in this system is CB1954 (see Scheme 7.6), which is activated by conversion to a 4-hydroxylamine derivative thus generating a potent DNA cross-linking agent (see Section 7.4.2) when administered together with NRH (see Structure 7.5). Interestingly, although activation has been shown to occur in some rat tumors, human cancer cells are inherently resistant to CB1954 because they cannot efficiently catalyze this conversion. Also, the expression of DT-diaphorase (NQO1), an enzyme known to be expressed in high concentrations in certain human tumors, appears to be related to NQO2 activity.

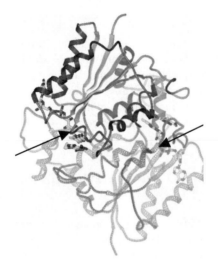

STRUCTURE 7.5 Molecular model showing the flavin-containing NQO2 oxidoreductase enzyme that catalyzes reduction of the prodrug CB1954 (bound in left-hand side of image; arrowed) in the presence of the electron donor NRH (bound in right-hand side of image; arrowed).

In preclinical studies, the presence of NRH has been shown to provide a very large (i.e., 100- to 3000-fold) increase in the cytotoxicity of CB1954 in both non-transfected human tumor cell lines and NQO2-transfected rodent cells. Other reduced pyridinium compounds have been shown to serve as cosubstrates for NQO2, and Knox and co-workers have identified key SAR features through a medicinal chemistry program. One derivative in particular, 1-carbamoylmethyl-3-carbamoyl-1,4-dihydropyridine (Structure 7.6), has been identified as a good cosubstrate for NQO2 but with several key advantages over NRH, such as greater stability and cellular penetration properties, along with the ability to potentiate the cytotoxicity of CB1954.

STRUCTURE 7.6 Structure of the synthetic NQO2 cofactor 1-carbamoylmethyl-3-carba-moyl-1,4-dihydropyridine.

Importantly, 1-carbamoylmethyl-3-carbamoyl-1,4-dihydropyridine is easy to synthesize, and Phase I clinical trials are now underway with patients being treated with both CB1954 and this cofactor to establish whether selective activation of the prodrug occurs in tumor cells.

7.5.3 ASPARAGINASE

Asparaginase (Cristanaspase™; Erwinase™) therapy takes advantage of the fact that certain types of tumor cells (in particular lymphoblastic leukemia cells) are unable to synthesize their own asparagine. Systemic administration of the enzyme aspara-ginase causes a significant reduction in systemic concentrations of asparagine, thus starving the cancer cells of this nutrient and leading to cell death while healthy cells continue to synthesize their own. This process is described in detail in Chapter 8 (Biological Agents).

7.6 PHOTOACTIVATED DRUGS (PHOTODYNAMIC THERAPY)

Photodynamic therapy (PDT) involves the administration of a nontoxic prodrug that can be activated selectively at the tumor site by the light of a specific wavelength. This general strategy has been in use for many years for the treatment of psoriasis using 8-methoxypsoralen (PUVA treatment) (Structure 7.7). This agent is relatively nontoxic until exposed to UV light, when it then cross-links DNA at thymine sites, causing distortion of the DNA helix with consequent toxicity toward the psoriatic cells.

STRUCTURE 7.7 Structure of 8-methoxypsoralen used in the phototherapy of psoriasis (PUVA treatment).

This prodrug photoactivation concept was extended into cancer therapy in the early 1990s, when it was discovered that porphyrin-type molecules are selectively taken up by some tumors. This led to the development of porfimer sodium (Photofrin™) and temoporfin (Foscan™), which are now used in the photodynamic treatment of various tumors (Structure 7.8 and Structure 7.9). After systemic administration, these drugs accumulate in malignant tissue and can then be activated by laser light to produce cytotoxic effects. Indications are presently limited to obstructing esophageal cancer and NSCLC, and to situations in which a tumor manifests near the skin surface (e.g., advanced head and neck cancers). However, recent progress in the development of surgical lasers with flexible optical fibers has allowed experimental use of these agents for the treatment of tumors in inaccessible places, such as parts of the GI tract and the ovaries. An intense nonlaser light source has also been developed for this type of therapy, and many new types of prodrugs are under investigation. However, one challenge to progress in broadening the use of this therapy is that the shorter wavelengths of light required to activate the present generation of photoactivated agents do not penetrate tissues very effectively. Therefore, new agents that can be activated at longer wavelengths are required so that deeper tissue penetration of light sources can be achieved.

7.6.1 Porfimer Sodium

Porfimer sodium (Photofrin™), introduced in the early 1990s for the treatment of esophageal cancer, is not a single entity but a dark reddish-brown mixture of oligomers formed by up to eight porphyrin units joined through ether and ester linkages (Structure 7.8). As part of a two-stage treatment process, it is administered by intravenous infusion over 3 to 5 minutes, followed by illumination of the tumor with laser light at 630 nm after 40 to 50 hours. Porfimer sodium is licensed for use in obstructing esophageal and endobronchial NSCLC that can be reached with laser endoscopy methods. It is also in use experimentally for other tumors of the skin and body cavities (e.g., stomach, colon, ovarian) that can be reached with flexible lasers.

STRUCTURE 7.8 Structure of porfimer sodium (Photofrin™).

The cytotoxicity and consequent antitumor effect of porfimer sodium are light and oxygen dependent. After intravenous injection, the agent clears from most tissues in 40 to 72 hours. However, organs of the reticuloendothelial system (including the liver and spleen), the skin and, most importantly, tumor tissue, retain the agent for longer periods of time. Therefore, selectivity occurs through a combination of preferential uptake and retention of the agent by the tumor and accurate delivery of the laser light, which induces a photochemical rather than a thermal effect.

At the molecular level, propagation of radical reactions, initiated after porfimer sodium absorbs light to form singlet oxygen, is thought to be responsible for the cellular damage caused by this agent. Hydroxyl and superoxide radicals, lethal to cells, are thought to form during subsequent radical reactions. Ischemic necrosis secondary to vascular occlusion, thought to be partly mediated through thromboxane A_2 release, can also cause tumor shrinkage. From a patient-care perspective, it is important to note that the necrotic reaction and associated inflammatory responses may develop over several days.

Side effects of porfimer sodium include severe photosensitivity, with sunscreens offering no protection. Although some individuals are photosensitive to porfimer sodium for 90 days or more after administration due to residual drug in all areas of the skin, in general, patients are advised to avoid bright indoor light or direct sunlight for at least 30 days after treatment. Therefore, careful monitoring of treated patients is required, and it is important to expose small test areas of skin to sunlight immediately after treatment to assess the degree of photosensitivity likely to occur. Constipation can also be a problematic but treatable side effect.

7.6.2 TEMOPORFIN

Temoporfin (Foscan™) is a second-generation photosensitizing agent for PDT introduced in the mid- to late-1990s (Structure 7.9). The main advantage of temoporfin over porfimer sodium is that it is a discrete chemical moiety rather than a polymeric mixture. It is licensed for the PDT of advanced head and neck cancer.

Temoporfin is administered by intravenous injection over at least 6 minutes. Side effects include photosensitivity (as with porfimer sodium, sunscreens offer no protection), and so exposure of eyes and skin to bright indoor light or direct sunlight should be avoided for at least 15 days after treatment. Other side effects include constipation, local hemorrhage, facial pain and edema, dysphagia, and possible scarring near the treatment site. Temoporfin is contraindicated in patients with porphyria or other diseases exacerbated by light. In addition, ophthalmic slit-lamp examination cannot be carried out for 30 days after administration.

7.7 BORON NEUTRON CAPTURE THERAPY (BNCT)

Boron Neutron Capture Therapy (BNCT) is a targeted biphasic approach to cancer therapy in which boron-10 (^{10}B)–enriched delivery agents are first administered intravenously and are taken up by tumors to varying extents. Once maximal tumor uptake has been achieved, the target area is then irradiated with low-energy neutrons (epithermal neutrons in the 1 to 10,000 electron volt energy range), which become

STRUCTURE 7.9 Structure of temoporfin (Foscan™).

thermalized at depth in the tissue and are captured by the ^{10}B atoms, with the resultant reaction producing α-particles (^{4}He) and lithium-7 (^{7}Li) ions that destroy the tumor tissue:

$$^{10}B + neutrons \rightarrow ^{13}B \rightarrow ^{7}Li + ^{4}He + 2.79 \text{ MeV}$$

These particles are energetic but have a limited range of less than 9 μm in tissue, thereby preferentially targeting the radiation to the tumor while sparing surrounding tissues. They also have high linear energy transfer (LET) characteristics compared to X-rays and gamma rays and, as a consequence, the level of tissue oxygenation is less important, which is an additional advantage if the tumors contain hypoxic cells.

A major advantage of BNCT over conventional radiotherapy is that it enables relatively large volumes of normal tissue to be irradiated with a reduced risk of adverse effects. This is because the presence of a neutron capture agent ensures that the radiation dose to the tumor is significantly higher than that to the surrounding normal tissue, despite the fact that the delivery of low energy neutrons cannot be focused. However, this advantage is dependent upon successful selective uptake of the neutron capture agent by the tumor. A major disadvantage is the lack of availability of expensive nuclear reactor or linear accelerator facilities (required to produce neutron beams) in most hospitals, and a paucity of effective BNCT delivery agents.

Experimental and clinical studies relating to the use of BNCT have largely, but not exclusively, focused on brain tumors— glioblastomas, in particular. High-grade gliomas are frequently inoperable and, without treatment, can prove fatal within a few months of diagnosis, with death being due to extensive proximal disease and local metastatic spread. Gliomas rarely metastasize via the blood and are only occasionally dispersed along cerebrospinal fluid pathways. Therefore, a treatment capable of controlling the primary disease and local metastatic spread could potentially result in patient cures. Unfortunately, high-grade gliomas do not respond well to conventional radiation or traditional chemotherapeutic agents. Enhanced radiotherapy methods, including hypoxic cell sensitizers and fast neutrons, have also proved ineffective. Therefore, BNCT has the potential to address this challenging clinical problem provided that suitable ^{10}B delivery agents can be found, which is an active area of research.

A number of boron delivery agents have either been studied or are in development. For example, borocaptate sodium ($Na_2B_{12}H_{11}SH$) has been shown to be effective in limited clinical trials in Japan (but in fewer than 150 patients). In these trials, patients were irradiated with thermal neutrons after the administration of BSH and, in those with relatively small superficial (less than 6-cm deep), high-grade brain tumors, 5-year survival rates approached 60%. This compared well with a 5-year survival rate of less than 3% in patients with comparable tumors treated with conventional radiotherapy. However, the use of thermal neutrons with limited penetration characteristics in tissue was a major limitation of this Japanese study.

More recently, much effort has focused on the development of more-deeply penetrating epithermal neutron beams. At present, nuclear reactor based epithermal beam facilities are available at the Brookhaven National Laboratory (BNL) in the U.S., at the European Joint Research Center (Petten) in the Netherlands, and at the University of Helsinki in Finland. In addition, the development of a new British epithermal beam facility has been funded at the University of Birmingham. This facility is unique in that it is based on a linear accelerator.

STRUCTURE 7.10 Structure of the BNCT agent *p*-boronophenylalanine (BPA).

Clinical BNCT studies using an epithermal beam have been undertaken at BNL using the delivery agent *p*-boronophenylalanine (BPA) dissolved in a fructose solution and administered intravenously (Structure 7.10). Preliminary results have been encouraging in terms of tumor responses observed, and no evidence of radioinduced brain necrosis has been observed.

BNCT also has the potential to treat tumors other than those of the central nervous system. For example, clinical trials in Japan involving the treatment of malignant melanoma using BPA as the neutron capture agent have provided encouraging results. The development of improved neutron capture agents could potentially enable a wider range of tumor types to be treated. Recently, a compound called *boronated porphyrin*, or BOPP, has been developed that accumulates in tumors in a manner similar to the PDT agents.

7.8 NOVEL DRUG DELIVERY APPROACHES

Effective drug delivery remains a challenge in the management of cancer. Existing drugs could be significantly more effective if techniques could be developed to deliver them selectively to the tumor site while avoiding healthy tissues. Therefore, there is a focus on the development of sophisticated targeted delivery systems that will not only supplement conventional chemotherapy and radiotherapy but may also prevent the occurrence of drug resistance. Knowledge and experience from areas such as nanotechnology, advanced polymer chemistry, and electronic engineering are being drawn upon to help develop these novel approaches. Examples from the areas of gene therapy, nanotechnology, novel polymers, and ultrasound are highlighted below.

7.8.1 GENE THERAPY

A significant research effort is underway to establish methodologies to deliver viral and other vectors to specific organs and tissues to enable therapeutic strategies such as GDEPT. There is also interest in the use of gene therapy strategies in combination with radiotherapy, including the use of radiation-sensitive promoters to control the timing and location of gene expression specifically within tumors. In particular, it is thought that gene therapy may enhance the effectiveness of limited-dose radiotherapy, which falls short of destroying a cancer. Yet another possibility is radioprotective gene therapy using transgenes coding for antioxidants that may ameliorate the effects of radiation-induced reactive oxygen species and could be used to spare normal tissues. Tumor-specific vectors leading to the expression of anti-angiogenic proteins are also of interest as a means to shut down blood flow specifically in tumors.

Therefore, not surprisingly, most research activity in the gene therapy area involves the development of have viral vectors and techniques to "wrap-up" DNA vectors in microparticles, such as liposomes or nanoparticles, to protect them from degradation by enzymes in the bloodstream and to target them to specific organs or tissues. In the latter case, the DNA must then be released, incorporated into the cancer cell genome, and efficiently expressed, which is also an active area of research.

Finally, it should be noted that the long-term safety of gene therapy techniques is a significant concern. In practice, it is highly unlikely that vectors can be targeted completely specifically to tumors, organs, or tissues, and some degree of collateral delivery to healthy cells is always likely. This could lead to cancer in these healthy tissues after a period of time, a phenomenon already observed in one clinical trial of children who received gene-replacement therapy for a noncancer disease. The viral vector used caused DNA to be added to the genome of healthy cells, and some patients developed a form of leukemia.

7.8.2 NANOTECHNOLOGY-BASED DRUG DELIVERY

Nanobiotechnology is a research area being applied to the improvement of drug delivery in various cancer therapies. It has been known for some time that encapsulation of cancer drugs in particles such as liposomes can modify their behavior after administration. Advantages of polymeric micellar drug delivery systems include: (1) long circulation time in the blood and stability in biological fluids; (2) appropriate size (10 to 30 nm) to escape renal excretion but to allow for extravasation at the tumor site; (3) simplicity in incorporating the drug compared to covalent bonding of the agent to a polymeric carrier; and (4) drug delivery that is independent of drug characteristics. Some micellar systems are dynamically stable because their solid-like cores dissociate slowly at concentrations below their critical micelle concentration. Others are not so stable and require additional stabilization that may be achieved, for instance, by cross-linking the micelle core. In a study of the pharmacokinetics and distribution of doxorubicin in micelles formed by drug-polymer conjugates, the micelles circulated much longer in blood than did free agents. Furthermore, the uptake of the conjugated drug by various organs proceeded much more slowly than that of free drug, and lower levels of conjugate were found in the heart, lungs, and liver than in the tumor. These findings were the basis for developing Caelyx™, a liposomal form of doxorubicin that is licensed for advanced AIDS-related Kaposi's sarcoma and for advanced ovarian cancer when platinum-based chemotherapy has failed (see Chapter 3). Due to the different pharmacokinetic profile of this liposomal preparation (i.e., accumulation at the tumor site), the incidence of cardiotoxicity is reported to be lower, as is the potential for local necrosis at the site of administration. A similar product, Myocet™, is licensed for use in combination with cyclophosphamide for metastatic breast cancer (see Chapter 3). Combined delivery of drugs and surfactants in micelles and liposomes, and the use of polymer-drug conjugates have also been studied as a means to overcome drug resistance by bypassing the P-Glycoprotein (P-Gp) pump system. The potential for liposomes and polymers to accumulate in tumors has also led to therapies such as PDEPT.

Various other long-circulating colloid drug delivery systems have been studied for use in cancer chemotherapy. Only a few known block copolymers form micelles in aqueous solution, and these include AB-type block copolymers (e.g., poly[L-amino acid]-co-polyethylene oxide). In particular, the PLURONIC family of ABA-type triblock copolymers having the structure PEO-PPO-PEO is well known, where PPO is polypropylene oxide, the hydrophobic central PPO block forms a micelle core, and the flanking PEO blocks form the shell or corona that protects the micelles from recognition by the reticuloendothelial system (RES).

A common structural motif of all these long-circulating systems, whether they are nanoparticles, liposomes, or micelles, is the presence of polyethylene oxide (PEO) on their surfaces. The dynamic PEO chains prevent opsonization and render the particles unrecognizable by the RES. This advantage has encouraged extensive research into the development of new techniques (ranging from physical adsorption to chemical conjugation) to coat particles with PEO. From a technological perspective, polymeric micelles formed by hydrophobic-hydrophilic block copolymers, with the hydrophilic blocks comprised of PEO chains, are very attractive drug carriers. These micelles have a spherical, core-shell structure, with the hydrophobic block forming the core of the micelle and the hydrophilic block or (blocks) forming the shell. They have promising properties as drug carriers both in terms of their size and architecture. The hydrophobic drug molecules partition inside the micelles, which are 15 to 35 nm in diameter. It is thought that these structures enter cells by phagocytosis or endocytosis. The drug is thus delivered inside the cells by local delivery or by fusion of the particle with the cell membrane, which destabilizes the micelle.

There is now significant ongoing research into new types of nanoparticles and the nanoencapsulation of drugs for targeted delivery to tumors of various organs both as a single therapy and in combination with other treatments, such as radiotherapy. In particular, there is growing interest in attaching entities such as antibody fragments to the surface of drug-containing nanoparticles in order to guide them specifically to a tumor, where they can release the active agent. Nanoparticles are also being used in gene therapy strategies as previously described. Finally, nanotechnology-based diagnostics can be combined with therapeutics, which is likely to be important in the future for the growing emphasis on the personalized management of cancer.

7.8.3 INTRACRANIAL DELIVERY

The improvement of cancer drug delivery to the brain is now a major area of research. One of the current limitations of the treatment of brain tumors is the lack of a suitable method to deliver therapeutic agents directly to the lesion. The challenge for systemic therapy is to develop methods to allow drugs to cross the blood–brain and brain–tumor barriers in order to allow higher concentrations to be obtained within the tumor bed. There are also opportunities in local drug delivery, and one commercially successful product of research in this area, Gliadel™, is based on local, controlled delivery of carmustine by a biodegradable polymer implanted at the tumor site after surgical resection (see Chapter 3). This allows effective concentrations of carmustine to reach any remaining tumor cells, a situation not usually achieved with systemic administration due to dose-limiting bone marrow toxicity. It also avoids some of the other adverse effects of carmustine, such as cumulative renal damage and delayed pulmonary fibrosis. Many other promising examples of both polymer and nonpolymer based drug delivery systems are under investigation.

7.8.4 ULTRASOUND TARGETING

The efficacy of cancer chemotherapy is often limited by the toxic effects of the current generations of drugs. One approach to this problem is to sequester a drug into a "package" that interacts minimally with healthy cells, keeping the drug contained until release at an appropriate time at the tumor site. Recently, an experimental method has been developed that involves ultrasound to achieve this objective. It involves systemic administration of a micellar drug carrier with a hydrophobic core containing an effective amount of an anticancer drug. Ultrasonic energy is then applied to the tumor site to release the drug from the hydrophobic core of the micelles. Examples of polymers studied include ABA triblock copolymers, AB diblock copolymers, mixtures of these, and mixtures of such polymers with PEGy-lated diacylphospholipids.

Ultrasound has been used extensively for both medical diagnostics and physical therapy, and is widely regarded to be safe when exposed to healthy tissues. Advantages of ultrasound include the fact that it is noninvasive, and that its energy can be controlled and focused easily with a capability of penetrating deep into tissues to predetermined depths. In particular, large amounts of ultrasonic energy can be accurately deposited deep into tumors. The depth of penetration and the shape of the energy deposition pattern may be controlled by varying the ultrasound frequency and the type and shape of the transducer. Optimal power densities at the target site may be obtained by adjusting the output power of the ultrasonic transducer.

Two possible mechanisms for the ultrasound enhancement of chemotherapy have been proposed. The first relates to enhanced drug release from micelles (as previously described). The other is associated with the apparent enhanced uptake of the micellar-encapsulated drug by cells. For example, there are literature reports of ultrasound-induced hypersensitization of anthracycline-sensitive cell lines, which would appear to support the enhanced drug-uptake mechanism. However, other experiments on the effect of low-frequency ultrasound pulse duration on drug uptake suggest that both mechanisms may work in concert.

Both low-frequency and high-frequency ultrasound have proved effective in triggering drug release from micelles. Low-frequency ultrasound is more effective but does not allow sharp focusing. In contrast, high-frequency ultrasound allows sharp focusing but does not penetrate as deeply into the interior of the body. Therefore, optimal design of ultrasound treatment will depend on tumor size and location. For example, for tumors 2 cm in diameter or larger, application of 100 kHz (or a lower frequency) ultrasound is feasible, with optimal power densities achieved by controlling output energy. For smaller tumors that are not so deep, a higher-frequency ultrasound can be used because it provides sharper focusing. In addition, power densities produced at the focal site by existing hyperthermia devices appear sufficient to cause drug release from micelles. Importantly, *in vivo* experiments have demonstrated improved survival rates for ovarian carcinoma–bearing mice treated with 3 mg/kg of doxorubicin in PLURONIC micelles delivered intraperitoneally in combination with ultrasound (1 MHz, 30 s, 1.2 W/cm^2) applied 1 hour after injection of the drug. In the same experiment, no effect was observed in untreated control mice or mice treated with 3 mg/kg of doxorubicin administered intraperitoneally in a physiological solution.

The current findings suggest that ultrasound targeting is a promising new area of research, particularly as high-frequency ultrasound is widely used in clinical practice for imaging purposes (though at much lower power densities than used in these experiments). The ideal scenario would be to ultimately combine imaging and therapeutic ultrasound transducer arrays in one instrument that could be used initially to image the tumor but, followed by automatic focusing of the ultrasound beam for therapeutic purposes. Finally, the combination of micellar drug delivery carriers and ultrasound is especially promising for treating MDR-positive tumors that do not respond to conventional treatment regimens.

FURTHER READING

Algire, G.H., and Chalkley, H.W. "Vascular Reactions of Normal and Malignant Tissue *In Vivo,*" *J. Nat. Cancer Inst.,* 6:73-85, 1945.

Denekamp, J. "Endothelial Cell Proliferation as a Novel Approach to Targeting Tumour Therapy," *Br. J. Cancer,* 45:136-139, 1982.

Folkman, J. "Tumor Angiogenesis: Therapeutic Implications," *NEJM,* 285:1182-86, 1971.

Folkman, J. "Angiogenesis in Cancer, Vascular, Rheumatoid, and Other Disease," *Nat. Med.,* 1:27-31, 1996.

Hahnfeldt, P., et al. "Tumor Development under Angiogenic Signaling: A Dynamical Theory of Tumor Growth, Treatment Response, and Postvascular Dormancy," *Cancer Res.,* 59:4770-75, 1999.

Hua, J., et al. "The Role of New Agents in the Treatment of Colorectal Cancer," *Oncology,* 66(1):1-17, 2004.

Hurwitz, H., et al. "Bevacizumab Plus Irinotecan, Fluorouracil, and Leucovorin for Metastatic Colorectal Cancer," *NEJM,* 350:2335-42, 2004.

Presta, L.G., et al. "Humanization of an Anti-Vascular Endothelial Growth Factor Monoclonal Antibody for the Therapy of Solid Tumors and Other Disorders," *Cancer Res.,* 57:4593-99, 1997.

Saaristo, A., et al. "Mechanisms of Angiogenesis and Their Use in the Inhibition of Tumor Growth and Metastasis," *Oncogene,* 19:6122–29, 2000.

Salgaller, M.L. "Technology Evaluation: Bevacizumab, Genentech/Roche," *Curr. Opinion Mol. Ther.,* 5(6):657-67, 2003.

Sheng, Y., et al. "Combretastatin Family Member OXI4503 Induces Tumor Vascular Collapse through the Induction of Endothelial Apoptosis," *Int. J. Cancer,* 111:604-10, 2004.

8 Biological Agents

8.1 INTRODUCTION

Biological agents are large macromolecular entities, usually proteins or glycoproteins, that fall broadly into four categories: biological response modifiers (BRMs), immunotherapy agents, enzymes, and vaccines. They are produced either by extraction from human cells or by genetic engineering processes.

BRMs include such agents as antibodies, interferons, and interleukins and may be used to control processes that facilitate the growth of tumors. They can also stimulate the immune system and make cancer cells more recognizable and susceptible to destruction. This can be achieved by enhancing the potency of immune system cells, such as macrophages, natural killer (NK) cells, and T cells. BRMs can also encourage cancer cells to adopt the growth patterns of normal cells; they can halt or even reverse processes that convert normal or precancerous cells into cancer cells. They may also be used as an adjunct to other cancer therapies. For example, they can help the body to repair or replace otherwise healthy cells that have been damaged or destroyed by chemotherapy or radiation, and help to prevent metastasis. Growth, colony-stimulating, and tissue necrosis factors are also considered to be BRMs, but they are still at an early experimental stage (see Chapter 9).

Immunotherapy, such as the bacillus Calmette-Guérin (BCG) bladder instillation, is designed to enhance or stimulate the immune system's response to tumor cells — in this case, locally for the treatment of primary or recurrent bladder carcinoma and for the prevention of recurrence following transurethral resection. The best example of an enzyme used in cancer chemotherapy is asparaginase, which is designed to interfere with the supply of a nutrient (asparagine) essential for tumor cells but not for healthy cells, and is used for the treatment of acute lymphoblastic leukemia (ALL).

Vaccines represent the fastest growth area for biological therapeutics, with worldwide revenues predicted to reach $6 billion by 2010. Of the more than 100 cancer vaccines currently in late-stage research, it is estimated that less than half will ever reach commercialization. However, although anticancer vaccine research is a highly specialized field that has been slow to develop, the recent success of clinical trials of the human papilloma virus (HPV) vaccine to prevent cervical cancer has stimulated great interest in this area.

Finally, it is worth noting that one of the major concerns when administering a biological agent is hypersensitivity of the patient to the protein or glycoprotein and the possibility of severe anaphylaxis.

8.2 BIOLOGICAL RESPONSE MODIFIERS (BRMS)

Interferons, interleukins, antibodies, and other agents that act through the immune system to treat cancer are known as biological response modifiers (BRMs). They work by modifying the interaction between the immune system and cancer cells in order to boost the body's own ability to deal with them. Growth, colony-stimulating, and tissue necrosis factors are also considered to be BRMs but these are in the early stages of development as therapeutic agents (see Chapter 9). This area of research is highly active and many more agents are likely to become available in the near future.

8.2.1 MONOCLONAL ANTIBODIES (MAbs)

Monoclonal antibodies (MAbs) can now be produced in large quantities to good manufacturing practice (GMP) standards, and there is a significant and growing interest in this area due to the success of signal inhibitors, such as trastuzumab (Herceptin™) and cetuximab (Erbitux™) (see Chapter 5), and the anti-angiogenic agent bevacizumab (Avastin™) (see Chapter 7), which target growth factor receptors on the cell surface (i.e., human epidermal growth factor receptors 2/neu, or epidermal growth factor receptor, in the case of trastuzumab and cetuximab, respectively or growth factors themselves (i.e., vascular endothelial growth factor in the case of bevacizumab). However, other antibodies, such as rituximab (MabThera™) and alemtuzumab (MabCampath™), are capable of causing lysis of B lymphocytes and so can be used to treat diseases such as chronic lymphocytic leukemia (CLL), advanced follicular lymphoma, and diffuse large B-cell non-Hodgkin's lymphoma.

8.2.1.1 Rituximab

Rituximab (MabThera™) is a mAb licensed for use in drug-resistant advanced follicular lymphoma and for diffuse large B-cell non-Hodgkin's lymphoma in combination with other anticancer drugs. It works by targeting B lymphocytes and causing them to undergo lysis. It should be used with caution in patients with a history of heart disease because it can exacerbate angina, arrhythmias, and heart failure.

One of the most common toxicities associated with rituximab is the infusion-related *cytokine release syndrome,* which is characterized by severe dyspnea and usually occurs during the first infusion. Other associated adverse effects include nausea, vomiting, fever, chills, allergic reactions, such as rash and flushing, and tumor pain. Transient hypotension can also occur during infusion. Premedication with an analgesic, an antihistamine and a corticosteroid can help with these side effects. More rarely, pulmonary infiltration and tumor lysis syndrome can occur.

8.2.1.2 Alemtuzumab

Alemtuzumab (MabCampath™) is similar to rituximab in that it is a mAb therapy that works by targeting and then lysing B lymphocytes. Its main use is in treating CLL patients who have either failed to respond to alkylating drugs or who have remitted for only a short period (i.e., less than 6 months) following treatment with fludarabine. As with rituximab, infusion-related cytokine release syndrome is a problematic side effect of alemtuzumab.

8.2.2 Interferons

Interferons are types of cytokines that occur naturally in the body. There are three major forms: alpha, beta, and gamma. Interferons were the first cytokines to be produced in the laboratory for the treatment of cancer, and interferon alpha is the most widely used for this purpose. Interferons can improve the way the immune system responds to cancer cells. For example, some interferons can stimulate macrophages, T cells, and NK cells, thus enhancing the antitumor activity of the immune system. They can also have a direct action on tumor cells, slowing growth or promoting differentiation into cells with more normal behavior.

The first interferon was discovered in 1957 by scientists at the National Institute for Medical Research in the U.K. Interferons are glycoproteins normally induced in response to viral infections and are usually only effective in the species in which they are produced. Although initially studied for their antiviral activity, in 1981 clinicians in Yugoslavia reported substantial improvements or even total remissions in head and neck cancers after human leukocyte interferon preparations were injected directly into the tumors. Despite the intense interest after this discovery, a considerable time elapsed before any interferons were developed for clinical use, the major difficulty being commercial production of pure interferons in sufficient quantities. Relatively large doses are required for treatment, and initially only minute amounts of varying purity were available from human tissue culture methods. However, in the early 1980s, significant advances were made in production techniques, including the development of recombinant DNA technologies that allowed biosynthetic interferons of high purity to be manufactured in clinically useful quantities.

Combinations of interferon alpha with a number of traditional anticancer agents, such as doxorubicin, cisplatin, vinblastine, melphalan, and cyclophosphamide, have been evaluated in patients with cancers including ovarian, cervical, colorectal, and pancreatic carcinomas with promising results. In the U.K., interferon beta is also available as interferon beta-1a (Avonex™, Rebif™) or as beta-1b (Betaferon™). It is licensed only for various stages of multiple sclerosis and not cancer.

8.2.2.1 Interferon Alpha

Interferon alfa has demonstrated limited antitumor activity in certain lymphomas and solid tumors, and has also been used to treat chronic hepatitis B and C. Three preparations (IntronA™, Roferon-A™, and Viraferon™) are licensed in the U.K., all administered by subcutaneous injection. IntronA™, which is interferon alfa-2b (rbe), is licensed for use in hairy cell leukemia, chronic myelogenous leukemia (CML), lymph or liver metastases of carcinoid tumors, follicular lymphoma, and maintenance of remission in multiple myeloma and as an adjunct to surgery in malignant melanoma. Roferon-A™, which is interferon alfa-2a (rbe), is used to treat AIDS-related Kaposi's sarcoma, recurrent or metastatic renal cell carcinoma, hairy cell leukemia, progressive cutaneous T-cell lymphoma, follicular non-Hodgkin's lymphoma, and CML, is also used as an adjunct to surgery in malignant melanoma. In addition, it is also used to treat chronic hepatitis B and C.

Viraferon™, which is another interferon alfa-2b (rbe), is licensed for treatment of chronic hepatitis B and C, rather than cancer. A polyethylene glycol-conjugated ("pegylated") derivative of interferon alfa-2b, known as peginterferon alfa-2b (Pegasys™, PegIntron™, and ViraferonPeg™), has also been introduced more recently. Pegylation increases the persistence of interferon in the blood, and this form is licensed for use in the treatment of chronic hepatitis C, ideally in combination with ribavarin.

The most common dose-related side effects associated with the interferons include influenza-like symptoms, lethargy, nausea, and anorexia. Other adverse effects include ocular side effects, depression (including suicidal behavior), myelo-suppression (particularly affecting granulocyte counts), cardiovascular problems (including hypotension or hypertension), nephrotoxicity, and hepatotoxicity. Less common events include hypersensitivity, thyroid abnormalities, hyperglycemia, alopecia, psoriasiform rash, confusion, coma, and seizures.

8.2.3 INTERLEUKINS

Interleukins are naturally occurring proteins found in the body. Several classes have been identified and studied for their role in immunomodulatory and inflammatory processes. For example, interleukin-1 has been shown to possess both direct and indirect antitumor effects, and has been investigated for its ability to protect bone marrow cells from the deleterious effects of radiation and chemotherapy. It can also release a cascade of hematopoietic growth factors.

Interleukin-2 has also been extensively studied, and a recombinant form of this (aldesleukin [Proleukin™]) is the only interleukin presently licensed in the U.K. for use in anticancer therapy. It is used to treat metastatic renal cell carcinoma; however, the response rate is less than 50%, and it causes capillary leakage leading to hypotension and pulmonary edema. Studies are also underway to determine whether interleukin-2 can enhance the efficacy of cancer vaccines.

8.2.3.1 Interleukin-2

Aldesleukin (Proleukin™), a recombinant interleukin-2 usually given by subcutaneous injection, is licensed for the treatment of metastatic renal cell carcinoma. It is now rarely given by intravenous infusion because of an association with capillary (vascular) leak syndrome (VLS), which can cause pulmonary edema and hypotension. Aldesleukin produces tumor shrinkage in a small proportion of patients, but it has not been shown to increase survival. The major common adverse effects are bone marrow, hepatic, renal, thyroid, and central nervous system (CNS) toxicity.

8.3 IMMUNOTHERAPY

In order to defend the body against attack by foreign (i.e., "nonself") invaders, such as bacteria and viruses, the body has a complex immune system that consists of networks of cells and organs that work closely together for a common cause. Scientists believe that the immune system is one of the body's main defenses against cancer and that it usually identifies and eliminates both precancerous and cancer cells. However, some cancer cells learn how to avoid detection by the immune system, and these cells proliferate to form malignant tumors. Hence, the risk of developing cancer should be greater in those whose immune system has broken down or is not functioning properly. Kaposi's sarcoma, common in patients with AIDS, is thought to be an example of this.

Based on this concept, one strategy to prevent or treat cancer involves stimulating or enhancing the immune system's responses to tumor cells. Agents and therapies designed to achieve this are known as *immunotherapies*. Nonspecific immunomodulating agents (as distinct from the more specific vaccines — see Section 8.5) are designed to broadly stimulate or indirectly support the immune system, a good example being BCG, which is used before or after surgical removal of a tumor to augment the treatment of superficial bladder cancer. More specific immunomodulating agents work by targeting key immune system cells and give rise to secondary responses, such as increased production of immunoglobulins and cytokines. So far, BCG is the only immunomodulating agent licensed in the U.K. for the treatment of cancer, although a number of new products are currently under development.

8.3.1 BCG BLADDER INSTILLATION

BCG is a live, attenuated strain of bacterium derived from *Mycobacterium bovis* that works by stimulating the immune system at a local level in the bladder. The two commercial products, ImmunoCyst™ and OncoTICE™, are derived from the Connaught and TICE strains of bacteria, respectively. BCG is licensed as a bladder instillation for the treatment of primary or recurrent bladder carcinoma, and for the prevention of recurrence following transurethral resection. BCG is contraindicated in patients with active tuberculosis, an impaired immune response, or other infections (e.g., human immunodeficiency virus, urinary tract infection).

Influenza-like syndrome, malaise, fever, hematuria, dysuria, urinary frequency, cystitis, and potential systemic BCG infection are some of the more common problems associated with effects of BCG. More rarely, hypersensitivity reactions (e.g., rash and arthralgia), renal abscess, transient urethral obstruction, bladder contracture, orchitis, and ocular disturbances may occur.

8.4 ENZYME-BASED THERAPIES

One successful enzyme-based therapeutic strategy involves the use of an enzyme to break down and reduce systemic concentrations of nutrients required by tumor cells. A selectivity for tumor cells is achieved because healthy cells are usually able to synthesize supplies of nutrients for themselves whereas, due to genetic abnormalities, tumor cells may be deficient in particular biosynthetic pathways and so become starved of one or more nutrients. The best known example of this strategy is the use of the enzyme asparaginase (crisantaspase [Erwinase™]), which breaks down the amino acid asparagine, thus lowering its concentration in blood and tissues.

8.4.1 ASPARAGINASE

The enzyme asparaginase (crisantaspase [Erwinase™]) is produced by *Erwinia chrysanthem* (a bacterial plant pathogen). It is a 133 kDa molecular weight tetrameric protein used almost exclusively in the treatment of ALL. The mechanism of action is based on the fact that these particular tumor cells synthesize lower levels of asparagine than that required for significant growth; as a result, they must obtain this amino acid exogenously. Healthy cells, on the other hand, can synthesize their own asparagine. The treatment involves intramuscular or subcutaneous administration of asparaginase, which reduces the concentration of asparagine in the body by converting it to aspartic acid and ammonia, thereby removing it from the protein synthesis cycle (Scheme 8.1). Whereas healthy cells rapidly synthesize their own supply of asparagine, the tumor cells eventually succumb to the reduced levels in their environment.

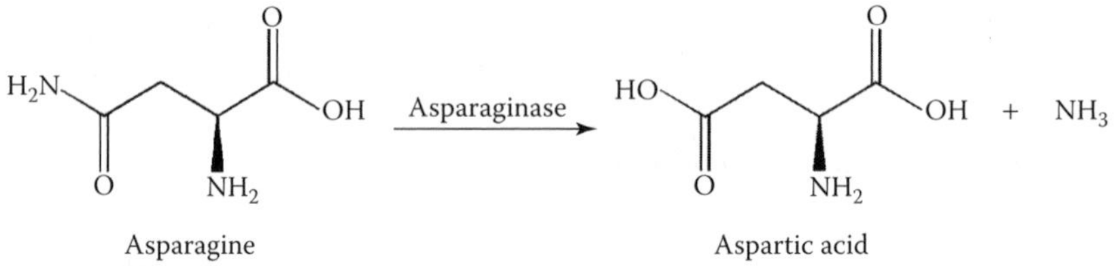

SCHEME 8.1 Cleavage of asparagine to aspartic acid and ammonia by asparaginase (crisantaspase [Erwinase]).

Side effects of asparagine include nausea, vomiting, CNS depression, and liver function and blood lipid changes. As crisantaspase is a large protein, facilities for management of anaphylaxis should always be available during administration. Resistance to the drug develops when the tumor cells begin to synthesize their own asparagine.

8.5 VACCINES

Specific immunologic approaches to cancer therapy (as distinct from nonspecific approaches — see Section 8.3) are growing in number and sophistication, and vaccines are at the forefront of this in both cancer treatment and prophylaxis. Vaccines are designed to boost an individual's natural defenses against tumor cells and have the potential to:

- prevent a cancer from forming
- stop an existing tumor from growing
- stop a tumor from returning after it has been treated
- eradicate tumor cells not affected by previous therapies

The cancer vaccine area has been very slow to develop and, until recently, the best known example of a cancer-related vaccine was one licensed for the prevention of infection with hepatitis B virus, which can cause liver cancer. However, a large number of new vaccines are presently in clinical development, with many providing impressive results (such as Merck's HPV-targeted vaccine that promises to significantly reduce the occurrence of cervical cancer). Based on these encouraging results, there is now a significant and growing interest in this area, and a vibrant market for cancer vaccines is predicted for the next decade.

8.5.1 THE IMMUNE SYSTEM

The immune system is partly constructed from a network of specialized immune cells. These cells are produced in the bone marrow from *stem cells,* which can produce a number of different types of immune cells. These immune cells circulate in the blood or lymphatic systems and also collect in the lymph nodes. Some immune cells have very specific functions, whereas others have more general (or nonspecific) functions. For example, T cells (T lymphocytes) and B cells (B lymphocytes) are types of specific immune cells, each type being activated by a single well-defined substance known as the T cell or B cell antigen, respectively. When a T cell or a B cell becomes activated by recognizing the corresponding antigen, it makes many identical copies of itself so that each copy can recognize the same antigen as the original T or B cell. Furthermore, there are two main forms of T cells. Cytotoxic T cells identify and kill cells that carry the antigen they recognize, whereas helper T cells release chemical messengers known as *cytokines* that recruit other types of immune cells to the attack site. Helper T cells also help cytotoxic T cells to function. The B cells make antibodies, with each cell making only one type of antibody directed against its specific antigen. In the same way that helper T cells help cytotoxic T cells to function, they also stimulate B cells to make antibodies.

The immune system also contains antigen-presenting cells (APCs), such as macrophages and dendritic cells. Their function is to patrol the body and to continuously sample their surrounding environment. They engulf any antigen-bearing "foreign" entities they come across (e.g., dead cells, debris, viruses, and bacteria) and then display small portions of the antigen on their surface. Dendritic cells are distinct by being more stationary. They monitor their environment from one location, such as the skin. Lymphocytes (T or B cells) that meet an APC can sense whether their specific antigen is present on its cell surface. If it is, then the lymphocytes become activated.

8.5.2 TYPES OF CANCER VACCINES

At present, there are two major categories of cancer vaccines. The first represents vaccines designed for prophylactic use (to prevent cancer), and the second represents vaccines used to treat existing cancers. Both T cells and B cells can be activated as part of an immune response for the purpose of either cancer prevention or treatment.

With preventive vaccines targeted against infectious agents that are known to cause cancer (e.g., hepatitis B and HPV), the activated B cells produce antibodies that bind to the infectious agents and interfere with their ability to infect cells. Because the agents must infect cells to transform them, this lowers the chance of the cancer occurring. For vaccines designed to treat cancer, antibodies specific for cell-surface antigens on cancer cells can target the antigens and, by a number of indirect mechanisms, cause the cell to die.

Part of the challenge of developing vaccines in either category is that the immune system has the almost impossible task of differentiating between healthy and tumor cells. To protect the body, the immune system must be able to ignore or "tolerate" healthy cells but recognize and attack abnormal ones. However, to the immune system, cancer cells differ from normal ones in only very small and subtle ways. Therefore, unfortunately, it usually tolerates them rather than attacking them. It follows that cancer vaccines must not only produce a robust immune response but should also activate the immune system sufficiently strongly to allow it to overcome its usual tolerance of cancer cells. Another reason why cancer cells may not stimulate a strong immune response is that they can develop ways to evade the immune system. For example, they may shed certain types of markers from their surfaces so that they become less "visible" to the immune system.

8.5.3 STRATEGIES FOR STIMULATING THE IMMUNE SYSTEM

Strategies that have been developed to stimulate immune responses against cancers include direct use of tumor antigens, enhancement of immunogenicity of tumor antigens, primed APCs, and anti-idiotype antibodies.

8.5.3.1 Direct Use of Tumor Antigens

This approach involves the identification of unusual or unique cancer-related molecules (i.e., tumor antigens) that appear on the surface of cancer cells but are rarely present on the surface of normal cells. These tumor antigens then form the basis of vaccines. More than 60% of research and development in the vaccine area is based on this approach. Apart from this being the most logical path to cancer vaccine development, other advantages include higher specificity, ease of production, lower cost of manufacture, and reduced levels of concern related to product contamination.

8.5.3.2 Enhancement of Immunogenicity of Tumor Antigens

A second general approach is to develop a vaccine capable of making tumor antigens more visible to the immune system. This can be done in several ways, including:

- Altering the structure of a tumor antigen slightly to make it look more "foreign," and then administering the altered antigen to the patient as a vaccine. One way to alter an antigen is to modify the gene needed to make it.

- Placing the gene known to be associated with a tumor antigen into a tissue or organ-specific viral vector and using the virus as a vehicle to deliver the gene to cancer cells (or to normal cells for prophylaxis purposes). Cancer cells infected with the viral vector will make much more tumor antigen than uninfected cells and should thus be more visible to the immune system. Cells can also be infected with the viral vector in the laboratory and then given to patients as a "vaccine."

- Placing genes for other proteins that normally help stimulate the immune system into a viral vector along with a tumor antigen gene.

8.5.3.3 Primed APCs

"Primed" dendritic cells or other APCs can be used as a component of vaccines. Dendritic cells can be primed in three ways:

- APCs can be exposed to tumor antigens in the laboratory and then injected back into patients. These cells are then primed to activate T cells.
- APCs can be infected with a viral vector that contains the gene for a tumor antigen.
- APCs can be treated with DNA or RNA that contains genetic instructions for the antigen. The APCs then make the tumor antigen and present it on their surface.

8.5.3.4 Anti-Idiotype Antibodies

Special types of antibodies known as *anti-idiotype antibodies* are known that have antigen-binding sites that mimic tumor antigens. These antibodies present tumor antigens in a different way to the immune system, thus stimulating B cells to make antibodies against the relevant tumor antigens.

8.5.4 Manufacture of Vaccines

Approximately 75% of the current cancer vaccine pipeline is comprised of generalized or "off-the-shelf" vaccines based on specific carbohydrates, proteins, or other easily replicated structures that are capable of being mass produced. The remainder consist of "personalized," vaccines (i.e., based on antigens harvested from an individual patient's tumor cells). This trend reflects the fact that generalized vaccines are simple to manufacture and commercially viable on a large scale. However, mainly due to their limited range of relevant antigen expression, they are associated with a higher rate of clinical failure than personalized approaches. To date, despite concerns relating to the complexity of manufacture and formulation, as well as product sterility and distribution problems, the "personalized" dendritic cell-based vaccine approach has provided the most convincing clinical evidence of efficacy. For example, Dendreon's Provenge™ is based on patient-specific dendritic cells loaded with a prostatic acid phosphatase, and has been the only vaccine to date to demonstrate a survival advantage in a randomized large-scale trial in prostate cancer, even though it failed to meet the primary endpoint of time-to-progression.

The vaccines can be made either from a patient's own tumor cells or antigens from another cancer patient's cells. If whole human cells are used as vaccines, they are usually treated with enough radiation to keep them from dividing (growing and multiplying), or enough to kill them. It is noteworthy that most tumors of a given type but from different patients share many antigens. When a patient's own tumor antigens or cells are used, the vaccine is called an *autologous* vaccine. When another patient's tumor antigens or cells are used, the vaccine is called an *allogeneic* vaccine.

Cancer vaccines may include additional ingredients known as *adjuvants* that help boost the immune response. These substances may also be given separately to increase a vaccine's effectiveness. Many different kinds of substances have been used as adjuvants, including cytokines, proteins, bacteria, viruses, and certain chemicals.

8.5.5 ACTIVITY OF VACCINES

Studies in laboratory animals have demonstrated that cancer vaccines show the greatest promise in situations where they are used to prevent a cancer from returning after the primary tumor has been eradicated by chemotherapy, radiation, or surgery. One interpretation of this finding is that the immune system is more likely to be successful when it has to deal with a smaller number of cancer cells. Conversely, attempting to shrink existing tumors using vaccine therapy is more problematic because the immune system is likely to be overwhelmed and ineffective when matched against a larger number of cancer cells.

At present, the most success has been achieved with preventative vaccines targeted toward causative viruses (e.g., hepatitis B and HPV). Experimental cancer vaccines as a treatment for advanced cancers have been far less successful and are presently recommended to patients only when all other therapies have been exhausted. Combination treatments are also being investigated, and many current studies are evaluating the co-administration of vaccines with other biological agents such as interleukin-2.

8.5.6 EXAMPLES OF VACCINES IN DEVELOPMENT

There are too many experimental cancer vaccines in early-stage clinical trials to discuss here in detail. Instead, some examples are listed below to provide insight into the range of clinical studies presently underway. It is important to note that, for vaccines, the promise observed in early-stage clinical trials, which often enroll only a small number of patients, is not always sustained in larger trials. For example, in one recent trial of a melanoma vaccine, the early findings suggested that the vaccine might help prevent melanoma from recurring in patients at high risk from this. However, in a subsequent larger trial that include approximately 750 patients who were at high risk for melanoma recurrence, high-dose interferon proved superior to the vaccine in preventing return of the disease.

8.5.6.1 Non-Hodgkin's Lymphoma

In a recent early-stage study in 20 patients, 18 who were given vaccinations against non-Hodgkin's lymphoma stayed in remission for 4 years on average. The vaccine contained a protein specific to each patient's tumor cells (i.e., an autologous vaccine) as well as two other substances to help boost the immune response. A large, randomized, Phase III trial of this vaccine is now underway.

8.5.6.2 Non-Small-Cell Lung Cancer (NSCLC)

In a recent Phase I/Phase II investigation, 3 out of 33 patients with advanced non-small-cell lung cancer had a complete remission of the disease and were still alive at least 3 years after treatment with an experimental vaccine therapy. The vaccine was prepared by adding the cytokine granulocyte–macrophage colony stimulating factor (GM-CSF) gene to each patient's tumor cells (i.e., an autologous vaccine).

8.5.6.3 Melanoma

An early-stage clinical trial showed that, when administered along with a melanoma peptide vaccine, an antibody that blocked the activity of a key immune-system regulatory molecule caused tumors to shrink in patients with metastatic melanoma.

8.5.6.4 Prostate Cancer

Results from a second multicenter Phase II study of GVAX vaccine, designed for the treatment of prostate cancer, were recently reported. The trial found that for 22 patients with advanced hormone-refractory metastatic prostate cancer who were receiving the highest dose, the final median survival was not less than 24.1 months, compared with 18.9 months for patients treated with Taxotere™ and prednisone. A Phase III clinical trial is presently underway of GVAX in combination with Taxotere and prednisone.

8.5.6.5 Cervical Cancer

In a recent clinical trial, a Merck vaccine (Gardasil™) that targets the HPV virus prevented virtually 100% of precancerous lesions that can lead to cervical cancer in women. Gardasil™ targets HPV types 16 and 18 which are associated with approximately 70% of cervical cancers, and HPV types 6 and 11, which are responsible for approximately 90% of genital warts. In the Phase III study, known as Future II, the vaccine prevented high-grade (i.e., the most serious) precancerous lesions and noninvasive cancers in 97% to 100% of women who received it. The women were first screened to make sure that they were not infected with HPV; they were then given either vaccine or placebo injections and monitored for approximately 2 years. More than 12,000 women between the ages of 16 and 26 enrolled in this randomized, prospective, placebo-controlled, double-blind study, which is just one component of ongoing Phase III trials that involve more than 25,000 males and females across 33 countries. Gardasil™ is expected to receive a licence in the near future, and there are ongoing discussions about the age at which it should be administered.

8.5.6.6 Head and Neck and Cervical Cancer

Lovaxin C™ is a *Listeria* vaccine, presently at the preclinical toxicology stage, that has been developed for the treatment of head and neck and cervical cancers. The pending clinical trial will be the first assessment of a *Listeria monocytogenes*–based live cancer vaccine in humans. The vaccine is designed to target the HPV-E7 antigen, which is expressed in both head and neck and cervical cancers. It produces an abnormally strong immune response to HPV-E7, and so activity in the clinic is anticipated.

8.5.7 CONCLUSION

In conclusion, cancer vaccine development is still in its infancy but has been buoyed recently by the success of vaccines against causative agents such as HPV. There is also a feeling among scientists that vaccines have an inherent advantage over other therapies in that they utilize the body's own finely tuned defense mechanism. Vaccines might also prove more effective when combined with other therapies.

FURTHER READING

Koutsky, L.A., et al. "A Controlled Trial of a Human Papillomavirus Type 16 Vaccine," *NEJM*, 347:1645-51, 2002.
Morse, M.A., et al. *Handbook of Cancer Vaccines*. Totowa, NJ: Humana Press, 2004.
O'Mahony, D., et al. "Non-Small-Cell Lung Cancer Vaccine Therapy: A Concise Review," *J. Clin. Oncol.*, 23(35):9022-28, 2005.

9 The Future

9.1 INTRODUCTION

Cancer research is funded not only by large pharmaceutical companies and smaller biotechs but also by such sources as government agencies, charitable organizations, and trusts and by special initiatives. Some people argue that, given the field of chemotherapy was initiated approximately 60 years ago with the discovery of the nitrogen mustards, the current lack of curative (or even highly efficacious) drugs and therapeutic strategies is a reflection of the complexities of the disease and the shortage of ideas for novel agents and therapies rather than a lack of funding for basic research. Considering the resources that have been channeled into cancer research worldwide since President Nixon declared his famous "War on Cancer" in 1971, the current lack of effective drugs and therapies would tend to support this view. However, the relatively recent discovery of imatinib (Gleevec™) has provided both the industrial and academic cancer research communities with a fresh impetus and enthusiasm for developing new types of so-called *molecularly targeted* agents. This mission has been spurred on by new and exciting developments in molecular biology, such as the human genome project, DNA arrays, and proteomics, which have already allowed many novel biological targets to be identified. This chapter describes some examples of these new areas currently under development. New research tools available to scientists in cancer research are also discussed, along with the rapidly developing area of chemoprevention.

9.2 NOVEL BIOLOGICAL TARGETS AND THERAPEUTIC STRATEGIES

There are many new approaches to cancer therapy, some at the very early research stage and others at later phases of development. These approaches range from novel drugs, biological agents and drug delivery systems, to gene targeting, gene therapy and, more recently, the possibility of targeting cancer stem cells. A selection of examples from each of these areas is described below.

9.2.1 NOVEL THERAPEUTIC TARGETS

9.2.1.1 RESISTANCE INHIBITORS

Because drug resistance can develop in most classes of agents, it provides a major obstacle to successful cancer chemotherapy. The various mechanisms of resistance available to the cancer cell can be categorized as non-class-specific, class-specific, or drug-specific. The main non-class-specific mechanism available to most cancer cells involves the induction of proteins such as multiple drug resistance protein (MDRP) and breast cancer resistance protein (BCRP), which can actively transport a variety of different classes of drugs out of a cell. This is a highly effective mechanism of resistance because it works irrespective of the chemical structures of the drugs. Similarly, a cancer cell can develop resistance by enhancing the metabolism of drugs from any mechanistic or structural class.

An example of class-specific resistance can be found within the DNA-interactive family of agents. After exposure to alkylating or cross-linking agents, cancer cells can increase DNA repair activity to develop resistance or, because all covalent DNA-binding agents are electrophilic, they can also increase the expression of glutathione transferase, which facilitates reaction of glutathione with the drugs to render them non-DNA-reactive.

Lastly, cancer cells can sometimes develop resistance to specific drugs. Instances of this can be found in the signal transduction family of agents. For example, cancer cells become resistant to imatinib (Gleevec™) by mutating the drug-binding site of the BCR-ABL kinase enough to block imatinib binding. However, in this case, the structure of other kinases is unaffected, and so the activity of other inhibitors targeted to different kinases will be maintained.

Agents capable of blocking any of these drug resistance pathways should not only enhance the action of the relevant anticancer drugs but should also allow them to be used for longer periods of time before resistance develops.

9.2.1.1.1   p-Glycoprotein Inhibitors

An early lead in this therapeutic strategy came from the observation that co-administration of verapamil (a calcium-channel antagonist) can enhance the activity of some anticancer drugs. Verapamil, a viscous, pale yellow oil (also known as dexverapamil hydrochloride), was first patented by Knoll in the early 1960s (Structure 9.1). It was the prototype calcium antagonist and was developed for use as an antihypertensive, antianginal, and antiarrhythmic (class IV) agent.

STRUCTURE 9.1   Structure of verapamil (* = chiral center).

It was eventually discovered that verapamil inhibits the glycoprotein efflux pump in multidrug resistant tumor cells, which explained its ability to enhance the activity of anticancer agents. Interestingly, both stereoisomers were found to inhibit the p-glycoprotein efflux pump, whereas its vasodilating activity resides primarily in the (S)-isomer. This led to a number of clinical studies in the mid- to late-1990s that involved using verapamil in combination with a number of anticancer drugs, including mitoxantrone, doxorubicin, and vinblastine. Unfortunately, at the maximally toler-ated daily dose (480 mg oral), it was not possible to achieve blood levels equivalent to the concentrations that provided an optimum cytotoxic potentiating effect *in vitro*. As a result, further clinical exploration of verapamil in this context was discontinued. However, this area of research is still active, and a number of pharmaceutical companies have novel p-glycoprotein inhibitors in various stages of research and development.

9.2.1.1.2 DNA Repair Inhibitors

The efficacy of covalent-binding DNA-interactive agents such as the nitrogen mus-tards and cisplatin is known to be reduced by the numerous repair mechanisms that normally protect DNA from damage by carcinogens, radiation, and viruses. DNA repair is mediated by a remarkable set of enzymes that first recognize the damage incurred (usually through a distortion of the DNA helix), and then signal to other enzymes to repair the lesion. A number of distinct repair pathways have now been identified, and their mechanisms range from simply removing an extra methyl group to completely excising the damaged segment of DNA and resynthesizing a new one. The pathways and mechanisms by which these enzymes work is an active area of research, and the first crystal structures of repair enzymes have only recently become available.

An interest exists in developing agents that inhibit DNA repair and, can thus be used clinically to enhance the efficacy of DNA-interactive anticancer drugs. How-ever, there is concern that healthy cells may become correspondingly more sensitive to the DNA-interactive agents during the treatment period, with no net gain in therapeutic index compared to use of the drug alone. There is also an associated risk that carcinogens and other sources of DNA damage (e.g., radiation and viruses) might be rendered more toxic during the treatment period. Some examples of the different classes of DNA repair inhibitors under investigation are described below.

9.2.1.3.1 DNA Alkyltransferase Inhibitors

Inhibitors have been developed against the DNA O6-alkyltransferase (ATase) enzyme, which repairs the O6-methylguanine adducts produced by temozolomide (Temodal™). PaTrin-2 (Patrin™, Lomeguatrib™), which is discussed in Chapter 3, emerged as one of the most potent compounds of its class and with a good toxicity profile. It is presently being investigated in crossover clinical trials with temozolo-mide in melanoma and colorectal cancers.

9.2.1.3.2 Poly(ADP-Ribose) Polymerase Inhibitors

An alternative approach is to target the mechanism that signals repair enzymes to accumulate at the site of a DNA lesion. The zinc-finger DNA-binding enzyme poly(ADP-ribose) polymerase 1 (PARP-1) becomes activated by sensing and binding to breaks mainly in single-stranded DNA, although it can also detect breaks in double-stranded helices. It then initiates DNA repair, especially after exposure to camptothecin derivatives, alkylating agents, or ionizing radiation. It does this by recruiting and then ADP-ribosylating nuclear proteins, such as the scaffold protein XRCC1, which can, for example, direct DNA polymerase b (polb) to replace a damaged nucleotide. Nuclear proteins associated with apoptosis may also be recruited. Therefore, the PARP-1 mediated poly(ADP-ribosyl)ation of nuclear proteins transforms DNA damage into signals that lead either to cell death or activation of the base-excision repair pathway. Interestingly, it appears that only the most abundant enzymes, PARP-1 and PARP-2 (both nuclear enzymes), are activated by DNA damage, although up to 16 other different types of PARP have been identified to date.

Since the discovery of PARP-1 about 40 years ago, many inhibitors have been developed both as potential adjuvants for use with DNA-interactive agents in anticancer therapies and as tools for investigation of PARP-1 function. There is now convincing biochemical evidence at the cellular level, supported by experiments involving the genetic manipulation of PARP-1 activity, demonstrating that PARP inhibition can induce an enhanced sensitivity to ionizing radiation, topoisomerase I inhibitors, and DNA-alkylating agents. Finally, it is worth noting the high degree of homology between the catalytic domains of PARP-1 and PARP-2, which means that most inhibitors will work on both enzymes.

4-Amino-1, 8-naphthalimide AG-104699

STRUCTURE 9.2 Structures of PARP inhibitors 4-amino-1,8-naphthalimide and AG-104699.

An example of a potent inhibitor of PARP that is used experimentally to study repair pathways in *in vitro* systems is 4-amino-1,8-naphthalimide (see Structure 9.2). This agent is 1000-fold more potent than one of the prototype lead PARP inhibitors, 3-aminobenzamide, and exhibits mixed-type inhibition with respect to substrate. A number of later-generation inhibitors have provided significant enhancement of the antitumor activity of temozolomide, topoisomerase inhibitors, and ionizing radiation in *in vivo* models; some of these, in combination with temozolomide, have led to complete tumor regression in model systems.

These promising preclinical results encouraged a number of pharmaceutical companies to take PARP inhibitors into clinical trials. For example, AG-104699, which is presently in Phase II (Pfizer), was developed from a benzimidazole-lead using a crystal structure–based design approach (Structure 9.2). KU-59436 is an orally available, potent inhibitor of PARP-1 and PARP-2 that is currently under development by KuDOS Pharmaceuticals. This compound, whose structure is still proprietary, leads to significant tumor growth delays in *in vivo* models of human colon carcinoma when administered in combination with temozolomide or irinote-can. KuDOS has also discovered that some PARP inhibitors have stand-alone activity against certain common tumors that lack working copies of BRCA1 or BRCA2 (tumor suppressor genes). This means that inhibition of PARP may benefit a specific subset of patients with mutated BRCA1 or BRCA2 genes that predisposes them to a number of early-onset tumors, particularly those of the breast, pancreas, prostate, and ovaries. However, in the short term, it is more likely that PARP inhibitors will be given in combination with agents such as temozolomide or irinotecan to improve the treatment of cancers including melanoma and glioma and colorectal and gastric cancers, respectively.

9.2.1.2 Telomerase Inhibitors

Telomeric DNA has emerged in the last decade as a novel anticancer target. Telo-meres are short, repeat DNA sequences at the ends of chromosomes (5′-TTAGGG-3′ [3′-chromosome end]) that protect sequence information near the chromosome ends from degradation and ensure complete replication of chromosome ends. Somatic cells have a finite lifespan because normal DNA polymerase is unable to fully replicate the ends of telomeric DNA due to its mechanism of action. Therefore, their telomeres progressively shorten with each successive round of replication until they become critically short, at which point they enter irreversible replicative senescence and ultimately apoptosis. This is thought to act as a type of "biological clock," leading to natural cell death when the telomeres have been depleted, a process that can be linked to aging.

Telomerase is a reverse transcriptase enzyme capable of adding telomeric repeats back on to the ends of chromosomes. Although the telomerase enzyme is not expressed in normal cells that have a defined life cycle, cancer cells can exploit this mechanism by expressing the enzyme and replacing telomeres to balance those lost during replication, thus attaining a state of immortality. This is supported by the

G-quartet   Folded G-quadruplex

STRUCTURE 9.3 Structures of a G-quartet and a folded G-quadruplex.

finding that more than 80% of different tumor types express telomerase. Proof-of-principle experiments using dominant-negative telomerase cells, antisense, or active-site inhibitors have shown that inhibition of telomerase in cancer cells can cause them to enter senescence and apoptosis.

One therapeutic strategy involves directly targeting the telomeric DNA substrate of telomerase, which can be induced to fold into a four-stranded DNA quadruplex structure, especially if stabilized by a small ligand. The quadruplex structure (see Structure 9.3), which is comprised of a series of stacked G-quartets, is not recognized by the template of telomerase and, therefore, cancer cells cannot add back telomerase sequences and eventually die. One advantage of this strategy is that it should be a highly selective therapy because normal cells do not usually utilize telomerase and so should be unaffected by the ligand.

(a)   (b)

STRUCTURE 9.4   (a) Structure of an example of an acridine-based G-quadruplex inhibitor; (b) molecular model of the inhibitor (in white) bound to the end of a G-quadruplex. (Figure kindly provided by Professor S. Neidle.)

Several types of ligands that can stabilize the G-quadruplex have been identified. Some have been derived by structure-based design and others by screening of individual compounds or libraries. The structure-based design approach has produced a class of trisubstituted acridine ligands have been developed by Neidle and co-workers (see Structure 9.4). The acridine chromophore was designed to selectively bind to the unique structural features of the G-quadruplex. Interestingly, structurally distinct G-quadruplex-binding ligands identified by high-throughput screening of compound libraries all share the same structural feature of extended planarity, a requirement for optimal interaction with G-quadruplex DNA. It is also worth noting that the acridine-based inhibitors do not appear to fulfill the conventional requirements for druglike features because they possess three cationic charges. However, they appear to be readily taken up by cells, possibly as a consequence of their hydrophobic aromatic rings.

All of the various types of G-quadruplex-stabilizing ligands in development show significant antitumor activity in *in vivo* models, particularly in combination with traditional cytotoxic agents. Provided they show antitumor activity with a reasonable toxicity profile in the clinic, one potential advantage of these agents is that they should be active across a broad range of tumor types due to their mechanism of action.

9.2.1.3 Epigenetic-Based Therapies

Heritable changes in gene function can occur without modifications to the DNA sequence itself. This area of study is known as *epigenetics,* a term that is sometimes used more broadly to describe the mechanisms involved in the development of an organism, such as gene silencing and imprinting. For decades the mechanism of heredity has appeared to be a relatively simple one that is coded and translated through the sequence of DNA. However, recent discoveries have highlighted how inherited changes in gene function can occur outside of this, through the modification of DNA or chromatin structures. These so-called "epigenetic changes" are present from birth to death and are involved in the first crucial steps that govern embryonic development, and also in influencing the expression or silencing of genes in epigenetic diseases. Researchers are now trying to understand how these epigenetic mechanisms interact with each other, how their disruption can lead to such conditions as cancer and mental retardation, and how drugs can be designed to treat these disorders.

It is now known that gene expression can be profoundly affected by changes in chromatin structure. Factors such as proteins that are required for the DNA transcription process, and hence gene expression, can only access the genes while the chromatin is "open" and not protecting that part of the genome. Conversely, these factors cannot reach the genome if the chromatin is "closed" (i.e., condensed), so the affected genes will remain "switched off." Epigenetic effects are usually mediated either by acetylation of histone proteins or by methylation of cytosines in genomic DNA. These chemical modifications lead to alterations in chromatin structure (i.e., open vs. closed), which may not only influence gene expression but also dictate heritable patterns of gene expression. Thus, research efforts in this area focus mainly on trying to establish relationships between chromatin structure, genomic instability, and cancer, and more specifically on elucidating the functions of DNA methylation, histone modification, and the protein complexes that sense and interpret these changes in chromatin.

Many heritable diseases result from mutations (DNA sequence changes) that stop gene expression, although some disorders are caused by epigenetic modifications that cause inappropriate gene silencing. Indeed, many cancers involve the epigenetic silencing of tumor suppressor and other genes that would normally control gene expression. However, inappropriate gene activation can also lead to diseases. For example, Burkitt's lymphoma is caused by overexpression of *myc*, the function of which is to promote cell proliferation. As might be anticipated, in most normal cells, the *myc* gene is located in an area of "closed" chromatin, as only a low expression level is required. However, in affected lymphocytes, the *myc* gene can move into a region of "open" chromatin through abnormal chromosomal rearrangements, thus leading to overexpression of the corresponding protein. This stimulates unregulated proliferation of lymphocytes and leads to the clinical symptoms of lymphoma. The main types of epigenetic modifications occurring in human tumor cells are DNA methylation and histone deacetylation. Patterns of methylation are put into place and then maintained by a family of enzymes known as DNA methyltransferases. There are four known human DNA methyltransferases — DNMT1, DNMT2, DNMT3A, and DNMT3B — which use a highly conserved catalytic mechanism to methylate cytosine residues in the context of genomic DNA. These enzymes engage in a variety of specific protein-protein interactions that presumably determine their functional specificity. For example, DNMT1 is sometimes associated with the replication machinery, thus implying a function in the maintenance of DNA methylation patterns, whereas DNMT3A can be associated with transcription factors, consistent with a role in *de novo* methylation. Furthermore, analysis of DNMT knockout cells suggests a significant level of cooperation between individual enzymes, thus introducing an additional level of complexity.

DNA methylation can inhibit the binding of control proteins, such as transcription factors, to promoters, thus directly switching off gene expression. However, the attraction of methyl-binding domain (MBD) proteins associated with histone deacetylases (HDACs) represents a more generalized effect. The chromatin structure can be completely changed by these enzymes, which chemically modify the histones. Acetylation of the chromatin causes it to "open" and become accessible to the

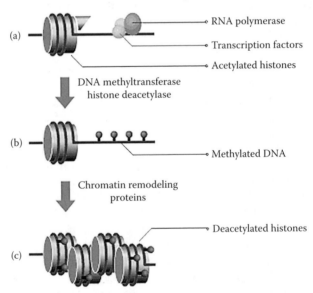

SCHEME 9.1 Epigenetic changes that can halt gene expression. (a) "Open" chromatin is characterized by histones with acetylated tails and nonmethylated DNA, which allow the recruitment of transcription factors and subsequent activity by RNA polymerase. (b) The activity of RNA polymerase may be blocked by the methylation of DNA (through DNA methyltransferase), which may directly inhibit the binding of transcription factors. Methylated DNA may also attract histone deacetylases enzymes, which are associated with methyl-binding domain (MBD) proteins that are initially recruited to methylated DNA. (c) Transcription factors cannot access the genome when the chromatin is in a "condensed" or "closed" state as a result of DNA methylation and histone deacetylation, and transcription cannot take place. The change in chromatin structure is often referred to as "chromatin remodeling."

necessary transcription factors, thus promoting gene expression. Conversely, gene silencing results from deacetylation of the histones, which causes condensation of the chromatin (see Scheme 9.1).

Human tumor cells frequently show altered patterns of DNA methylation, particularly at CpG islands, which are DNA sequences rich in CpG dinucleotides often found close to gene promoters. Methylation within islands appears to be associated with transcriptional repression of the linked gene. Genes involved in all facets of tumor development and progression can become methylated and epigenetically silenced. Re-expression of such silenced genes could lead to suppression of tumor growth or sensitization to anticancer therapies.

An important distinction between genetic and epigenetic changes in cancer is that the latter might be more easily reversed through therapeutic intervention. New therapeutic strategies are emerging based on reversing the DNA methylation process and inhibiting histone deacetylation. These approaches are being supported by the development of novel methods to rapidly screen and evaluate DNA methylation and histone acetylation patterns in the human genome. Furthermore, identifying epigenetic alterations in a precancerous lesion could lead to the discovery of biomarkers that may add to the knowledge of risk assessment and early detection, and potentially provide molecular targets for chemopreventive intervention. Examples of approaches to new therapies, including the inhibition of DNA methylation and histone deacetylation, are described below.

9.2.1.5.1 DNA Methyltransferase Inhibitors

Genomic DNA methylation patterns in cancer cells are characterized by two distinct features. First, repetitive sequences such as centromeric satellites tend to be hypomethylated, which has been linked to chromosome abnormalities. Strongly reduced levels of DNA methylation have been shown to cause genomic instability and concomitant tumorigenesis in studies involving transgenic mice. The second feature is that CpG islands in the promoter regions of various genes become hypermethylated and the corresponding genes silenced. Hypermethylation of tumor suppressor genes has been observed in many types of tumor cells and is generally assumed to be functionally equivalent to genetic loss-of-function mutations.

The possibility of reversing epigenetic mutations has generated interest in the development of DNA methyltransferase inhibitors. The prototypical inhibitor 5-azacytidine (5-AzaC; Mylosar™) has shown promising response rates in myelodysplastic syndrome patients and has recently been approved for therapeutic use (see Structure 9.5). It is a pyrimidine nucleoside analog that was originally isolated as an antitumor antibiotic from *Streptoverticillium ladakanus*. A chemical synthesis was reported in 1964, and it was evaluated in the clinic for acute nonlymphocytic leukemia in the late 1980s. However, 5-AzaC and related analogs become incorporated into genomic DNA, an inhibitory mechanism closely associated with their high cytotoxicity. Therefore, there is interest in discovering other classes of noncytotoxic inhibitors. A few compounds are known that reduce DNA methylation in human cells but are nonspecific inhibitors of DNA methyltransferases. In 2005, Lyko and co-workers used a molecular modeling approach based on a three-dimensional representation of the catalytic domain of

5-Azacytidine (5-AzaC; Mylosar™)   RG108

STRUCTURE 9.5 Structures of 5-AzaC (Mylosar™) and RG108.

human DNMT1 to produce the novel agent RG108 (see Structure 9.5). The model, used for an *in silico* screen of a small-molecule database, identified the structure of RG108, and it was subsequently synthesized. RG108 efficiently blocks DNA methylation both in a cell-free *in vitro* system and in human cancer cell lines. Furthermore, a drug-dependent demethylation and reactivation of epigenetically silenced tumor suppressor genes has been demonstrated. Therefore, RG108 appears to be an interesting lead molecule with fundamentally novel characteristics, particularly with regard to its lack of cytotoxicity at therapeutic concentrations and its ability to demethylate and reactivate tumor suppressor genes without affecting the methylation of centromeric satellite sequences. Therefore, it should prove useful for the experimental modulation of epigenetic gene regulation, which is likely to grow in importance as a novel cancer therapy in the near future.

9.2.1.5.2 Histone Deacetylase Inhibitors

Transcription in cells is controlled by a number of mechanisms, including the degree of acetylation of lysine residues in the histone N-terminal tails. This is controlled by two families of enzymes, known as histone acetyltransferases (HATs) and histone deacetylases (HDACs), which dictate the pattern of histone acetylation and deacetylation associated with transcriptional activation and repression, respectively (see Scheme 9.1).

HATs, of which there are more than five classes, catalyze the addition of an acetyl group to the lysine residues of histone N-terminal tails, thereby masking the positive charge on these residues. This reaction causes a decreased affinity of the histones for DNA (e.g., chromatin "opening"), which is generally associated with transcriptional activity. Deacetylation by HDACs removes these charge-neutralizing acetyl groups, thus allowing the histone protein to become more positively charged. This causes an increased affinity of the histones for DNA and leads to transcriptional repression (e.g., chromatin "closing"). HATs and HDACs do not work independently but exist in multiprotein complexes, where they work together to maintain fine control of transcriptional activation and repression. There is also growing evidence that HDAC enzymes play a role in other biological processes, such as microtubule structure and function, and the cell cycle.

The initial interest in developing HDAC inhibitors was based on the hypothesis that they should cause general up-regulation but might exert a selective effect on tumor cells by activating genes that these cells manage to down-regulate during the course of their transformation (e.g., tumor suppression genes). Three classes of HDACs exist (classes I, II, and III), and inhibitors that act on classes I and II are known. A significant research effort is now underway to identify novel inhibitors that can selectively target specific classes of HDACs, or even individual enzymes within a class. HDAC inhibitors are sometimes known as *chromatin remodeling agents* because they work by changing the shape of chromatin, the nuclear material consisting of DNA, the histones, and associated proteins.

Several structural classes of HDAC inhibitors exist: hydroxamates (e.g., trichostatin A [TSA], suberoylanaline hydroxamic acid [SAHA]; see Structure 9.6); cyclic tetrapeptides (e.g., FR901228, apicin); aliphatic acids (e.g., sodium/phenyl butyrate, valproic acid); benzamides (e.g., MS-275); and ketones (e.g., trifluoromethyl ketones, α-ketomides). Valproic acid was an early lead molecule that was shown to inhibit class I HDACs with more efficiency than it did class II. SAHA is one of the best-known HDAC inhibitors (Structure 9.6). It was synthesized as a more potent version of an old differentiation agent originally investigated in the clinic by Dr. Paul Marks (who was then president of Memorial Sloan-Kettering Cancer Center in New York) and was found to be an HDAC inhibitor in 1998. In 2001, Marks and three colleagues founded the company Aton Pharma (recently purchased by Merck) to further develop SAHA.

Trichostatin A   Suberoylanilide Hydroxamic Acid (SAHA)

STRUCTURE 9.6   Structures of Trichostatin A (TSA) and suberoylanilide hydroxamic acid (SAHA).

A large number of studies in various tumor cell lines have demonstrated that the biological effects of HDAC inhibitors include growth arrest, activation of differentiation pathways, and induction of apoptosis. For example, it has been shown that some HDAC inhibitors can reverse the malignant phenotype of lung cancer cells growing *in vitro*. It is also known that up-regulated HDAC expression in lung cancer cells growing *in vitro* can lead to unresponsiveness to tumor growth factor beta, which normally inhibits epithelial cell proliferation, thus suggesting that HDAC inhibitors may be useful in the treatment of lung cancers. Interestingly, some HDAC inhibitors, such as the experimental agent FR901228, have overcome both MDRP- and BCRP-mediated chemotherapy resistance in small-cell lung cancer cell lines.

More than eight HDAC inhibitors are presently in Phase I or Phase II clinical trials, including SAHA (Aton Pharma/Merck), depsipeptide (Fujisawa), phenylbutyrate (Elan), LAQ824 (Novartis), PXD101 (TopoTarget), MS-275 (Schering), pyroxamide (Aton Pharma), and MGCD0103 (MethylGene). The U.S. FDA has already granted orphan drug status to SAHA, which is presently in Phase II clinical trials for a variety of cancers, including cutaneous T-cell lymphoma. Although there is a possibility that HDAC inhibitors with standalone clinical activity in some tumor types will emerge, it is likely that most will be used in combination with established cytotoxic drugs.

Before HDAC inhibitors entered the clinic, many skeptics argued that drugs such as these that work by remodeling chromatin would be associated with severe side effects because they might trigger a generalized gene transcription cascade. However, HDAC inhibitors appear to be safe in the clinic, producing no major adverse events.

9.2.1.4 Antimetastatic Agents

A primary tumor is often not the direct cause of death of a patient because it can be removed by surgery and treated with radiotherapy or chemotherapy. Rather, secondary tumors (metastases) usually lead to death because they become too dispersed throughout the body to make further treatment, particularly surgery, possible. Therefore, there is interest in identifying agents capable of inhibiting the mechanism by which tumor cells move around the body, either via the blood or the lymph system, to establish themselves in new locations.

It is known that tumor enlargement, invasion, and metastasis are complicated multistep processes that involve tumor cell proliferation, degradation of the extracellular matrix by proteolytic enzymes, the migration of cells through basement membranes to contact the circulatory systems (blood or lymph), migration of the tumor cells to distant metastatic sites and, finally, growth of a new tumor (or tumors). The matrix metalloproteinase (MMP) family of zinc endopeptidase enzymes is now known to play an important role in the proteolytic degradation of the extracellular matrix proteins important for normal tissue remodeling and certain pathologic conditions. More than 20 mammalian MMPs are now documented, ranging from the well-known stromelysin, gelatinase, and collagenase enzymes to the recently identified membrane-type MMPs. There are five classes of human MMPs, consisting of more than 20 structurally related members that are categorized based on their substrate specificities and primary structures. These include collagenases (MMP-1, -8, and -13), gelatinases (MMP-2 and -9), stromelysins (MMP-3, -7, -10, -11, and -12), membrane-type MMPs (MMP-14, -15, -16, and -17; also known as MT1-MMP, MT2-MMP, MT3-MMP, and MT4-MMP), and nonclassified MMPs. Nearly all the components of the extracellular matrix, including proteolytic resistant proteins such as fibrillar collagen and elastin, can be degraded by these proteinases.

In humans, trophoblast invasion, skeleton and appendage development, ovulation, and mammary gland involution require finely regulated MMP activity. Furthermore, loss of control of MMP activity can have serious consequences and is associated with such problems as degradation of myelin-based protein in neuroinflammatory diseases, loss of aortic wall strength in aneurysms, tissue degradation in gastric ulceration, increased matrix turnover in restenotic lesions, opening of the blood–brain barrier after brain injury, destruction of cartilage and bone in rheumatoid arthritis and osteoarthritis, and tissue breakdown and remodeling during invasive tumor growth, angiogenesis, and metastasis.

As anticipated, many cancer cells overexpress various MMPs, facilitating their infiltration of healthy tissues and eventual metastasis. For example, tumor cells taken from head and neck, colon, lung, breast, prostate, ovarian, osteosarcoma, and pancreatic cancers have been demonstrated to overexpress MMPs in comparison to matched healthy cells. In particular, MMP-2 and MMP-9, which are closely associated with tumor growth, angiogenesis, and metastatic potential, are commonly reported to be overexpressed in tumor tissues. On this basis, it was hypothesized that suppression or inhibition of MMP enzymes might significantly reduce tumor growth, angiogenesis, invasion, and metastasis. Early studies centered on MMP-2 and MMP-9 because they are so frequently overexpressed in tumors cells.

The earliest MMP inhibitor prototypes were synthesized in the early 1980s, but clinical evaluation of MMP inhibitors as a class took place only more recently. One of the reasons for this delay was that, although a number of inhibitors with high *in vitro* potency were initially discovered, it proved difficult to find clinical candidates that could be given orally and that would be suitable for clinical trials. The active site of MMP enzymes contain a zinc (II) ion, and so a common property of all MMP inhibitors is the presence of a zinc binding group (ZBG) capable of chelating the zinc, such as carboxylic or hydroxamic acid groups, or a sulfhydryl (Scheme 9.2). Most of the MMP inhibitors currently under clinical investigation contain a hydroxamic acid group, which has proved particularly effective at binding to zinc.

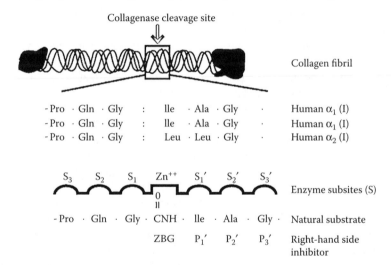

SCHEME 9.2 Diagram of the collagenase cleavage site of collagen which provided a model for the design of the first generation of matrix metalloproteinase (MMP) inhibitors.

The designs of the first MMP inhibitors were based on the known amino acid sequence of the collagenase cleavage site on collagen. This strategy involved the synthesis of short peptides with an attached ZBG that mimicked part of this amino acid sequence. Through these studies, it was established that the most potent inhibitors mimicked the sequence to the right of the cleavage site and incorporated a hydroxamic acid group as the ZBG (see Scheme 9.1). Inhibitors designed by this approach (e.g., Batimastat™; see Structure 9.7) had broad activity across the different

STRUCTURE 9.7 Structures of BB-2516 (Marimastat™), BB-94 (Batimastat™), and FYK-1388.

families of MMP enzymes but had little activity against other types of metallopro-teinase enzymes, such as enkephalinase or angiotensin-converting enzyme.

Further structure activity relationship studies showed that a certain degree of enzyme selectivity could be achieved by introducing larger P1'-substituents (see Scheme 9.2), which enhanced inhibition of stromelysin-1 and the gelatinases while reducing activity toward matrilysin and fibroblast collagenase. It was also found that certain combinations of substituents gave compounds such as Marimastat™, which is broad spectrum and orally active (Structure 9.7). The improved bioavailability of Marimastat is not fully understood but might be due to protection from peptidases by its unique pattern of substituents, reduced first-pass metabolism, or improved absorption.

In general, due to the problems encountered in obtaining inhibitors with good oral activity, with some exceptions, there has been a trend away from peptidic-based compounds in recent years. For example, Bayer discovered that fenbufen, a known anti-inflammatory agent, possesses modest inhibitory activity against gelatinase-A, and a series of analogs with greater potency than fenbufen was prepared. A number of novel nonpeptidic compounds have also been discovered by screening synthetic and natural product libraries. Through this means, tetracycline-type derivatives with MMP inhibitory activity and hydroxamic acid derivatives structurally similar to Batimastat™ have been obtained from natural product screens. However, synthetic compound libraries have provided the most unique new inhibitors, and Novartis has identified a novel nonpeptidic inhibitor of stromelysin-1 (see Structure 9.8), which after further elaboration led to the orally active experimental agent CGS 27023A.

In recent years, X-ray crystal structures of inhibitor-MMP complexes have pro-vided structural information that has allowed a more rational approach to inhibitor design. Structural information has confirmed the MMP binding modes of inhibitors such as Batimastat, and structural differences between the active sites of different MMPs have been explored. For example, it is now known that the most significant differences between the various MMPs relate to the size and shape of their S1' pockets (see Scheme 9.2). With the exception of matrilysin and fibroblast collagenase, in which the S1' subsite is relatively shallow, most MMPs have a deep pocket that reaches into the enzyme's core. This variation of S1' size for different MMPs explains why the incorporation of large P1' groups provides a degree of selectivity. This observation has been used by Agouron to design the nonpeptidic hydroxamic acid–containing inhibitor, AG-3340. Roche, also using a structure-based approach, has developed an orally active compound Ro 32-3555, which has selective inhibitory activity for col-lagenases instead of gelatinase-A or stromelysin-1 (Structure 9.8).

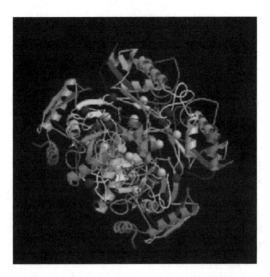

STRUCTURE 9.8 Structure of the stromelysin MMP enzyme MMP-12.

Based on these research programs, several compounds are now in clinical development, including AG3340, Ro 32-3555, D-2163, and the broad spectrum CGS-27023A, which is currently in Phase II and Phase III trials. Also, the novel broad-based inhibitor FYK-1388 (see Structure 9.7) is generating significant interest because it inhibits MMP-1, -2, -3, -7, -9, -13, and -14 and is particularly active against MMP-2 and -9, especially compared to Marimastat. Interestingly, *in vivo* experiments have suggested that the antitumor activity of FYK-1388 is not associated with a tumor cytostatic effect but can probably be accounted for by a reduction in tumor-induced angiogenesis through the blockade of several MMPs, and the inhibition of invasion through the extracellular matrix and associated blood vessels. Finally, it is worth noting that Marimastat itself is still being studied in a number of clinical trials as a broad-spectrum agent.

In terms of adverse reactions observed with MMP inhibitors, it is striking that patients in several clinical trials of agents such as Marimastat and Batimastat have suffered musculoskeletal effects, most likely linked to an effect on collagenase or other structural proteins. Although collagen turnover in adults is slow (approximately 50 to 300 days), MMPs are crucial for the maintenance of load-bearing tissues such as tendons, joints, and bones. Significantly, the side effects have included pain in the hands and shoulders, and generalized pain and tenderness of the joints.

Interestingly, in *in vivo* models, the experimental agent FYK-1388 shows significant antitumor activity at much lower doses than for previous generations of MMP inhibitors, such as Batimastat and marimastat, and with no sign of toxicity. Therefore, it is hoped that FYK-1388 will generate fewer musculoskeletal effects in clinical trials.

9.2.1.5 Heat Shock Protein (HSP) Inhibitors

Proteins are involved in nearly every aspect of cellular function, and producing proteins with the correct amino acid sequence is only the first step for a cell. Equally important, most proteins, which are usually very large and complex molecules, need to be folded into exactly the right three-dimensional shape before they can function properly. Any proteins that are not folded up properly are identified and tagged for degradation by the proteasomes.

The molecules that carry out the folding process are proteins themselves and are known as *chaperone molecules* or *heat shock proteins* (HSPs, or Hsps). One particular protein of this type, HSP90, is important because it is a master protein that controls a series of other HSPs. It is crucial for the folding of several "client proteins," many of which are highly relevant to cancer. For example, client proteins include mutated p53 (important in most cancers), Bcr-Abl (important in acute lymphoblastic leukemia), human epidermal growth factor receptor 2 (HER2)/neu (important in breast cancer), Raf-1, ErbB2 and other kinases, Cdk4, c-Met, Polo-1, Akt, telomerase hTERT, and steroid hormone receptors.

Because cancer cells typically have a large number of mutated proteins that might not fold properly, they often compensate by overexpressing HSP90. The result is that even mutated proteins are folded sufficiently well to avoid disposal by the proteasomes, which allows cancer cells to survive. Therefore, blocking the activity of HSP90 should lead to a reduction in folding activity in all cells, with more client proteins being destroyed by the proteasomes. However, because cancer cells have a higher dependence on HSP90 than normal cells (because they have more mutated proteins to deal with), researchers recognized that HSP90 should be a good target for drug intervention. Furthermore, as it is involved in the correct functioning of so many oncoproteins and pathways (including signal transduction and transcription), inhibition of HSP90 should block multiple oncogenic pathways in cancer cells, a preferable strategy compared with targeting a single point of vulnerability. Therefore, this type of "combinatorial blockade" of oncogenic targets should inhibit all of the hallmark traits of malignancy (see Chapter 1), and has the potential for broad-spectrum clinical activity across multiple cancer types. However, the risk of serious adverse effects was a concern because HSP90 is also involved in normal cell functioning.

In the late 1960s, the Upjohn Company discovered geldanamycin (Structure 9.9), a natural product isolated from *Streptomyces hygroscopicus,* that has the appropriate shape to interact in a "pocket" of the N-terminal domain of HSP90, thus

STRUCTURE 9.9   Structures of Geldanamycin, 17-AAG, 17-DMAG, and Radicicol.

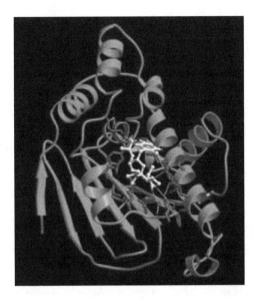

STRUCTURE 9.10 X-ray crystal structure of a complex of geldanamycin (center, in white) and HSP90. The side chains of geldanamycin are shown interacting in a well-defined deep pocket of HSP90 whose surface is covered with residues known to be conserved across different HSP90-expressing species.

blocking its ability to assist in the folding of proteins (Structure 9.10). Geldanamycin is a benzoquinone ansamycin, a member of a family of natural products originally attributed with weak antibiotic activity but later discovered to have potent antitumor activity. A natural product with similar activity, radicicol, was also isolated from the mycoparasite *Humicola fuscoatra* (see Structure 9.9).

Recent X-ray crystallographic studies have revealed the structure of the geldanamycin-binding domain of HSP90 (see Structure 9.10). The structure reveals a pronounced 15 Å–deep pocket formed by residues 9-232 that is highly conserved across species. Geldanamycin binds into this pocket, adopting a compact structure similar to that of a polypeptide chain in a turn conformation. This binding mode, along with the pocket's similarity to other substrate-binding sites, suggests that its probable function is to bind a portion of the client protein substrate and participate in the conformational maturation and refolding reaction.

Unfortunately, although geldanamycin was active in preclinical studies, it turned out to be a poor candidate for clinical trials due to its *in vivo* toxicity, instability, and poor water solubility. Therefore, during the 1980s, the National Cancer Institute (NCI) worked on several geldanamycin analogs in an effort to develop nontoxic, more water-soluble versions of the drug. Eventually, in 1992, 17-allylamino-demethoxygeldanamycin (17-AAG) was produced (in collaboration with Kosan), and this analog has been undergoing Phase I and Phase II clinical trials since 1999 (see Structure 9.9). 17-AAG binds specifically to HSP90 in a similar manner to geldanamycin itself but with a lower affinity. Therefore, it is surprising that 17-AAG and geldanamycin have similar potencies in tumor cells *in vitro*. Because 17-AAG had a better toxicity profile than geldanamycin, it was clearly a preferred clinical candidate and is presently being investigated in several trials. Preliminary results indicate that the target dose of 17-AAG can be achieved without dose-limiting toxicity. In particular, it is considered less hepatotoxic than geldanamycin and has been administered at doses of up to 450 mg/m^2/week. The most common side effects observed with 17-AAG include anorexia, nausea, and diarrhea (all of which can be dose-limiting). Less-common side effects include hepatotoxicity, fatigue, and blood dyscrasias (e.g., thrombocytopenia, anemia). A schedule dependency has been observed, and multiple daily dosing appears to increase the overall toxicity of 17-AAG, even with prolonged intervals between dosing cycles.

In the late 1990s, a further analog, 17-dimethylaminoethylamino-demethoxy-geldanamycin hydrochloride (17-DMAG), was identified by the NCI in collaboration with Kosan (see Structure 9.9). 17-DMAG has excellent bioavailability, is widely distributed to tissues, and is quantitatively metabolized much less than 17-AAG. It is presently being evaluated in the clinic for the treatment of both lymphomas and solid tumors.

Finally, it is worth noting that results from preclinical studies suggest that geldanamycin and its analogs have synergistic activity with several other traditional cytotoxic agents and inhibitors of signal transduction pathways. Therefore, if these agents are successful in the clinic, they may ultimately be used in combination therapies.

9.2.1.6 Thioredoxin Reductase Inhibitors

The thioredoxin reductase enzyme is a key component of cellular metabolism, and there is currently interest in developing inhibitors on the basis that a selective effect on tumor cells might be achievable because some have a faster rate of metabolism compared to healthy cells. The texaphyrins are aromatic pentadentate ligands that belong to a general class of compounds referred to as "expanded porphyrins" that have been developed as thioredoxin inhibitors by Pharmacyclics. The central core of a texaphyrin can form highly stable complexes with lanthanide series cations such as Gd(III) and Lu(III). Two metallotexaphyrins, motexafin gadolinium (Xcytrin™ [G-Tex]) and motexafin lutetium (Lt-Tex), are presently in clinical development (see Structure 9.11).

Motexafin gadolinium is designed to target cancer cells via this unusual and novel mechanism. Once inside a cancer cell, the compound disrupts metabolism and energy production by inhibiting thioredoxin reductase. Through a process known as "futile redox cycling," it produces free radicals toxic to the cancer cell via stimulation of the production of hydrogen peroxide and other reactive oxygen species. This results in redox stress, a situation in which tumor cells become susceptible to oxidative damage. Clinical studies have suggested that both primary and metastatic tumors selectively accumulate motexafin gadolinium, and that it remains inside the tumor tissue for many days.

Magnetic resonance imaging (MRI) can be used to follow the course of treatment with motexafin gadolinium because the metal atom in its center (gadolinium) is

Motexafin gadolinium (G-Tex): X = Gd (Xcytrin™)
Motexafin lutetium (Lt-Tex): X = Lu

STRUCTURE 9.11 Structure of motexafin gadolinium (Xcytrin™) and motexafin lutetium.

paramagnetic. MRI imaging in clinical studies has confirmed that motexafin gadolinium is selectively taken up by cancer cells in a number of different tumor types.

This drug, which is administered by intravenous injection, has reached Phase III clinical evaluation in patients with advanced renal cell cancer that is recurrent. Clear benefits have been observed, including stable or regressive disease with relatively mild side effects. Side effects include fatigue, headache, nausea, and blistering of the fingers. It has also been evaluated in Phase III trials as an MRI-detectable radiation enhancer, and clinical studies are planned for both motexafin gadolinium alone and in combination with different anticancer agents in other tumor types, such as brain metastases from lung cancer, head and neck tumors, and glioblastomas, as well as multiple myeloma, chronic lymphocytic leukemia (CLL) and lymphomas.

Motexafin lutetium, which accumulates in both tumor tissue and atherosclerotic plaque, is being evaluated in the clinic as a photosensitizer for use in the photo-dynamic treatment of tumors, and as a treatment for age-related macular degeneration and atherosclerosis.

9.2.1.7 HDM2-p53 Binding Inhibitors

The transcription factor protein p53, often referred to as the "guardian of the genome," is activated in response to cell stress (e.g., low oxygen levels, heat shock, DNA damage), which prevents further proliferation of the stressed cell by promoting cell cycle arrest or apoptosis. Therefore, it plays a pivotal role in regulating the growth of many tumors and in their response to cytotoxic agents. For example, tumors with mutated p53s generally respond poorly to cytotoxic agents, and there is a loss of apoptosis in affected cells. The role of p53 as a tumor suppressor is supported by the observation that approximately 50% of human tumors have a mutated or nonfunctional version of the gene. The MDM2 protein, a key negative regulator of p53 that is overexpressed in many human tumors, functions by binding to p53 and targeting it for proteasomal degradation (Structure 9.12). Therefore, one therapeutic strategy involves the development of small ligands that can inhibit the interaction of MDM2 with p53 in tumor cells containing the functional protein, thereby maintaining its levels and consequent tumor suppressor activities. This approach has been validated *in vitro* in various tumor cell lines by demonstrating that an anti-MDM2 monoclonal antibody (Mab) or antisense oligonucleotide (targeted to down-regulate MDM2 production) leads to p53 accumulation followed by cell cycle arrest or apoptosis. Such inhibitors may also be useful as chemopreventive agents.

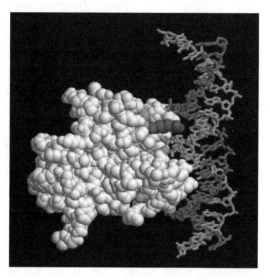

STRUCTURE 9.12 Model of the p53 core domain (mostly white) bound to DNA.

The discovery of small molecules capable of blocking the interaction of two proteins such as MDM2 and p53 has been a significant challenge. However, the potential of this approach was demonstrated several years ago using synthetic peptides that were shown to have potent inhibitory activity towards the MDM2-p53 interaction. Furthermore, once a crystal structure of the p53-MDM2 complex was obtained, key interactions involved in its inhibition could be studied. In particular, a hydrophobic cleft in the MDM2-recognition surface was identified.

A number of inhibitors of the MDM2-p53 interaction have been found by screening chemically diverse libraries using enzyme-linked immunosorbent and fluorescence-based cell-free assays. One of the best-known classes was discovered by Roche and is referred to as the "nutlins," named after the location of the research team in Nutley, New Jersey (see Structure 9.13). The most active of the nutlins, nutlin-2 and nutlin-3, have *in vitro* potencies (IC_{50}s) in some cell lines in the order of 0.14 μM and 0.09 μM, respectively. Biochemical studies have confirmed that the

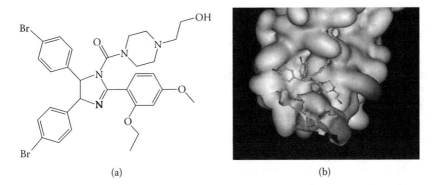

(a)   (b)

STRUCTURE 9.13 (a) Molecular structure of Nutlin-2, an MDM2-p53 inhibitor. (b) View of the crystal structure of the complex of Nutlin-MDM2 showing one of the bound nutlin molecules lying in a hydrophobic cleft of the protein. (Kindly provided by Professor Stephen Neidle.)

nutlins induce the expression of p53-regulated genes and exhibit potent antiproliferative activity through the induction of apoptosis in cancer cells expressing wild-type (i.e., functional) p53, but not in cells with mutated p53. At least one nutlin analog, nutlin-3, has acceptable pharmacokinetic properties and has demonstrated significant antitumor activity in human tumor xenografts in nude mice, reducing tumor growth by 90% at a dose of 200 mg/kg. Therefore, there is growing interest in this class of agent in both academic and industrial laboratories, with continuing attempts to identify further suitable anticancer drug leads.

STRUCTURE 9.14 Structure of the natural product chlorofusin.

Lead molecules have also emerged from natural sources. Chlorofusin is a novel fungal metabolite isolated from a *Fusarium* species (see Structure 9.14). It was identified as a moderately potent p53-MDM2 inhibitor during a screen of 5300 microbial extracts. This high-molecular-weight agent (1363 Da) is comprised of two components, the chlorofusin cyclic peptide (shown on the right hand side of Structure 9.14), and the smaller chromophore (shown on the left), which is a relative of the azaphilone family of natural products. Chlorofusin has been shown to interact with the p53-binding pocket of MDM2, thus blocking the approach of p53. Interestingly, the cyclic peptide fragment has been synthesized by Searcey and co-workers but does not bind to MDM2 or have any biological activity. Synthesis of the chromophore, whose absolute stereochemistry is yet to be reported, is presently underway and will answer the question of whether the MDM2-binding activity resides in the chromophore alone or only when it is joined to the cyclic peptide. A number of other natural products, including some members of the chalcone family, have also been shown to possess MDM2-p53 inhibitory effects.

In silico screening of virtual libraries has also been used in an attempt to identify novel inhibitors. For example, one such study identified a sulfonamide (NSC 279287) from the NCI database that was subsequently shown to enhance p53-dependent transcriptional activity at a low micromolar level in an MDM2-overexpressing cell line.

9.2.1.8 Radiosensitizing Agents

There has been longstanding interest in agents that can potentiate the effect of radiotherapy by modifying the slope of the radiotherapy dose-response curve. This strategy is partly based on the discovery that hypoxic tumors tend to be resistant to both radiotherapy and chemotherapy, and that oxygenation can enhance the antitumor effect. *In vitro* experiments in tumor cell lines demonstrating the effect of oxygen enhancement, led the way in the 1970s to clinical trials of the use of hyperbaric oxygen in combination with radiotherapy. Since then, there has been a search to find agents that can mimic the effect of oxygenation. Such agents need to reach the tumor in adequate concentrations and to have predictable pharmacokinetics for synchronization with radiation treatment. A successful radiosensitizer that could be administered with each session of radiotherapy would need to be minimally toxic itself and cause no enhancement of radiation toxicity. However, despite extensive research, an ideal agent has not yet emerged, although several compounds are presently in clinical trials.

Through the years, various chemical agents have been studied that are capable of producing radiosensitization of hypoxic cells *in vitro*. One of the best-known classes of agent developed for this purpose is the nitroimidazoles, which produce sensitization by mimicking the presence of oxygen. Two nitroimidazole agents in particular, misonidazole and pimonidazole, have been extensively studied since the late 1970s (see Structure 9.15). They have been combined with irradiation for the treatment of a number of different cancers in widespread clinical trials.

Misonidazole (USAN)   Pimonidazole hydrochloride

STRUCTURE 9.15 Structures of Misonidazole (USAN) and Pimonidazole hydrochloride.

Misonidazole and pimonidazole work, at least *in vitro,* by virtue of their high electron affinity. This induces the formation of free radicals and depletes radioprotective thiols, thereby sensitizing hypoxic cells to the cytotoxic effects of ionizing radiation. This allows single-strand breaks in DNA to occur, with subsequent inhibition of DNA synthesis followed by cell death. However, despite the promising *in vitro* results, the combination of misonidazole and radiation in several clinical trials (including a number of Phase III trials) since 1978 has failed to demonstrate improved survival. Furthermore, the clinical usefulness of misonidazole and a close analog (desmethylmisonidazole) was limited by severe but reversible toxicities, including nausea and vomiting and, more seriously, neurotoxicity (peripheral neuropathy).

Many other classes of agents have been investigated in search of therapeutically useful compounds. For example, it has been known for some time that the sensitivity of cells and tissues to radiation may be reduced by exposure to prostaglandins. Therefore, researchers have tried to obtain therapeutic gain by administering nonspecific prostaglandin inhibitors, such as nonsteroidal anti-inflammatory drugs (NSAIDs), prior to radiotherapy. This approach has not proved successful, as the NSAIDs raise the sensitivity of both normal and tumor cells with no net therapeutic gain. However, there has recently been renewed interest in selectively sensitizing tumors to radiation while sparing healthy tissue with the discovery of some highly selective inhibitors of the inducible cyclooxygenase 2 (COX-2) enzyme, which do not act on the constitutive COX-1.

A number of agents, many based on the misonidazole structure, are still being evaluated in clinical trials. These include efaproxiral, etanidazole, fluosol, and nimorazole. Given recent advances in the way radiation is administered (e.g., highly accurate computer-controlled confocal beams), the emergence of an efficient radiosensitizing agent could significantly improve this treatment modality.

Finally, it is worth noting that, in preclinical *in vitro* studies, agents such as misonidazole can be shown to enhance the antitumor effects of some traditional cytotoxic drugs, such as cyclophosphamide. This combination has also been evaluated in the clinic, where it was found to be well tolerated. In particular, the pharmacokinetics of cyclophosphamide were not affected by misonidazole, and its myelotoxicity was not potentiated.

9.2.2 NEW BIOLOGICAL AGENTS

9.2.2.1 Growth Factors

Growth factors, or *cytokines,* are proteins that influence cell growth and maturation. Recombinant technology has allowed the production of large amounts of cytokines, and several are being evaluated in clinical trials. For example, hematopoietic growth factors have found use in counteracting the myelosuppressive side effects associated with many anticancer agents. Granulocyte colony stimulating factor (G-CSF) and granulocyte-macrophage colony stimulating factor (GM-CSF) boost the circulating number of neutrophils, eosinophils, and macrophages by inducing inflammation. It has been shown that some tumor cells possess receptors for these CSFs, and a number of clinical trials are ongoing. Erythropoietin is naturally produced by the body in response to hypoxia and has been used to treat certain types of malignant anemias. Recombinant technology has now been applied to the production of erythropoietin, and a number of clinical trials are presently underway.

Inhibiting certain growth factors can lead to useful antitumor activity, and known inhibitors include octreotide (Sandostatin™), which is already in clinical use (see Chapter 6), and suramin (a polysulfonated naphthylurea). These two agents are analogs of somatostatin, a naturally occurring growth hormone. Octreotide is administered subcutaneously and is useful in controlling symptoms, but does not always lead to tumor mass reduction. Suramin binds proteins extensively due to its polyanionic nature, and early clinical trials were hampered by severe toxicities (renal and liver dysfunction, adrenal insufficiency, and peripheral neuropathy). However, more recent clinical trials in breast and prostate cancer patients with suitable dosage adjustments have produced promising results. The clinical evaluation of lymphokine, a glycoprotein produced by lymphocytes and natural killer cells, is also under investigation.

9.2.2.2 Tumor Necrosis Factor

Tumor necrosis factor (TNF) is a glycoprotein produced by macrophages, monocytes, and natural killer cells that is partly responsible for tumor cell lysis. Phase I and Phase II studies, in which the agent was administered intravenously or intramuscularly, have recently taken place but with mixed results. One problem is ensuring that a high enough concentration of TNF reaches the tumor site. However, this must be balanced against the drug's adverse effects, which include hypotension and cardiotoxicity. Transient fever and hematological disturbances have also been observed in some patients, although these are reversible on cessation of treatment. Studies on the potential clinical applications of TNF are continuing.

9.2.2.3 Immunomodulation

9.2.2.3.1 Thalidomide

There is significant interest in discovering and developing agents that can stimulate the immune system to attack and kill tumor cells. Ironically, thalidomide, a drug that became infamous in the 1960s for its teratogenic effects, is gaining much attention as a lead molecule in this new therapeutic area.

Thalidomide is a chiral glutamic acid derivative first reported in 1957 by Chemie Grünenthal. It is very poorly soluble in water and exists as a racemic mixture of two enantiomers. In 1957 thalidomide was approved in various European countries as a sedative and became widely used in the rest of the world, specifically marketed to pregnant women. By late 1961, the association between thalidomide and fetal abnormalities had become recognized. Although it was withdrawn commercially at this time with historic compensatory payments, thalidomide has continued to be investigated for the treatment of other conditions, including erythema nodosum leprosum, AIDS wasting, nonmicrobial aphthous ulcers of the mouth and throat

STRUCTURE 9.16 Structures of the two enantiomers of thalidomide.

(associated with AIDS), resistant or relapsing multiple myeloma, and other malignancies.

Thalidomide is used today mainly for the treatment of multiple myeloma. Although its main pharmacological activity is immunomodulatory, the precise mechanism of action in multiple myeloma has not been fully established. Apart from a direct effect on myeloma cell growth and survival, other suggested mechanisms include an effect on interleukins and interferon-γ, inhibition of cell surface adhesion molecules that help leukocyte migration, changes in the ratio of CD8+ (cytotoxic T cells) to CD4+ (helper T cells) lymphocytes, suppression of the production of TNF-α, and a possible effect on angiogenesis. Interestingly, in some circumstances thalidomide can also exhibit partial immunostimulatory as well as anti-inflammatory effects.

The most commonly observed adverse reactions in humans include somnolence, constipation, and polyneuropathy, leading to painful tingling and numbness. Furthermore, hypersensitivity can occur, accompanied by such symptoms as maculopapular rash and possibly fever, tachycardia, and hypotension.

Finally, it is noteworthy that thalidomide is administered as a racemic mixture. However, the individual enantiomers differ from the racemate, each having higher water solubility by a factor of approximately three. There is also indirect evidence that enantiomerically pure thalidomide is absorbed more readily and undergoes faster hydrolytic cleavage. However, both *in vitro* and *in vivo* studies have shown that, under physiological conditions, there is very rapid racemization, so that whichever enantiomer is administered, a similar equilibrium mixture will quickly be established. The (+)-(R)-form is the predominant enantiomer at equilibrium in a ratio of 1:1.6.

Given the interesting clinical results being obtained with thalidomide, a significant research effort is presently underway to fully establish its mechanism of action and to develop more efficacious and less toxic analogs.

9.2.3 NOVEL DRUG DELIVERY SYSTEMS

A number of new drug delivery systems that have reached the clinic, such as polymer and ultrasound-based systems, were described in Chapter 7. However, there are many other novel delivery systems at much earlier stages of development. A few examples of these are described below.

9.2.3.1 Near-Infrared-Activated Nanoshells

A new form of targeted cancer therapy based on photothermal technology is being developed by researchers at Rice University (Houston, TX, U.S.A.) in collaboration with Nanospectra Biosciences, Inc. It is a noninvasive treatment that uses nontoxic gold nanoshells in combination with near-infrared light that passes harmlessly through soft tissue. The near-infrared light, which is just outside the visible spectrum, is used to raise the temperature of the nanoshells that have accumulated selectively in the tumor, thereby destroying the cancer cells with heat. In theory, healthy tissue should not be damaged, and antitumor efficacy for the treatment has been demonstrated in *in vivo* models.

The multilayered nanoshells, invented by Naomi Halas (Rice University) in the 1990s, consist of a silica core covered by a thin gold shell, and are approximately 20 times smaller than a red blood cell. The composition, shape, and size of the nanoshells all affect their unique optical characteristics. In particular, the thickness of the gold shell and the size of the core can be varied so that they will respond to a specific wavelength of light when heated. For this anticancer treatment, the nanoshells are "tuned" to raise their temperature in response to near-infrared light.

An important part of this therapy, the accumulation of nanoshells in the tumor, is based on the observation that, when administered systemically, small particles such as the nanoshells tend to accumulate in tumors. This phenomenon, known as the enhanced permeation and retention (EPR) effect, is thought to be caused by the poorly developed blood vessels in tumors that have enhanced vascular permeability. Thus, small particles leak out and accumulate in the interstitial spaces within the tumor. Limited macromolecular recovery via post-capillary venules and poor lymphatic drainage also play a role in the accumulation process (see Chapter 7). After the nanoshells have been allowed to accumulate, near-infrared light is then applied to the skin above the tumor site. This is converted into heat by the nanoshells, which kills neighboring cancer cells while not injuring nearby healthy tissue due to the highly localized effect of the heating.

Therapeutic efficacy has been established in *in vivo* models after systemic administration of nanoshells followed by a 6-hour pause to allow accumulation at the tumor site. A narrow beam of near-infrared light from a laser was then applied to the skin above each tumor, after which time the surface temperature at these sites was observed to increase significantly. No temperature increases were observed in light-treated nontumor locations or in control experiments in which laser treatment alone was used, suggesting that, as anticipated, the nanoshells had accumulated in the tumors. Crucially, all the tumors in the nanoshell, and light-treated models had disappeared within 10 days.

Future clinical trials will demonstrate whether these encouraging preclinical results can be translated into therapeutic gain in humans. Also, any potential long-term toxicities of nanoshells in humans need to be identified and investigated.

9.2.3.2 Radioactive Glass Beads (Microspheres)

A new experimental treatment for inoperable liver cancer patients has been developed by researchers at Northwestern Memorial Hospital (Chicago, IL, U.S.A.) and is based on miniature radioactive glass microspheres (i.e., beads). The microspheres are 15 to 35 μm in diameter and carry high doses of radiation (yttrium-90) to the liver to kill tumor cells. This is a completely novel way to deliver radiation to liver tumors, and can improve quality of life and prolong survival for liver cancer patients.

The procedure involves storage of the microspheres inside a nuclear reactor for several days to generate the radioactive yttrium. Then, using an X-ray-based imaging technique, a radiologist places a small catheter into the hepatic artery (i.e., the artery supplying blood to the liver). Millions of the radioactive microspheres are then injected into the hepatic artery, where they are allowed to irradiate the liver until they lose their radioactivity (approximately 12 days). The encouraging results obtained from this approach are due to the ability of the beads to deliver much larger doses of radiation to the tumor than can be achieved by external radiotherapy techniques. Also, despite the high level of radiation achieved in the liver, healthy tissue is not damaged because the radiation penetrates only 2.5 mm into surrounding tissues.

This therapy is a useful addition to current treatments for liver cancer because, although surgical removal of a liver tumor offers the best intervention, only 15% or less of cancer patients are suitable for surgery, usually because their tumors are too far advanced at the time of diagnosis or because of other medical considerations. Systemic chemotherapy is still the first-line treatment for most liver cancer patients but unfortunately it is not very effective. Therefore, this procedure, which can be carried out in an outpatient setting, offers the possibility of effective treatment without the usual side effects of nausea, weakness, hair loss, and radiation burns. So far, very encouraging results have been obtained from clinical trials with hundreds of liver cancer patients in various countries, including Thailand, Singapore, Hong Kong, New Zealand, Australia, and Canada. Thus, it is likely that this treatment will become more widespread in the future. It is also possible that this approach of delivering radioactive microspheres through a catheter could be applied to other organs, including the kidneys and brain. Future variations on the theme of this technology could utilize glass beads modified to carry and then release chemotherapeutic agents rather than radioactivity. This is an especially exciting possibility because the present approach (using radioactive glass beads) has the disadvantage that the clinic must be located near a suitable nuclear reactor, which is expensive to install and operate and may limit uptake of the procedure in the future.

9.2.4 NUCLEIC ACID TARGETING

Once a gene has been identified as important for survival and growth of a tumor cell, it is theoretically possible to develop agents that can either target the gene itself or the equivalent messenger RNA (mRNA), thus blocking production of the corresponding protein while leaving the transcription of all other genes intact. These strategies are known as *antigene* and *antisense* approaches, respectively, and should

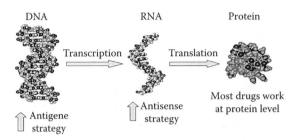

SCHEME 9.3 Diagram illustrating that, whereas most drugs work at the protein level, the antigene and antisense strategies intervene at the DNA and RNA levels, respectively, to down-regulate or block production of protein.

produce agents of low toxicity because only the "faulty" gene in the cancer cells should be affected (Scheme 9.3).

However, for this approach to work, *sequence selectivity* is a crucial issue. It has been calculated that it may be necessary to actively recognize between 15 and 20 base pairs of DNA or RNA in order to selectively target one gene in the entire human genome (approximately 30,000 genes) (Scheme 9.4). Many suitable gene targets have been identified during the past decade, and now await the development of effective gene-targeting technologies.

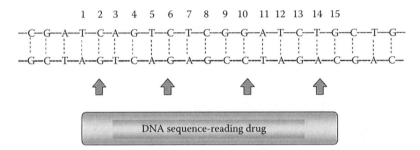

SCHEME 9.4 Diagram illustrating that, based on statistical calculations, a DNA sequence–reading drug is probably required to recognize 15 DNA base pairs or more in order to target a single gene in the entire human genome.

A number of approaches to producing molecules with DNA- or RNA-recognition properties have been proposed and investigated during the past two decades. However, despite intense research activity, few (with the exception of antisense oligonucleotides) have progressed to the clinic and commercialization. *Ribozymes* are a special family of oligonucleotides that target and bind to mRNA but then induce cleavage. Zinc finger proteins (ZFPs) have been developed that can recognize and bind to specific DNA sequences, and GeneICE™ is a technology involving the use of synthetic oligonucleotides to locate a particular sequence of DNA and then attract HDAC enzymes that shut down gene expression. Various decoy technologies have also been developed that attempt to intercept DNA-binding transcription factors using small DNA-like fragments that prevent them from interacting with their intended target. Finally, RNA interference (RNAi) is a relatively new technology that provides effective down-regulation of genes by introducing short, double-stranded RNA fragments (siRNA) into a cell. However, with the exception of an antigene approach using low-molecular-weight ligands, all of these strategies involve macromolecules which, although they may work well *in vitro,* provide significant drug delivery challenges on translation to *in vivo* models or the clinic. The most important gene-targeting strategies are described below.

9.2.4.1 Antigene Strategy

One advantage of the antigene strategy is that, in theory, inhibition of a given gene can prevent the formation of numerous copies of the corresponding RNA transcript and protein. One antigene approach utilizes oligonucleotides that interact with double-stranded DNA to form a so-called "triple helix." The oligonucleotide, which is typically 15 to 20 base pairs in length, aligns in the major groove and is held in place mainly by hydrogen-bonding interactions with the DNA bases. Although much effort has been put into this area, two major problems remain. First, recognition of DNA is presently limited to runs of cytosines or thymines, and this level of selectivity is insufficient to successfully target clinically important gene sequences. Second, in practice, oligonucleotides do not make ideal drugs because they suffer from a number of problems, including instability and poor pharmacokinetic and cellular penetration characteristics. Significant efforts have been made to stabilize these oligonucleotides against chemical or enzymatic degradation through chemical modifications to the backbone phosphate groups, and these efforts have been largely successful. However, attempts to improve the pharmacokinetic and cellular penetration characteristics have been less successful, with most research directed towards incorporation of the oligonucleotides into nanoparticles such as liposomes. It should be noted that strategies to enhance the stability of oligonucleotide-DNA complexes have also been investigated. For example, one approach has involved tethering oligonucleotides to intercalating or alkylating moieties in order to anchor them in the major groove with higher affinity.

(a)   (b)

STRUCTURE 9.17 (a) Structure of an example of a Dervan hairpin polyamide. (b) Diagram showing details of the DNA recognition process with imidazole and pyrrole heterocycles (shown as circles) favoring guanine and cytosine bases, respectively. β-Alanine linkers (shown as diamonds) are placed between blocks of two heterocycles (and on the tail) to maintain synchronization of the heterocycles with DNA bases and to allow the presence of an intervening AT base pair.

In the small-molecule area (e.g., less than 1500 Da molecular weight), agents have now been produced that can recognize more than 10 base pairs of DNA, and this remains an active area of research. For example, the original hairpin polyamides of Dervan and co-workers are polymers of eight-membered heterocycles that fold back on themselves to form hairpin structures that bind in the minor groove of DNA. The intriguing aspect of hairpin polyamides is that the five-membered heterocycles can be chosen to specifically recognize individual DNA base pairs, with pyrroles and imidazoles favoring cytosine and guanine bases, respectively (see Structure 9.17). A β-alanine linker is also included between each set of two heterocycles to ensure that they correctly align with individual DNA bases. However, although it has been possible to design hairpin polyamides with impressive selectivity for certain sequences as demonstrated with naked DNA *in vitro,* their cellular and particularly nuclear membrane penetration characteristics have proved problematic and none, as yet, have been taken into development.

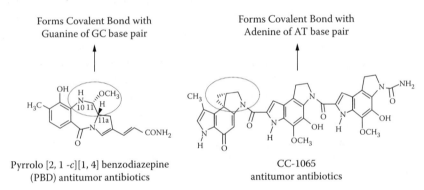

Forms Covalent Bond with
Guanine of GC base pair

Forms Covalent Bond with
Adenine of AT base pair

Pyrrolo [2, 1 -*c*][1, 4] benzodiazepine
(PBD) antitumor antibiotics

CC-1065
antitumor antibiotics

SCHEME 9.5 The PBD and CC-1065 families of antitumor antibiotics bind in the minor groove of DNA and form covalent bonds highly specifically with guanine and adenine DNA bases through their carbinolamine and cyclopropane functionalities, respectively (circled). One approach to gene targeting involves combining the G- and A-reactive components of these molecules in novel gene targeting entities to obtain both GC- and AT-recognition properties.

Other approaches to the design of antigene agents that interact in the minor groove of DNA are currently underway based on natural products that are known to possess DNA sequence-recognition properties. For example, the pyrrolobenzodiazepine (PBD) antitumor antibiotics produced by *Streptomyces* species bind specifically and covalently to the guanine base of purine-guanine-purine (preferably AGA) sequences, and thus can be used as components of longer molecules that target wider DNA motifs (Scheme 9.5). The other significant advantage of using natural products such as PBDs is that they readily penetrate both the cellular and nuclear membranes of tumor cells, presumably because they have evolved as chemical weapons and so have been naturally selected for their ability to reach the genome. Experimental gene-targeting agents based on the PBDs have now been produced that reach up to 20 base pair spans while retaining their ability to penetrate cellular and nuclear membranes with great ease. Furthermore, extended PBD molecules of this type have been shown to inhibit transcription *in vitro* in a sequence-selective manner and are now at the stage of translation into *in vivo* models. Similarly, extensive work has been carried out on the CC1065 family of antitumor antibiotics, which also bind in the minor groove of DNA but, in contrast to the PBDs, target adenine bases (see Scheme 9.5). Research is now underway to combine components of both PBD and CC1065 molecules in order to produce gene-targeting agents capable of recognizing both GC and AT base pairs in stretches of DNA through covalent bond formation.

9.2.4.2 Antisense Strategy

An antisense oligonucleotide is a relatively short length of chemically modified single-stranded nucleic acid (e.g., 10 to 20 base pairs) with a sequence complementary to a region of target mRNA. Hybridization with the RNA then interferes with the process of translation for their particular protein. However, a similar set of problems to those described for antigene oligonucleotides exists, such as poor stability, pharmacokinetic, and cellular penetration characteristics. This is highlighted by the fact that the first antisense product to be given FDA approval for marketing in the late 1990s (i.e., Fomiversen [ISIS-2922] to treat cytomegalovirus-induced retinitis) has to be administered locally by injection directly into the eye, thus avoiding the problems associated with systemic delivery. A further consideration is that the antisense strategy may not prove as efficient as the antigene approach, in which inhibition of just one copy of the gene should prevent numerous copies of mRNA from being produced.

Despite disappointments with the clinical evaluation of antisense technology in the mid-1990s when unexpected toxicities and a general lack of efficacy were observed, there is currently a resurgence of interest and a number of agents are presently in clinical trials. For example, oblimersan sodium (Genasense™ [Genta Inc.]), an antisense agent targeted to the Bcl-2 oncogene, is being extensively evaluated in clinical trials. In early 2004, the FDA approved the new drug application (NDA) for Genasense™, allowing its evaluation in combination with dacarbazine in chemotherapy-naïve patients with advanced melanoma. This was a significant event, as it was the first NDA approved for systemic antisense therapy and was also the first for an agent that promotes chemotherapy-induced apoptosis. More importantly, it was the first new agent to be developed for advanced melanoma in 30 years and, for this reason, was given "priority review" status by the FDA.

The mechanism of action of Genasense™ involves the inhibition of expression of the protein Bcl-2, which is often overexpressed in tumor cells and is responsible for blocking chemotherapy-induced apoptosis. Therefore, Genasense™ was designed to improve the activity of currently used cytotoxic agents in cancers such as melanoma, in which Bcl-2 is overexpressed. The clinical trials, which have focused on pretreatment with Genasense™ before administering a standard chemotherapeutic agent, have either been completed or are ongoing in cancer types, including multiple myeloma, CLL, lung cancer, prostate cancer, melanoma, and acute myeloid leukemia in combination with chemotherapeutic agents such as docetaxel, paclitaxel, irinotecan, imatinib, rituximab, fludarabine, cyclophosphamide, gemtuzumab ozogamicin, cytosine arabinoside, and dexamethasone, and also radiation, Mabs, and immunotherapy.

Similarly, GEM-231 (Hybridon Inc.), an antisense agent targeted to the R1 subunit of protein kinase A is being evaluated against a number of solid tumors in combination with docetaxel, paclitaxel, and irinotecan. A further example is AVI-4126 (AVI Biopharma Inc.), an antisense agent targeted to the *c-myc* oncogene that completed successful Phase I studies in 2005 and is presently being evaluated in combination with a number of chemotherapeutic agents for the treatment of solid tumors, including those of the prostate.

With regard to adverse events, a number of reactions to the intravenous injection of backbone-modified antisense oligonucleotides have been observed. The most serious adverse reactions can arise from so-called "off target" events that occur when an antisense oligonucleotide binds to other RNAs of similar sequence, thus leading to unpredictable side effects. However, side effects such as complement activation and prolongation of blood coagulation have also been observed.

Finally, research is underway to discover low-molecular-weight ligands (e.g., molecular weight less than 1000 Da) that might interact with mRNA in a sequence-selective manner, thus avoiding the pharmaceutical problems (e.g., stability and delivery) apparent with antisense oligonucleotides.

9.2.4.3 Ribozymes

Discovered in the early 1980s by Thomas Cech and Sidney Altman, who shared the 1989 Nobel Prize in Chemistry, ribozymes are similar to antisense oligonucleotides in that they target and bind to complementary sequences of mRNA. However, in contrast to antisense oligonucleotides, once bound to their target RNA, ribozymes can elicit cleavage of the bound mRNA fragment by invoking an enzyme known as RNase H (Scheme 9.6). Also, unlike antisense oligonucleotides, the unique base-pair sequence of a ribozyme allows it to adopt the shape of either a *hairpin* or a *hammerhead,* which gives rise to the main way of categorizing ribozymes.

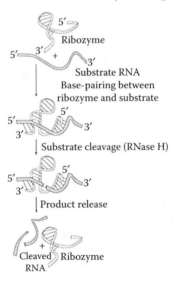

SCHEME 9.6 Mechanism of RNA cleavage by a hairpin ribozyme.

Ribozyme agents have been in development since the mid-1990s but, so far, no clinical candidates have emerged. In general, ribozymes suffer from the same stability and delivery problems as antigene and antisense oligonucleotides.

9.2.4.4 Zinc Finger Proteins (ZFPs)

Zinc Finger Proteins (ZFPs) are novel, two-component, semisynthetic proteins produced by Sangamo Inc. that are based on the structure of transcription factors. They are designed to regulate gene expression by binding sequence-selectively in the minor groove of DNA. ZFPs are comprised of a *recognition* domain (consisting of multiple zinc fingers) that identifies a specific DNA sequence in the minor groove and then binds to it. A second *functional domain* adds a specific biochemical activity to the protein, such as cleavage, activation, or repression (Scheme 9.7). Each zinc finger is produced by phage display technology and is selected to bind specifically to three consecutive base pairs of DNA (i.e., a base pair triplet). Two or more individual zinc fingers can then be joined to make up the recognition domain, which will bind to six or more base pairs (in multiples of three). Through careful choice of the zinc fingers used and careful chemical modification of the critical amino acid contacts of the ZFPs with functional groups in the DNA minor groove, novel proteins with the ability to recognize distinct base pair sequences in any given gene can be produced.

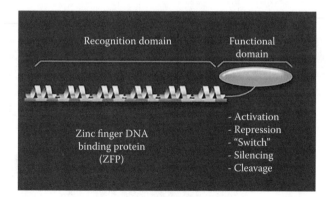

SCHEME 9.7 ZFPs are comprised of a *recognition domain* that identifies and then binds to a defined sequence of DNA and a *functional domain* that adds a specific biochemical activity to the protein such as cleavage, activation, or repression.

A range of different functional domains can be used to engineer ZFPs so that they can modulate gene expression by acting as an activator or a repressor. Furthermore, a more advanced ZFP has been produced that possesses a functional domain with a "switch" component that enables a small druglike molecule to regulate the activity of the ZFP and thus the expression of the target gene. It has been proposed that these types of ZFPs could be used in *regulatable gene therapy*, where they might potentially allow more precise dosing and control over the duration of exposure to the therapy. A further application is to add the functional domain of a DNA-cleaving enzyme (i.e., a restriction endonuclease) to the DNA recognition element of a ZFP. Such proteins could be used as tools to help replace "faulty" DNA sequences in mutated genes with corrected sequences, thus restoring gene function.

However, although ZFPs have proved useful in experiments using naked DNA, or in *in vitro* experiments for purposes such as target validation where the ZFPs can be microinjected or electrophoresed into cells, the main problems preventing their development as therapeutic agents are the usual ones associated with proteins including poor *in vivo* stability and pharmacokinetics, and a lack of ability to penetrate cellular and nuclear membranes. As with antigene and antisense oligonucleotides, there may be ways to overcome these problems in the future by, for example, incorporating ZFPs into nanoparticles that can accumulate in tumors and can deliver the ZFPs into cells by processes such as endocytosis. However, even though this approach could overcome the barrier of the cellular membrane, it is not clear how ZFPs would then cross the nuclear membrane to reach the nucleus.

9.2.4.5 GeneICE™ Technology

GeneICE™, a technology introduced by Cronos Therapeutics Ltd., makes use of a normal biological process (i.e., chromatin condensation) to block transcription of specific genes. The GeneICE™ agent is designed to interact with the target gene at a specific sequence and causes down-regulation by modifying the chromatin structure that governs gene expression.

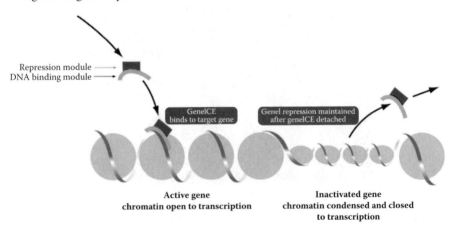

SCHEME 9.8 Mechanism of gene silencing via the GeneICE™ technology.

A GeneICE™ agent consist of two components, the first being a specific *DNA-binding module* based on an oligonucleotide that binds selectively to the gene to be silenced. This is coupled to a *repressor module,* which is a peptide fragment with the ability to recruit HDAC complexes to the gene. Deacetylation of the histones associated with the gene then leads to localized chromatin condensation that "closes" the DNA-histone complex and blocks transcription (Scheme 9.8).

Unfortunately, because this technology is based on oligonucleotides, it suffers from the same problems as the oligonucleotide-based antigene, antisense, and ribozyme approaches previously described with regard to stability, pharmacokinetics, and cellular and nuclear penetration. In addition, it is also unclear whether the localized chromatin condensation process is specific enough for the transcription of neighboring genes to be unaffected.

9.2.4.6 Decoy Strategies

Decoy strategies involve flooding a cell with short, oligonucleotide-type constructs (i.e., decoys) that resemble the DNA-binding site of a particular transcription factor protein. A decoy molecule then binds selectively to the transcription factor and prevents it from binding to its usual DNA sequence, thus inactivating the factor and switching the gene off or on, depending on whether the factor normally inhibits or promotes transcription. One advantage of this strategy is that other transcription factors that recognize different DNA sequences should be unaffected. However, a disadvantage is that any given transcription factor will usually control more than one gene, thus potentially reducing the selectivity of therapies based on this approach.

Decoys are normally made from short stretches of RNA or DNA bases joined together and chemically customized to have a longer half-life in plasma. One example of this approach, recently pioneered by Genta Inc., involves use of a decoy to target the cyclic adenosine monophosphate response element–binding protein (a transcription factor). In preclinical studies, inactivation of this protein has proved to be selectively toxic, with a higher potency towards cancer cells than normal cells. As a result, lead compounds are currently being optimized for clinical evaluation. However, it should be noted that, because decoys are usually based on strands of chemically-modified nucleic acids, they suffer from the same problems as the previously discussed oligonucleotide-based approaches in terms of poor stability, pharmacokinetics, and cellular and nuclear penetration characteristics.

9.2.4.7 RNA Interference (RNAi)

RNA interference (RNAi) is a recently identified phenomenon in which the introduction of short, dsRNA fragments into a cell can down-regulate or ablate production of the protein corresponding to the gene associated with the sequence of the dsRNA introduced. In fact, this is a natural process that occurs in all cells and is thought to act as protection from RNA viruses. Essentially, the cell responds to the introduction of extraneous dsRNA by destroying all internal mRNA of the same sequence. This phenomenon was first observed in the *Caenorhabditis elegans* worm and later in drosophila, trypanosomes, and planaria. The post-transcriptional gene silencing observed in plants is also thought to be related to an RNAi mechanism.

In the natural pathway, dsRNA present in a cell after infection is recognized by the cell as foreign. First, it is bound by the enzyme *Dicer*, which cuts it into nucleotide duplexes, each containing 21 base pairs and known as siRNAs. These, in turn, become bound by a composite of proteins known as the RNA-induced silencing (RISC) complex. One strand of the siRNA duplex is then stripped off, leaving the remaining "complementary" strand associated with the protein complex. The RISC complex, using adenosine triphosphate (ATP) as an energy source, then locates and

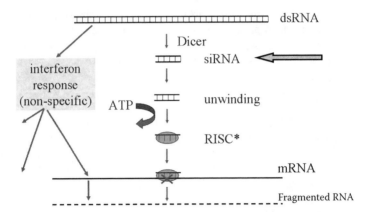

SCHEME 9.9 Mechanism of gene silencing via RNAi. dsRNA is first cut into shorter fragments of 21 bases (i.e., to siRNA) by the enzyme *Dicer*, and these fragments attach to a RISC protein complex, after which one strand is stripped off to leave the complementary strand. The RISC complex then uses this single-stranded RNA as a template to locate and bind mRNA containing the matching sequence. Once a match has occurred, the mRNA is then enzymatically degraded.

binds the target RNA using its complementary siRNA strand as a template. Only those RNAs that can form a complete set of Watson-Crick base-pairs with the siRNA complementary strand will bind, thus ensuring that only the specific target RNA is bound and destroyed. Once the RISC complex correctly matches up its complementary siRNA strand to its target RNA, the latter is cleaved and destroyed enzymatically (Scheme 9.9).

Potentially, this natural phenomenon can be utilized to down-regulate viral RNA, normal cellular RNA, or a disease-related RNA. The process can be initiated by introducing into the cell a 21 base-pair dsRNA (i.e., a siRNA) with a sequence matching the mRNA of the gene intended for down-regulation or ablation. siRNAs are remarkably potent and only a few dsRNA molecules per cell are required to produce effective interference. The reason for this is that a single dsRNA molecule can "mark" hundreds of target mRNAs for destruction before it is spent. In addition, the RISC complexes work catalytically with many mRNA molecules being destroyed by one complex. If the target RNA is a viral genome, then viral replication will be blocked. Similarly, if the target RNA is disease-related, then protein synthesis by that gene will be prevented and the disease may be ameliorated.

However, as with the antigene and antisense-type approaches that utilize oligonucleotides, the major problem with developing a therapy based on this technology relates to the difficulties associated with delivering dsRNA constructs to cells. In fact, the problem is even more severe than for delivering antigene or antisense oligonucleotides, because siRNAs are double stranded and hence more fragile than single-stranded oligonucleotides. Therefore, although the RNAi phenomenon has great promise for the *in vivo* down-regulation or ablation of genes critical for the survival of cancer cells, most current therapeutic strategies are based on gene therapy approaches in which a viral construct is used to deliver a fragment of DNA that can be inserted into the host genome. This insertion is then transcribed to a length of single-stranded RNA that is designed to fold back on itself to produce a double-stranded structure containing the critical targeting sequence. However, as with all gene therapy approaches, the problem is that a viral vector typically delivers the nucleic acid fragment to all cells of the body, with the consequent risk of collateral genomic damage. Therefore, as with antigene and antisense approaches, the first therapeutic applications of this technology are likely to be in cancers in which the siRNA can be delivered topically so that systemic drug delivery problems can be avoided. For example, at least one clinical trial using a topical method to deliver an anti-HPV siRNA to the cervix to treat cervical cancer is currently underway being discussed.

9.2.5 GENE THERAPY

The treatment of cancer by gene therapy has been discussed for many decades, but no successful methods have yet emerged despite intense research activity in this area. By far, the most significant problem to overcome is delivering a new or replacement gene specifically to tumor cells. It is feared, with good reason, that should the new genes be inadvertently delivered to healthy cells, then these may themselves transform into cancer cells at a later time due to their modified genome.

The most common form of gene therapy discussed and researched involves replacement of a mutated gene in a tumor cell, such as a mutated tumor-suppressor p53 gene, with a working copy. Usually, this is attempted using an organ-specific viral vector. For example, a genetically engineered hepatovirus can be used to deliver a new gene to cells in the liver. Adenoviruses have also been extensively investigated for this purpose. However, this approach has two main problems. First, in terms of efficacy, experiments in both *in vivo* models and humans have shown that viral vectors rarely infect all cells in a tumor mass, and so the treatment is only partially effective. Second, in terms of safety, organ-specific viral vectors are rarely completely specific for the target organ, and collateral infection of healthy cells can occur. This is regarded as a serious problem because previous clinical trials in children aimed at replacing a faulty metabolism-linked gene showed that some developed leukemia after the gene therapy, which was thought to be caused by collateral genomic damage of healthy cells. Despite these problems, much research is still underway to develop highly specific vectors. Other types of delivery systems, some involving nanoparticles, are also being investigated.

Other more sophisticated approaches, such as gene-directed enzyme prodrug therapy, are also under development. In this therapy, a viral vector is used to deliver a gene to cancer cells that codes for an enzyme not usually expressed in human cells, such as bacterial nitroreductase or carboxypeptidase (see Chapter 7). Once the tumor cells express the enzyme, an enzyme-sensitive prodrug is administered systemically and is converted to an active anticancer drug only at the tumor site. However, the success of this approach clearly depends on the specificity of the viral vector for tumor cells, and collateral infection of healthy cells could lead to adverse toxicities.

Despite these problems, gene therapy remains an attractive approach to cancer treatment and is still a highly active area of research. Therefore, it is likely that advances in tumor-specific gene delivery systems will emerge in the future.

9.2.6 Cancer Stem Cells

There are two competing theories of tumor development and metastases. According to one theory, all cells making up a tumor are basically identical and equally likely to divide to maintain tumor growth or to metastasize to form new tumors. The other theory maintains that only a small number of select cells from each tumor, known as *cancer stem cells,* have the ability to start new tumors. It is argued that these stem cells are highly robust and resilient to drugs and radiation. Therefore, the second theory may explain why a tumor returns after the original (i.e., the primary tumor) has been carefully removed by surgery (i.e., if one or more stem cells are left behind) and also why tumors are often so resistant to chemotherapy and radiation (i.e., it is the stem cells that survive and then repopulate the tumor mass).

The concept of cancer stem cells has been around since the 1950s. Only recently has evidence started to accumulate that supports the hypothesis regarding cancer stem cells. For example, a decade ago, it was shown that when cancer cells were harvested from patients with leukemia and implanted into rodents, only a select number of blood cells from the harvested population could successfully grow. Similarly, for solid tumors, it was more recently demonstrated that when tumor cells from human breast cancers are implanted into mice, approximately only 1 in 100 successfully form tumors. Therefore, it is now thought that cancer stem cells may play a role in most (or perhaps all) types of cancers (both solid and blood-related), and a significant research effort is now underway to identify tumor stem cells in all the various tumor types, including skin, lung, pancreatic, ovarian, and prostate cancer. A great deal of research into the origin of cancer stem cells is also ongoing. One possibility under investigation is that cancer stem cells may derive from mutated versions of the normal stem cells that sustain all of the various tissues of the body. From a clinical perspective, current thinking on how tumors spread, and the best way to treat different tumor types, could be changed by acceptance of the concept of cancer stem cells. For example, it may become evident that many currently used chemotherapy regimens kill most cells constituting a tumor, although the stem cells remain intact, thus allowing a relapse to occur after a period of time. This might not only ultimately change the way chemotherapy and radiotherapy are used, but could also bring about changes in the drug discovery process, with researchers designing or searching for novel agents that specifically target cancer stem cells rather than the main population of tumor cells for any given cancer type.

Normal stem cells in healthy organs and tissues have two important characteristics. The first is that they are *immortal,* meaning they can keep dividing indefinitely by a process known as "self-renewal." The second is that they are unspecialized, but their descendants can mature into all of the different cell types that form the various organs and tissues that make up the entire organism. Interestingly, once the maturation process has occurred, the stem cell descendants can rapidly divide but for only a limited number of cycles. To date, the best-defined normal stem cells originate in the blood system. This is based on work from the 1980s, when researchers identified and characterized mouse stem cells in murine bone marrow. It was established that, remarkably, just one type of hematopoietic stem cell produces both the various white blood cells of the immune system and the red blood cells. Thus, theoretically, the entire blood system of the animal can be reconstituted from a single hematopoietic stem cell. This finding led to a hunt for human leukemic stem cells which, when transplanted into immune-deficient mice, might produce human leukemic cells. Through these studies, it was discovered that, when leukemic blood cells from cancer patients were implanted into mice, approximately 1 in 1,000,000 cells had the ability to regenerate leukemia cells, and so these were considered to be cancer stem cells. When characterized, these cells were found to carry distinguishing CD34 proteins on their surfaces, thus indicating their similarity to normal hematopoietic stem cells (which also display CD34) but dissimilarity to other blood cells (which do not display CD34). Furthermore, it was discovered that the human leukemic stem cells lacked the protein known as CD38, which is usually displayed on the surfaces of most other leukemia cells.

Since these earlier experiments, stem cells have been discovered in all types of blood cancers, most recently in multiple myeloma. In this disease, plasma cells (that make antibodies) build up in the bone marrow, eventually causing its destruction. Researchers had suspected for some time that specialized cells (e.g., stem cells) might give rise to the numerous cancerous plasma cells formed because normal plasma cells divide very infrequently. It is now known that a subset of immune cells known as *B cells* are the stem cells in multiple myeloma. B cells have the ability to self-renew and can also develop into the mature plasma cells characteristic of this disease. This observation also supported earlier studies that proved that, during normal blood development, B cells produce plasma cells.

Further work in the 1990s demonstrated the presence of stem cells in many types of solid tumors. For example, it was discovered that a small number of cells in a population of testicular cancer cells had stem cell characteristics because they carried a protein on their surfaces also known to be carried by immature fetal cells. This discovery prompted researchers to separate populations of human breast cancer cells based on the proteins displayed on their surfaces. These different populations were then implanted into the mammary tissue of immune-deficient mice to establish whether growth would occur. One population of cells displaying the surface protein CD44 but lacking other proteins was found to have stem cell properties, and implanting as few as 100 cells reliably generated a tumor. In comparison, implantation of many thousands of unsorted breast cancer cells was required to form a tumor. Most importantly, it proved possible to isolate the same CD44-carrying stem cells from new tumors and reimplant them into other mice to reliably produce further tumors.

Researchers have also identified cells with stem cell–like properties in children's brain tumors. Initially, it was discovered that the protein CD133 is displayed on the surface of stem cells in healthy human brain tissue. This allowed separation of brain tumor samples into populations based on their surface proteins, and led to the identification of a small population (found in a variety of brain tumors) expressing CD133. These cells were grown *in vitro* and had the remarkable ability to differentiate into the same types of brain cells as found in the original tumor. Other tumor cells (i.e., non-stem cells) from the same tumor eventually stopped growing. Interestingly, CD133-expressing cells isolated from pediatric brain tumors have been transplanted into the brains of newborn rats, where they proliferate for up to a month, migrate throughout the brain, and produce nerve and other types of brain cells

However, it is still not entirely clear whether stem cells are differentiated cells that have acquired stem cell characteristics or whether they are normal stem cells that are behaving inappropriately. The latter is an appealing hypothesis because it has been thought for some time that normal stem cells are on the brink of transformation into other types of cells, including tumor cells. Another important consideration is that the immortal nature of stem cells means that the endless reproduction cycle allows unlimited time for tumor-promoting genetic mutations to build up. The fact that normal stem cells are highly robust and usually contain more efficient efflux pumps compared to the ones that some cancer cells use to protect them from chemotherapeutic agents may explain why many cancers are resistant to radiation and drugs.

The discovery of cancer stem cells has also influenced opinion on how tumors metastasize to other parts of the body. Metastasis used to be likened to an evolutionary process in which a small population of cells within the primary tumor would progressively accumulate genetic changes that would eventually empower them to spread to other tissues. However, based on current knowledge of stem cells, an alternative hypothesis idea is that cells from a primary tumor routinely spread throughout the body, but a metastatic tumor forms only when a relatively rare stem cell reaches a distant location.

Therefore, in summary, it is likely that the rapidly growing knowledge-base surrounding cancer stem cells will influence the way researchers search for new drugs in the future. For example, so far, the emphasis has been mainly on proliferation, and researchers have traditionally screened for novel agents that inhibit the growth of dividing tumor cells growing in vitro. However, it is now known that stem cells do not usually divide rapidly (i.e., they are quiescent), and only occasionally produce descendants that then rapidly reproduce. Therefore, it may prove necessary to devise new screening methods capable of identifying agents selectively toxic toward cancer stem cells. At least one pharmaceutical company (OncoMed Inc.) is exploiting these new observations and ideas in an attempt to discover novel drugs with selective activity against cancer stems cells, and this area of research is likely to grow in the future.

9.3   NEW RESEARCH TOOLS AND METHODOLOGIES

Since the dawn of chemotherapy in the mid-1940s, many incremental advances in biological and chemical technologies have been made and then applied to elucidating the molecular basis of cancer and to the discovery and development of new drugs and therapeutic strategies to prevent or treat it. In the early 2000s, these advances led to the discovery and subsequent commercialization of imatinib (Gleevec™), which is widely heralded as the first anticancer drug to be rationally designed based on a detailed knowledge of the molecular pharmacology of the cancer cell. Arguably, this represented a landmark advance in the design and discovery of new anticancer agents, contrasting with the relatively empirical approaches that had characterized the previous decades. Today's cancer researchers now have a number of powerful scientific tools and methodologies at their disposal, and some examples of the more important ones are outlined in the following subsections.

9.3.1 Gene Hunting and DNA Sequencing

Advances in molecular biology allow new genes to be identified and then sequenced with great rapidity. Modern methods of DNA sequencing are based on the technology and sequencing machines developed for the Human Genome Project during the 1990s. A remarkable number of cancer-related genes, such as the breast-cancer related BRCA1 and BRCA2, have been discovered and identified during the last decade. Hunting for new cancer-related genes (oncogenes) is now a major activity by both commercial and academic research groups throughout the world, and the list of known oncogenes is rapidly expanding.

The genes are normally discovered by applying modern molecular biology techniques to the study of cancer-related biochemical pathways in cancer cells growing *in vitro*. Once a new cancer-related gene has been identified, then its importance to the survival of the cell, and hence its value as a potential therapeutic target, can be validated using techniques such as RNAi to knock out the gene followed by observation of the effect. This may be followed-up with studies in *in vivo* models (e.g., knock-out mice), in which the effect of down-regulating or ablating the gene on tumor growth can be investigated. Once it has been established that the new oncogene is a viable therapeutic target, it can be sequenced and the corresponding protein produced in pure form for further studies through standard genetic engineering methodologies (e.g., transfection of *Escherichia coli*). It may then be possible to establish the structure of the protein by techniques such as X-ray crystallography and high-field nuclear magnetic resonance (NMR). After this, novel therapeutic agents can be designed to interact with the protein to block its function or perhaps intervene at the RNA (i.e., antisense) or gene (i.e., antigene) level.

9.3.2 Genomics and Proteomics

In their broadest sense, the terms *genomics* and *proteomics* refer to the recently developed technologies that allow gene expression and protein production, respectively, to be studied in living cells and tissues, including tumors. Genomics studies use DNA arrays (or "DNA microchips"), which have large numbers of cDNA fragments on their surfaces with sequences corresponding to different genes in the human genome. This technology was pioneered by Affymetrix Inc., whose latest generations of chips contain cDNAs corresponding to more than 50,000 genes (e.g., the Affymetrix Human U133+2.0 array contains cDNAs for 54,000 genes). Gene expression is studied by extracting the total population of mRNAs from cells and using this to produce fluorescently labeled cDNA probes using a reverse transcriptase enzyme. These probes are passed over the DNA array so that those corresponding to genes being expressed hybridize with the cDNAs on the chip and thus produce a signal at locations where a match occurs. Because the concentration of every cDNA reflects the amount of each corresponding mRNA species produced by the cell, the degree of fluorescence at each spot represents the expression level of that particular gene. Using suitable controls, suitable software will then be used to convert the

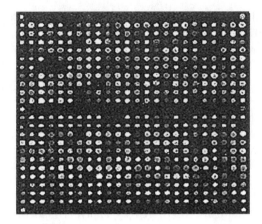

FIGURE 9.1 Software-generated image of a DNA array experiment showing the fluorescently labeled cDNA probes (generated from the cellular mRNA) bound to the cDNAs attached to the chip. Each spot represents a single gene, and the software color-codes the degree of fluorescence with green, yellow, and red to represent up-regulation, normal regulation, and down-regulation, respectively (compared to controls).

degree of fluorescence at each spot into a color code such as green, yellow, or red to represent overexpression, regular expression, or underexpression, respectively (Figure 9.1). However, for more comprehensive analyses, most software packages also produce accurate numerical values for the fluorescence intensities at each spot.

Although a useful and powerful technique, DNA array technology has two major drawbacks. First, the fact that a cell expresses a particular gene and generates the corresponding mRNA does not necessarily mean that it will go on to make the corresponding protein. For example, due to RNA splicing, one gene can lead to more than one type of protein, meaning that DNA array studies may give a false impression of which proteins are being expressed. The second major problem is "information overload." Quite often, a very large number of genes will be up- or down-regulated in a single experiment, and it may be difficult to ascertain which changes are relevant to the study in hand. This has led to the term *gene networks,* a phrase used to describe the concept that single genes or even single gene families are rarely up- or down-regulated in isolation, but instead usually affect a vast number of other genes and pathways. This makes the results of DNA array experiments difficult to interpret and has led to the relatively new area of research known as *systems biology,* which is the study of how one biochemical pathway in a cell affects other pathways.

Proteomics studies complement genomics in that they are designed to directly measure protein production rather than mRNA production, thus countering some of the known criticisms with DNA array technologies (e.g., RNA splicing). Proteomics studies normally involve extraction of the entire population of proteins from a cell, followed by their separation using techniques such as two-dimensional electrophoresis. Each protein is then identified using mass spectrometry.

An important point to note is that these two technologies (i.e., genomics and proteomics) are complementary and, as research tools, can be used experimentally to identify genes that are underexpressed or overexpressed in cancer cells compared to normal cells. Furthermore, they can be used to study tumor biopsy material taken from cancer patients, although this can be problematic because tumors are usually heterogeneous, which can lead to concerns that more than one cell type may be contributing to a result. More importantly, these techniques can be used to observe the effects of experimental anticancer agents on gene expression both in tumor cells growing *in vitro* and in biopsy material from clinical trials.

However, both genomic and proteomic techniques are expensive, particularly because experiments usually have to be carried out in triplicate to achieve statistical significance. Another problem is that both mRNA and protein expression can be highly dependent on factors such as cell cycle and the age of a cell. Furthermore, if studying the effect of a novel anticancer agent on a particular cell type, mRNA and protein expression can be highly dependent on the dose of the agent used and the time course of exposure. For example, for a cytotoxic agent, it may be possible to observe up- or down-regulation of specific biochemical pathways at lower concentrations or exposure times, whereas at higher concentrations or exposure times, only pathways associated with apoptosis may predominate. All of these factors combine to make DNA array and proteomic studies difficult to plan and expensive to carry out, and the results often difficult to interpret.

9.3.3 PROTEIN STRUCTURAL STUDIES

Thanks to advances in NMR and X-ray crystallography instrumentation and techniques, it is now possible to elucidate the three-dimensional structure of proteins to a high resolution (i.e., less than 1.5 Å), provided they can be isolated from cells, purified (for NMR and X-ray), and crystallized (for X-ray). Even if a protein cannot be crystallized, techniques have now been developed that will allow lower resolution structural information to be obtained (e.g., 5 to 10 Å, although sometimes better than 5 Å is possible if the sample is studied in two dimensions only). For example, membrane proteins are notoriously difficult to crystallize because they are often partly buried in the cellular membrane. However, cryoelectron microscopy methods have been developed that can provide useful two-dimensional maps of many membrane proteins.

Once the structure of a cancer-related protein has been established, its coordinates can be used to generate a three-dimensional model in one of the many molecular modeling software packages available, which can then be used for drug design purposes (e.g., SYBYL™ from Tripos Inc.). For example, such a model can be used for *in silico* screening whereby large virtual libraries of molecules (including chemical company catalogs) can be searched in order to find examples of molecules that fit various pockets and clefts of the protein. These molecules can then be obtained commercially or synthesized and used as leads for the development of novel inhibitors of function of the protein.

9.3.4 CHEMICAL TECHNOLOGIES

During the last two decades, various new chemical methodologies have been developed that allow large compound collections (i.e., libraries) to be produced for use in high-throughput screening operations. Compound collections are commercially available from companies specializing in this area and are usually provided in a multiwell plate format, with an accurately known amount of each compound per well. However, pharmaceutical companies and other research organizations may have their own compound collections synthesized in-house and built up over many years.

In general terms, libraries are either *diverse* or *focused*. Diverse libraries are collections of molecules of widely different structures, which can be useful for *random screening* against a target to find initial hits. A more focused library can then be based upon the particular structure(s) identified from the diverse library screen, which is essentially a means of early lead optimization. Within these two main categories, the libraries may be further classified as *parallel* or *combinatorial*. Parallel libraries tend to explore systematic changes of functional or chemical groups at one or more positions on a fixed chemical skeleton, and so can be made in few synthetic steps and are often used as focused libraries. The synthesis of parallel libraries has been made significantly more efficient through the development of new technologies and instruments, such as the microwave reactor and chemical work-station, which allow multiple compounds to be made very rapidly, and automatically and in high yield. Because these libraries are typically relatively small (approximately 20 to 100 compounds), individual members are usually purified to greater than 95% (and preferably greater than 99%) by techniques including preparative high-pressure liquid chromatography (HPLC).

Combinatorial libraries became popular with the pharmaceutical industry in the 1990s, typically comprising larger numbers of molecules, or mixtures of molecules, often of diverse structure, although the technique can also be used to make more focused libraries depending on the types of building blocks used. With combinatorial techniques, molecular fragments are joined together in a sequentially random fashion to provide large numbers of molecules (sometimes more than 1 million) within a short time. The use of a large number of chemical fragments of various structures connected in different ways leads to libraries of molecules of widely different three-dimensional shapes (i.e., of high molecular diversity). The key to success with combinatorial technologies is some form of tracking or tagging system that is used to ensure either that each new compound has a unique physical position within a two-dimensional array (or multiwell plate) or contains a traceable "label," which may take a variety of forms ranging from chemical to radiofrequency tags. These libraries can then be passed through high-throughput screens and the structure of any one compound (or compound in a mixture) providing a "hit" can be traced through the tag or by the positional history of the synthesis. The process of working out the structure of a hit in a combinatorial library is referred to as *deconvolution*. After identification, the active molecule can then be resynthesized on a larger scale for further evaluation. The main problem with combinatorial libraries is that, because of the nature of the technology, compounds often have relatively low purity (e.g., less than 90%), which can lead to false-positives and false-negatives. Therefore, there is a trade-off between the size and diversity of the library versus the lower purity of individual compounds or mixtures of compounds. For this reason, the pharmaceutical industry has moved away from combinatorial libraries, preferring to use externally purchased or in-house compound collections of known purity.

It should be noted that libraries of plant, bacterial, and fungal extracts are also commercially available and can be useful to obtain initial hits in screens. One disadvantage is that the extracts often consist of complex mixtures of ingredients, and a significant effort may then be required to identify the individual molecule or molecules responsible for the hit. However, one over-riding advantage is that natural product extracts typically contain compounds rich in structured diversity (and particularly stereochemistry) that have been produced through millions of years of evolution (see Section 9.3.7).

9.3.5 SCREENING METHODOLOGIES

During the last decade significant advances have been made in screening technologies, mainly based on the robotic handling of multiwell plates. Provided a robust biochemical assay can be developed for the interaction of an inhibitor or agonist with an enzyme, receptor or other protein, then large libraries of compounds can be rapidly and automatically screened with minimum human intervention. Assays of this type are usually light-based with a "hit" leading to the emission of light of a certain wavelength, which can be easily monitored and quantitated by automated detection systems. Bioinformatics systems are required to record, manipulate, and report the vast amount of data produced by these screens. As discussed above, the compound libraries used for screening are usually provided on multiwell plates either commercially or from in-house collections. These screening technologies have also been applied to searches for molecules that interact with DNA (i.e., antigene) and RNA (i.e., antisense).

9.3.6 *IN VIVO* MODELS

Dramatic developments in *in vivo* models have occurred during the last two decades. For example, a mouse strain known as Oncomouse™ is commercially available that has various (according to the model) oncogenes incorporated into its genome. These mice typically succumb to the oncogene-related tumor within a defined and reproducible period of time, and so the model can be used to study the relationship of the particular gene with the course of the disease, and also the effect of novel therapeutic agents on the time-to-disease, lifespan, and other factors. Similarly, *knock-out mice,* in which key genes have been deleted from their genomes, have been introduced. These models can be used to study, for example, the effect of deleting tumor suppressor genes.

A more recently introduced *in vivo* model involves an advance on the traditional human tumor xenograft, in that the transplanted tumor is transfected to express green fluorescent protein (GFP), a protein from the jellyfish *Aequorea victoria* that fluoresces green when exposed to blue light. Highly sensitive scanners have been developed that allow the entire mouse to be imaged, and thus tumor size and position can be measured based on the distribution and intensity of the fluorescence visualized through the skin and tissues of the mouse (even through the skull). This provides extremely precise measurements of tumor growth or shrinkage, thus allowing the effect of novel agents and therapies to be evaluated with great precision. However, the main advantage is that the scanning operation can be carried out on living mice without causing undue stress. Thus, unlike traditional human tumor xenograft experiments, the same animal can be subjected to multiple measurements, allowing tumors to be accurately tracked in one individual. Most importantly, the technique can additionally be used to observe the formation and growth of metastases in any part of the mouse, as these also express GFP.

9.3.7 SOURCES OF NEW LEAD MOLECULES

As previously described, modern methods of anticancer drug discovery involve identification and purification of a novel biological target (i.e., an enzyme or receptor), followed by its employment in a high-throughput screen against libraries of diverse compounds from either commercial or in-house sources. However, one of the main drawbacks of libraries of synthetic compounds is that even molecules from so-called diverse libraries occupy only a relatively small volume of possible three-dimensional space. Conversely, plants, bacteria, and fungi are rich sources of organic molecules, which generally occupy a much greater proportion of three-dimensional space because they produce highly complex molecules rich in stereochemistry developed through millions of years of evolution. This may help explain why such a high proportion of anticancer agents in clinical use today have their origins in natural products (Table 9.1).

TABLE 9.1

Examples of Anticancer and Chemopreventive Agents Derived from Natural Sources and Their Mechanisms of Action.

Drug	Mechanism of Action
Vinblastine, Taxanes	Spindle Inhibitors
Bryostatin	Protein Kinase C
Calicheamicin, Bleomycin	DNA Cleavage
Doxorubicin	DNA Intercalation
Ecteinascidin-743, PBDs	DNA Alkylation
Mitomycin C	DNA Cross-linking
Camptothecin/Topotecan	Topo I Inhibition
Podophyllotoxin/Etoposide	Topo II Inhibition
Flavopiridol/Roscovitine	CDK Inhibition
Combretastatins	Vascular Targeting
Porfimer	Photodynamic Therapy (PDT)
Herceptin	Antibody
Biologicals	Interleukins, Asparaginase
Sulforaphane, flavones etc.	Chemopreventive

It follows that there should be a greater chance of obtaining hits in high-throughput assays if natural product extracts are screened rather than libraries of synthetic compounds. However, with the upsurge of interest in parallel and combinatorial libraries in the 1990s, the desirability of using extracts of plant, bacterial, and fungal materials waned due to issues relating to their availability and stability, and especially the fact that they are usually complex mixtures. When a hit was obtained, it was necessary to separate the active agent from the complex mixture, identify its structure (including stereochemistry), and then either synthesize it or otherwise extract larger amounts from the original material, which may not have been available in useful amounts. Another perceived problem was that false-positives and false-negatives could be obtained in a screen due to the presence of other agonists or antagonists in the complex mixture.

However, technological developments have solved some of these problems and have now made the screening of natural product extracts more attractive. For example, a number of companies now supply natural product extracts in a multiwell plate format and guarantee the availability of more raw material in bulk should a hit be obtained from the screen. More importantly, *bioassay-guided fractionation* technology has revolutionized the separation and identification of hits in natural product extracts. The instrumentation involves an automated HPLC system directly coupled to a bioassay. An extract that shows activity in an initial screen is separated into fractions (usually broad ones at first) by the HPLC system (normally semipreparative or preparative), and a sample of each fraction is sent automatically to the bioassay. Active fractions are then sent back to the HPLC column and refractionated (into narrower fractions), each one again being sent to the bioassay. This process may be repeated for a number of cycles (usually fully automatically) until the narrowest fraction that contains the biological activity is obtained. At this stage, the purity may be sufficient to identify an active compound by techniques such as NMR or MS. If not, then only limited further purification should be required to achieve structural identification.

It is also worth noting a resurgence of interest in *ethnopharmacology* (or *ethnopharmacy*), the study of plant materials or other natural products that have been used by a local community as a traditional cure for a particular disease. The assumption is that these materials may contain an active constituent that can be identified, purified, and obtained in large amounts (by extraction or synthesis) for further studies. One recent example of this approach is the identification of a novel agent potentially useful for the treatment of leukemia and nonmelanoma skin cancers. Researchers at the University of Birmingham (U.K.) and Peplin Ltd. (Brisbane, Australia) discovered a new agent (known as PEP005 or 3-Angelate) in the sap of a weed (petty spurge, milk weed, *Euphorbia peplus*) traditionally used for treating corns and warts, on the basis that constituents apparently able to control cell growth and cause cell death may also have useful antitumor activity. The isolated agent has been shown to be 100-fold more cytotoxic toward certain tumor cells *in vitro* than healthy cells and is believed to work by activating protein kinase C which triggers apoptosis.

Given these developments, it is possible that more novel anticancer agent leads will be derived from natural product sources in the future.

9.4 CHEMOPREVENTIVE AGENTS

The prophylactic use of tamoxifen and related selective ER modulators (SERMs) by women to reduce the risk of breast cancer was discussed in Chapter 6. There is ongoing interest in identifying SERMs with improved toxicity profiles that could be taken safely over long time periods. With more than 10 million new diagnoses of cancer worldwide each year, there is also interest in other types of agents that may help to reduce the risk of cancer developing (i.e., *cancer chemopreventive agents*). Some experts have suggested that as much as two thirds of all cancers may be initiated and tumor growth hastened through lifestyle factors, including poor diet. It is now known that some diets may contain beneficial compounds capable of reducing cancer risk. For example, the results of a number of studies suggest that diets low in fat but rich in vegetables and fruits may contribute to a reduction in risk of developing various types of tumors. These findings are supported by observations that the incidence of bowel and colon cancers is significantly greater in the U.S. and most of Europe than in Asian countries, such as India and China, where diets are much higher in vegetables, fruits, and fiber but lower in fat (see Figure 9.2).

FIGURE 9.2 Fresh fruits on a market stall commonplace in Bangkok, Thailand (left), and a shop window in Seattle, WA, displaying chocolates (right). Asian diets, which are rich in fruit, vegetables, and fiber but low in fat, may explain the lower rate of bowel and colon cancer in these countries compared with rates in Northern America and a large part of Europe.

Based on these observations, specific food products have been investigated in an attempt to identify novel compounds that might reduce cancer risk and could be taken routinely in purified form as a prophylactic measure. This has given rise to the area of *chemoprevention,* which is rapidly growing in importance. The compounds in fruits and vegetables that can act as chemopreventive agents are known as phytochemicals (i.e., natural plant chemicals), and many thousands of compounds of this type may exist in nature, although very few have so far been isolated and subjected to proper scientific scrutiny. The most studied food items in the chemopreventive area include cruciferous vegetables, dietary fiber, garlic, ginger, grapes and red wine, green tea, honeybee propolis, hot peppers, olives and olive oil, soy food, tomatoes, and turmeric. Some drug substances are also thought to reduce cancer risk and are being studied. For example, it has been proposed that aspirin may reduce the risk of breast and bowel cancers through COX-2 inhibition, and clinical trials are underway to confirm this.

Various mechanisms have been proposed to explain the action of chemopreventive agents, which has given rise to two main categories: *anti-initiators* (or *blocking*) and *anti-promotional* (or *suppressing*) agents. Anti-initiators are claimed to selectively enhance the metabolism of some carcinogens, thus reducing the body's exposure to them. In particular, they may block DNA damage resulting from carcinogens (or resulting free radicals) by detoxifying the carcinogens via induction of Phase 1 and 2 enzymes. Alternatively, they may inhibit the metabolic activation of certain carcinogens that require this form of activation by selectively inhibiting Phase I enzymes. Anti-initiators may also work by increasing the expression of cytoprotective genes in human cells, including those that produce drug-metabolizing enzymes such as glutathione (GSH) transferase and NAD(P)H:quinone oxidoreductase 1 (NQO1) as well as antioxidant genes such as glutamate cysteine ligase modifier (GCLM) subunits and glutamate cysteine ligase catalytic (GCLC) genes, which help in the synthesis of GSH.

Anti-promotional agents may act directly on a variety of molecular targets, such as hormone receptors, mitogen-activated protein kinase, protein kinase C, second messengers, transcription factors such as NFkB and AP-1, cell cycle progression pathways, tumor suppressor genes, and enzymes such as ornithine decarboxylase and COX lipoxygenase. The existence of compounds capable of neutralizing carcinogens in the GI tract or preventing their absorption is also possible, although none has yet been identified. Another possibility is that, in the future, some compounds will be identified that can enhance the repair of DNA lesions caused by the interaction of carcinogens with the genome.

Although highly attractive as a concept, there are numerous practical problems with identifying and evaluating potential chemopreventive agents. One problem with the scientific validation of chemoprevention is that clinical trials need to be conducted over many years because, for example, cancer in old age might be associated with carcinogens ingested early in life. Second, it is notoriously difficult to accurately record food and drink intake in individuals over long time periods due to forgetfulness or a lack of truthfulness (especially in overweight individuals). A third problem is that it is difficult to devise *in vitro* screens for chemopreventive agents. For example, although it is possible to screen for compounds that enhance metabolism, there is no guarantee that faster metabolism will lead to prevention of carcinogenesis, and some carcinogens are actually activated by metabolism. Because of difficulties such as these, the chemoprevention has remained an underfunded fringe area of research for many years and is only recently beginning to attract the attention of funders and established scientific investigators. Hopefully, the growing momentum and application of firm scientific and statistical principles (e.g., properly conducted clinical trials) will help bring the study of chemopreventive agents into the mainstream during the next decade.

Nonetheless, a number of lead compounds (or families of lead compounds) of diverse structure have been identified through studies carried out to date, and a selection of these is discussed in the following subsections.

9.4.1 RESVERATROL

Resveratrol is a member of the phytoalexin family of compounds biosynthesized in plants during periods of environmental stress, including adverse weather conditions and attack by insects, animals, or pathogens (Structure 9.18). It has been found in more than 70 plant species, examples of which include peanuts, mulberries, red grapes, white hellebore, and fescue grass. Resveratrol is also the active ingredient in the powdered root of the Japanese knotweed, an Asian folk medicine known as "Kojo-Kon."

Red grapes are one of the best-known sources of resveratrol. Although resveratrol is not found in the flesh of the grape, 1 g of fresh grape skin contains approximately 50 to 100 µg of resveratrol. It follows that red wine is also a good source of resveratrol, where it is present at a level of approximately 1.5 to 3.0 mg/L, depending on the type of wine. Interestingly, researchers have suggested that resveratrol may be partly responsible for the ability of red wine to lower the level of blood cholesterol and fats. It has been further suggested that this may help to explain the "French Paradox," an observation that those consuming a Mediterranean-type diet (including red wine) appear to have a reduced risk of heart disease despite consuming higher levels of saturated fat (e.g., French cheeses).

STRUCTURE 9.18 Structure of resveratrol.

There is some evidence that all phases of cancer, including initiation, promotion, and progression, can be retarded by resveratrol. It can also boost levels of the Phase II drug-metabolizing enzyme quinone reductase, and is known to have anti-mutagenic and antioxidant activity. The increase in quinone reductase activity may be especially significant because this enzyme is known to detoxify carcinogens, thus decreasing exposure of the genome. All these physiological effects suggest that resveratrol may help to prevent the initiation process, the irreversible first phase of tumor formation. This is backed by experimental results showing that the development of precancerous lesions in mouse mammary glands produced by treatment with a carcinogen can be inhibited by the systemic administration of resveratrol with no toxic effects.

Resveratrol has other pharmacological effects, which suggest that, in addition to its anti-initiation activity, it may also inhibit other phases of cancer progression. For example, it inhibits hydroperoxidase and COX enzymes, suggestive of an anti-promotion effect. It also has antiplatelet aggregating and anti-inflammatory activity and can cause differentiation of human promyelocytic leukemia cells.

Because resveratrol is free from toxicities and is already ingested by the public in food and drink products, future research to establish dose levels and schedules may lead to purified resveratrol being taken by healthy individuals on a routine basis to reduce the risk of cancer. A number of clinical investigations are presently underway, and the NCI has been involved in studies using resveratrol obtained from the roots of a tree native to Peru. The potential benefits of this agent in coronary heart disease are also being investigated.

9.4.2 PHYTOESTROGENS

Phytoestrogens are polyphenolic members of a family of naturally occurring compounds (i.e., polyphenols) found in all plants. They occur in particularly high levels in such foods as soybeans, hops, grains, beans, and cabbage. Their chemical structure is similar to the mammalian hormones estrogen and estradiol, and so they have estrogenic activity but at a much lower level (i.e., 500-fold to 1000-fold less) than estradiol (Structure 9.19). Therefore, it has been suggested that phytoestrogens exert an antiestrogenic effect by binding to and then effectively blocking estrogen receptors while exerting only a weak estrogenic effect themselves. Suppression of normal estrogenic activity through this mechanism in estrogen-responsive tissues, such as breast tissue, may thus reduce the risk of cancer. Phytoestrogens may also have estrogen-independent actions on other cellular targets.

Isoflavone structure   Estradiol   Genistein: R = OH
Daidzein : R = H

STRUCTURE 9.19 Structures of the isoflavone ring system, estradiol (for comparative purposes) and the phytoestrogens genistein and daidzein.

Phytoestrogens are present in the stems, flowers, roots, and seeds of legumes, where they act as part of a plant's defense mechanism against microorganisms and fungal infections. They also attract specific nitrogen-fixing soil bacteria by acting as molecular signals transmitted from the roots of leguminous plants. Lignans, coumestans, and isoflavones are the main classes of phytoestrogens. Commercially, the isoflavones are receiving much interest at present, with the soybean providing the greatest quantities of raw material. The most important isoflavones in soy are genistein and daidzein. Daidzein has been isolated from red clover as well, and has specific protein kinase inhibition properties. In the U.K. diet, lignans are present in most fiber-rich foods and form an important source of phytoestrogens.

Cancer (breast and prostate cancer in particular), heart disease (antioxidant activity), osteoporosis, menopause, cognitive function and diabetes are major points of focus for the potential health benefits of phytoestrogens. Much interest has been shown in their potential anticancer activity, spurred on by the observation that Japan, whose population consumes large quantities of phytoestrogens, has one of the lowest breast cancer rates in the world. Similarly, there is a very low incidence of prostate cancer among Asian men. Interestingly, when members of these groups migrate to Northern America or Europe, their risk of developing cancer increases to a level similar to the general population of their newly adopted country, suggesting that an environmental factor such as diet may be important. The dietary consumption of phytoestrogens is known to be 30 times higher in Japan than in the U.K. In 1994, the total isoflavone intake of Japanese residents was estimated to be 150 to 200 mg/day, although more recent studies have suggested it might be lower (32 to 50 mg/day). However, even this lower estimate is insightful because it has been established experimentally that an intake of 50 mg/day can achieve isoflavone levels in the plasma of between 50 to 800 ng/ml, which is thought to be sufficient to contribute towards lowering breast cancer rates. Less is known about the possible impact of phytoestrogens in decreasing prostate cancer rates. So far, human clinical trials have produced no significant findings, although data from *in vivo* studies suggest a causal link.

Due to their positive (but weak) estrogenic activity, phytoestrogens have been advocated as a substitute for estrogen replacement therapy to prevent or reduce the symptoms of menopause. This concept is supported by the observation that menopause appears to be much less of a problem for women in countries where high amounts of soy are consumed. For example, in Japan, postmenopausal women reportedly suffer much less from night sweats and hot flashes than their U.S. and U.K. counterparts.

In summary, although the evidence available to date is insufficient to warrant the recommendation of specific dietary changes, it is highly likely that isoflavones and lignans ingested in the diet play a major role in the prevention of a number of different types of cancer. In particular, there is compelling evidence for a protective effect against prostate and breast tumors through maintaining a high intake of vegetables, legumes, fruits, and grains, and this has led major cancer research charities and governments in the U.K. and U.S. to recommend a daily intake of five portions per day of fruit or vegetables to protect against cancer. Not surprisingly, this is an active area of research, and the results of future studies should identify the particular food types and components responsible.

9.4.3 CURCUMIN

Curcumin (turmeric yellow) is the bright yellow ingredient in turmeric, a spice used in curry dishes (Structure 9.20). It has been known since the early 1990s that curcumin can slow the growth of tumors and arrest angiogenesis. In 2003, researchers discovered that curcumin can irreversibly inhibit aminopeptidase N (APN), an enzyme that promotes tumor invasiveness and angiogenesis. APN is a membrane-bound, zinc-dependent metalloproteinase that breaks down proteins at the cell surface, thus helping tumor cells invade the space of neighboring cells. This discovery was made when researchers at Sejong University (Seoul, Korea) were screening libraries to find inhibitors of APN as potential anti-angiogenic agents. After screening

STRUCTURE 9.20 Structure of curcumin.

approximately 3000 molecules, curcumin proved to be one of the most potent inhibitors. Through a combination of surface plasmon resonance experiments and *in vitro* enzyme assays it was established that curcumin's inhibition of APN is direct and irreversible. Although the exact mode of binding is not yet established, it has been postulated that the two α,β-unsaturated ketone moieties may covalently link to nucleophilic amino acids in APN's active site.

This finding suggests the possibility of developing other more potent APN inhibitors based on the structure of curcumin that might be taken orally as a pro-phylactic measure with minimum side effects. Curcumin itself is presently being evaluated in Phase I clinical trials to establish its pharmacokinetics and possible use in the prevention and treatment of colon cancer.

9.4.4 SULFORAPHANE

Sulforaphane is one of a group of compounds known as *isothiocyanates* found in broccoli, cabbage, cauliflower, kale, other cruciferous vegetables, and some fruits (Structure 9.21). It is thought to be partly responsible for the observation that populations consuming large quantities of vegetables and fruits in their diets appear to have a reduced risk of developing a number of different cancer types. Although most research to date has been conducted in *in vivo* models, a number of clinical studies are presently underway.

STRUCTURE 9.21 Structure of L-sulforaphane.

Sulforaphane was isolated and identified in 1992 by researchers at Johns Hopkins University (U.S.A.), who showed that it is a potent and selective inducer of Phase II detoxification enzymes. It is generated from its precursor form, sulforaphane glu-cosinolate (SGS), which functions as an indirect antioxidant. Unlike direct anti-oxidants, such as vitamins C and E, and beta-carotene, SGS does not directly neutralize free radicals. Instead, it boosts the activity of Phase II detoxification enzymes, which act as a defense mechanism by triggering broad-spectrum antioxidant activity, thus leading to the neutralization of free radicals before they can cause cellular damage, especially mutations to the genome that may lead to cancer. More importantly, in contrast to direct antioxidants that neutralize radicals on a one-for-one basis and are destroyed themselves in the process, indirect antioxidants such as sulforaphane are longer lasting, triggering an ongoing process that may continue for days after it has left the body.

In vivo experiments have shown that sulforaphane can block the formation of mammary tumors in rats if administered before treatment of a potent carcinogen. For example, in some experiments, the number of rats developing tumors could be reduced by as much as 60%, whereas in other studies, the number, size, and growth rate of tumors could be significantly decreased. It has also been shown that the formation of premalignant lesions in rat colons can be inhibited by sulforaphane, and that it can induce cell death in human colon carcinoma cells growing *in vitro*. These results suggest that the induction of apoptosis may be an important component of the chemopreventive properties of sulforaphane, in addition to the stimulating effect on detoxifying enzymes.

Extensive studies have established that different varieties of broccoli plants differ significantly in the quantities of sulforaphane they contain. There are also differences between fresh and frozen broccoli plants, and the concentration of SGS decreases as plants grow older. The greatest enzyme-inducing activity is found in young plants such as 3-day-old broccoli sprouts, which can contain as much as 50 times the concentration of SGS as mature, cooked broccoli. Not surprisingly, most commercially available broccoli is extremely variable in SGS concentration. However, some entrepreneurial growers have produced young, high-SGS yielding broccoli sprouts of consistent quality. Such plants are grown under carefully controlled and standardized conditions to maintain consistently high concentrations of SGS. Compared to mature, cooked broccoli, one such patented variety known as BroccoSprouts™ contains approximately 20-fold higher concentrations of SGS.

9.4.5 COX-2 INHIBITORS

There is evidence that the cyclooxygenase enzymes COX-1 and COX-2, and their related biosynthetic pathway, may play a role in both carcinogenesis and cancer progression. These enzymes catalyze the first step in the synthesis of prostaglandins from arachidonic acid. COX-1 is expressed constitutively in most tissues and seems to be responsible for the synthesis of prostaglandins that control normal physiologic functions. In contrast, COX-2 is not detected in most normal tissues, although it is induced by a variety of mitogenic and inflammatory stimuli that result in the enhanced synthesis of prostaglandins in both inflamed and neoplastic (i.e., tumor) tissues.

A number of lines of evidence suggest that COX-2 is a potentially interesting target for cancer prevention and treatment. For example, it is commonly overexpressed in both premalignant and malignant tissues. The most specific data supporting a cause-and-effect relationship between COX-2 and carcinogenesis have resulted from genetic studies in which multiparous female transgenic mice engineered to overexpress human COX-2 in their mammary glands developed mammary gland hyperplasia, dysplasia, and metastatic tumors. Also, transgenic mice overexpressing COX-2 in their skin can be shown to develop epidermal hyperplasia and dysplasia. Furthermore, consistent with these findings, COX-2 knock-out mice have a reduced incidence of skin papillomas and intestinal tumors.

In addition to the genetic evidence implicating COX-2 in carcinogenesis, numerous pharmacological studies have suggested that COX-2 is a viable therapeutic target. In particular, selective inhibitors of COX-2 reduce the formation, growth, and metastases of experimental tumors, and decrease the number of intestinal tumors in familial adenomatous polyposis (FAP) patients. However, the antitumor effects of selective COX-2 inhibitors may reflect mechanisms operating in addition to inhibition of COX-2. For example, recent studies of the expression of COX-2 in human prostate cancer have provided conflicting results; several have suggested that COX-2 is commonly overexpressed in prostate cancer, whereas others have found low levels of expression or none at all. However, even if COX-2 overexpression proves to be uncommon in prostate cancer, selective COX-2 inhibitors could still be useful if the COX-2-independent effects are proven to be clinically important.

Rofecoxib (Vioxx™)   Celecoxib (Celebrex™)   Aspirin

STRUCTURE 9.22 Structures of rofecoxib (Vioxx™), celecoxib (Celebrex™), and aspirin.

In this context, researchers recently investigated whether selective COX-2 inhibitors such as rofecoxib (Vioxx™) and celecoxib (Celebrex™) (Structure 9.22) possess COX-2-independent antitumor properties. Androgen-responsive (LNCaP) and androgen-insensitive (PC3) human prostate cancer cell lines were used in these studies because they both express only COX-1 and not COX-2. Clinically achievable concentrations of the selective COX-2 inhibitor celecoxib of 2.5 to 5.0 μmol/L were found to suppress the growth of both cell lines, whereas rofecoxib, a more potent selective COX-2 inhibitor, had no effect on cell growth at similar concentrations. Celecoxib appeared to exert its inhibitory effect on cell growth through induction of G1 cell cycle blockade and a decrease in the synthesis of DNA. Interestingly, it also showed activity in a PC3 human tumor xenograft at plasma levels (2.37 to 5.70 μmol/L) comparable to concentrations showing activity *in vitro*. It inhibited tumor growth in a dose-dependent manner, with the highest dose leading to a 52% tumor volume decrease and a 50% reduction in both microvessel density and cell proliferation. However, no measurable reduction in the intratumoral concentration of prostaglandin E2 was observed. In other *in vitro* and *in vivo* studies, celecoxib was shown to suppress cyclin D1 levels. Taken together, these results strongly suggest that celecoxib possesses significant COX-2-independent antitumor properties and may be active even in tumors that do not express significant amounts of COX-2.

In 2004, rofecoxib was withdrawn from the market by its manufacturer (Merck) based on unfavorable data from a 3-year "chemopreventive" clinical trial in patients with FAP. The trial, known as "APPROVe" (i.e., Adenomatous Polyp Prevention), involved patients with a medical history of colorectal adenomas. It was placebo-controlled, prospective and randomized, and attempted to evaluate the ability of rofecoxib to prevent recurrence of colorectal polyps. Unfortunately, patients on rofecoxib showed an increased relative risk of cardiovascular events, including heart attack and stroke, after 18 months of treatment compared with patients taking a placebo and this led to the drug's withdrawal. This was considered an ironic turn of events by observers of the pharmaceutical industry, given that Merck had developed rofecoxib as a novel anti-inflammatory agent, for which it was primarily marketed. However, the cardiovascular problems associated with celecoxib are currently thought to be compound- rather than class-specific, and the safety of celecoxib has been affirmed by Pfizer. All other COX-2 inhibitors are now under careful surveillance by the FDA, MHRA (U.K.), and other regulatory bodies, and this is also impacting newer agents, such as Novartis' Prexige™ and Merck's Arcoxia™.

Finally, there is some evidence that aspirin (see Structure 9.22), being a COX-1 and COX-2 inhibitor, may also possess chemopreventive activity. A number of studies are presently underway following patients who take low-dose (75 mg) aspirin for its antiplatelet activity to see whether they have a lower rate of bowel cancer than the general population after a number of years of treatment.

9.4.6 ANTIOXIDANTS

Antioxidants are commonly described as cancer chemopreventive agents, especially in the lay press, on the basis that they scavenge free radicals and prevent damage to cellular constituents, including the genome. In particular, vitamin C (ascorbic acid) (see Structure 9.23) has been recommended for many years as a chemopreventive agent. However, a recent retrospective study in patients suggests that it provides no benefit. White button mushrooms, which contain the antioxidant ergothioneine (see Structure 9.24), are also claimed to have cancer chemopreventive properties.

9.4.6.1 Ascorbic Acid (Vitamin C)

An authoritative analysis published in *The Lancet* in 2004 concluded that there is no evidence that vitamins (including vitamin C) help prevent the occurrence of common gastro-intestinal (GI) cancers. (see Structure 9.23)The study included more than 170,000 people at high risk for GI cancers and pooled 20 years of research data. However, despite this result, some experts still believe that antioxidants play a role in chemoprevention in certain individuals, and other studies are still ongoing.

STRUCTURE 9.23   Structure of ascorbic acid.

9.4.6.2   Ergothioneine

Ergothioneine, a constituent of white button mushrooms, the type most commonly consumed in the U.S. and Europe, has been proposed as an antioxidant "vitamin" with cancer chemopreventive and cardiovascular properties (Structure 9.24). White mushrooms are the best known source of ergothioneine, containing about 4 times more than chicken liver (previously considered the best source) and about 12 times more than wheat germ, also a good source of the nutrient. Even exotic mushrooms such as shiitake, oyster, and maitake contain about 40 times more of the antioxidant than wheat germ. A 3-ounce serving of button mushrooms (the amount usually served on a hamburger) contains approximately 5 mg of ergothioneine.

STRUCTURE 9.24   Structure of ergothioneine.

Ergothioneine was first discovered in 1909 in the sclerotia of the ergot fungus, *Claviceps purpurea*. It has since been found in blood, semen, and various mammalian tissues (principally the liver and kidneys), and was first synthesized in 1951. It is thought to work as a chemopreventive agent by helping the body to eliminate free radicals, which have been associated with both heart disease and cancer.

9.4.7   OLIVE OIL (THE "MEDITERRANEAN DIET")

It has been speculated for many years that the so-called "Mediterranean Diet" has a number of health benefits, including reduced risk of heart disease and cancer. In particular, red wine, tomatoes, and olive oil have all been suggested as likely beneficial components of this diet. Recently, two components of olive oil, oleic acid and oleocanthal, have been identified as likely chemopreventive components. These are described below along with lycopene, the proposed chemopreventive constituent of tomatoes.

9.4.7.1   Oleic Acid

An article in *Annals of Oncology* (2005) suggests that ingesting olive oil may reduce the risk of developing breast cancer. This finding was based on a study in which oleic acid (Structure 9.25), the main mono-unsaturated fatty acid component of olive oil, was found to prevent overexpression of the oncogene HER2/neu, which is known to be a significant factor in the development and management of certain types of breast tumors. Overexpression of HER2/neu is seen in about 20% of breast cancers and is associated with an unfavorable clinical outcome and resistance to chemotherapy.

STRUCTURE 9.25   Structure of (9Z)-9-octadecenoic acid (oleic acid).

Using human breast cancer models expressing high levels of HER2/neu, it was found that oleic acid suppresses overexpression of the oncogene, suggesting that dietary supplementation with this fatty acid could play a role in the prevention or management of this disease. Furthermore, oleic acid was found to make HER2/neu overexpressing cancer cells more sensitive to the effects of the Mab treatment trastuzumab (Herceptin™). At clinically relevant concentrations of trastuzumab, oleic acid and trastuzumab were found to act synergistically to down-regulate HER2/neu expression. Therefore, it is possible that oleic acid may also be beneficial in combination with other therapies directed at HER2/neu, and that dietary counseling could help delay or prevent trastuzumab resistance in this type of breast cancer.

9.4.7.2   Oleocanthal

It was reported in *Nature* in 2005 that oleocanthal, the dialdehydic form of (−)deacetoxy-ligstroside aglycone ("oleo" for olive, "canth" for sting, and "al" for aldehyde), is present in newly pressed extra-virgin olive oil (see Structure 9.26). The authors also noted that oleocanthal produces a marked stinging sensation in the throat, comparable to that obtained by tasting a solution of ibuprofen, an NSAID (see Structure 9.26). Working on the assumption that a shared pharmacological activity may explain these similar perceptions, the researchers isolated (−)oleocanthal from different premium olive oils and found that throat irritation intensity correlated positively with oleocanthal concentration. These results were confirmed by synthesizing and examining both isomers of oleocanthal to rule out the effects of minor contaminants. On the basis that it was shown 40 years ago that the bitterness of certain compounds can correlate with their pharmacological activity, the researchers went on to show that oleocanthal mimics the pharmacological effects of ibuprofen, a potent modulator of inflammation and pain.

Like ibuprofen, oleocanthal was shown to significantly inhibit the COX-1 and COX-2 enzymes which catalyze steps in the biochemical inflammation pathways starting from arachidonic acid. Both enantiomers of oleocanthal were found to cause

STRUCTURE 9.26 Structures of oleocanthal and ibuprofen.

dose-dependent inhibition of COX-1 and COX-2, with the (–) isomer slightly more active than the (+), but had no effect on lipoxygenase *in vitro*. This discovery suggested that protection against certain diseases might be obtained by the long-term ingestion of olive oil, due to the ibuprofen-like COX-inhibiting properties of oleocanthal. Calculations suggested that if 50 g of extra-virgin olive oil containing up to 200 µg/ml of oleocanthal is ingested per day, of which 60% to 90% is absorbed, then this approximates to an intake of up to 9 mg/day. Although this dose is relatively low, corresponding to about 10% of the ibuprofen dosage recommended for adult pain relief, it is established that regular, low doses of aspirin (e.g., 75 mg/day), another COX-inhibitor, confers cardiovascular health benefits. Ibuprofen has also been associated with inhibition of blood platelet aggregation, a reduction in the risk of developing certain cancers, and a decrease in COX-independent secretion of amyloid-$\beta42$ peptide in a mouse model of Alzheimer's disease.

Therefore, a Mediterranean diet rich in olive oil is believed to confer a number of health benefits, some of which seem to overlap with those attributed to NSAIDs. The discovery of COX-inhibitory activity in a component of olive oil offers a possible mechanistic explanation for this link, although as discussed in Section 9.4.5, confirmation of a direct link between COX-inhibition and an anticancer or chemopreventive effect has not yet been obtained.

9.4.7.3 Lycopene

Lycopene is an unsaturated open-chain carotenoid, also known as (*all-trans*)-lycopene, that provides fruits such as pink grapefruit, rosehip, watermelon, guava, and especially tomatoes with their red coloration (Structure 9.27). It is present in higher concentrations in ripe fruit, particularly in tomatoes (1 kg of fresh ripe tomatoes can yield as much as 0.02 g of pure lycopene). Although first isolated in impure form in 1910, its structure was not elucidated until the late 1920s. However, due to the

STRUCTURE 9.27 Structure of lycopene.

complexity of the molecule, particularly in terms of the number of double bonds and their relative configurations, it was not synthesized until 1950.

Lycopene is a powerful carotenoid antioxidant that neutralizes free radicals, thus preventing them from damaging cells and particularly their genomes. The accumulation of lycopene in body tissues is higher than for any other carotenoid, and it is deposited in high concentrations in the skin, colon, liver, prostate gland, and lungs. Epidemiological studies show that a high intake of lycopene-containing fruits can reduce the risk of developing certain types of tumors. For example, a study by Harvard's School of Public Health and Medical School on the diets of more than 47,000 men over a 6-year period found that, out of 46 vegetables and fruits evaluated for a possible association with health, only tomato-based foods that contained the highest amounts of lycopene showed any measurable effect. For example, a relationship was identified between a decreased risk of prostate cancer and a greater than average consumption of tomato-based foods, with resulting higher blood levels of lycopene. As a result of this observation, further studies are now looking at the possible role of lycopene in decreasing the risk of developing tumors of the lung, bladder, cervix, skin, and breast. In addition, it is thought that the habitual intake of tomato products by Italians may contribute to their low risk of developing GI cancers, and this hypothesis is being tested in further studies.

Interestingly, the results of some investigations suggest that any type of heat-based processing of tomatoes and tomato products may increase the bioavailability of lycopene by converting it to a form that is more easily absorbed by the body. It has also been shown that lycopene from tomatoes that have been processed into ketchup, paste, sauce, or juice is also absorbed more efficiently by the body. Other ongoing studies suggest that lycopene may be linked to a decreased risk of developing diseases associated with serum lipid oxidation and also macular degenerative disease.

9.4.8 THE KEAP1 GENE

The KEAP1 gene has recently received much attention in relation to chemoprevention. It has been proposed that down-regulation of KEAP1 (e.g., through antigene or antisense approaches) could represent a unique form of cancer chemoprevention. Anti-initiators (or blocking agents) work by increasing the expression of cytoprotective genes in human cells, including those that produce drug-metabolizing enzymes such as GSH transferase and NQO1, as well as antioxidant genes such as GCLC and GCLM subunits, which help to synthesize GSH. Therefore, to be chemoprotective, such chemical agents must induce a degree of reduction-oxidation (redox) stress in the cells, which can be potentially damaging. Thus, genes that are regulated in this manner contain antioxidant response elements (AREs) in their promoters, and their transcription is stimulated in response to the ARE-mediated recruitment of a complex that contains nuclear factor erythroid 2 p45-related factor 2 (NRF2). Thus, NRF2 accumulates in the nucleus in response to redox stress and after treatment with chemopreventive blocking agents.

The redox-sensitive kelch-like ECH-associated protein 1 (KEAP1) is an NRF2-specific adaptor protein for the CULLIN3–ROC1 ubiquitin ligase. Under homeostatic conditions, it promotes the proteasome-mediated degradation of NRF2 and thereby keeps ARE-containing genes switched off. However, under conditions of natural redox stress, or after treatment with chemopreventive blocking agents, KEAP1 no longer fulfills this role, NRF2 is stabilized, and the ARE-containing genes are switched on. Therefore, researchers have suggested that specifically reducing KEAP1 levels using a gene-targeting technology such as antisense or antigene should activate transcription of the ARE-containing genes in the absence of stress.

To test this hypothesis, a 21-nucleotide duplex siRNA was designed that, after transfection into human keratinocytes, successfully knocked down KEAP1 mRNA to less than 30% of normal levels. This up-regulated NRF2 and increased the levels of NQO1, GCLC, and GCLM, as well as aldo-keto reductase 1C1/2 and GSH, in the cells. Therefore, notwithstanding the delivery problems associated with siRNA, these results suggest that down-regulation of KEAP1 is a potentially valuable method for inducing the preadaptation of human cells to oxidative stress, thus potentially leading to a chemopreventive effect.

FURTHER READING

Al-Hajj, M., et al. "Therapeutic Implications of Cancer Stem Cells," *Curr. Opinion Gen. Devel.,* 14:43-47, 2004.

Beauchamp, G.K., et al. "Phytochemistry: Ibuprofen-Like Activity in Extra-Virgin Olive Oil," *Nature,* 437:45-46, 2005.

Borgatti, M., et al. "Decoy Molecules Based on PNA-DNA Chimeras and Targeting Sp1 Transcription Factors Inhibit the Activity of Urokinase-Type Plasminogen Activator Receptor (uPAR) Promoter," *Oncol. Res.,* 15:373-83, 2005.

Cassidy, A. "Dietary Phytoestrogens: Potential Anti-Cancer Agents?" *Brit. Nutr. Foun. Bull.,* 24:22-30, 1999.

Chen, C. and Kong, A.N. "Dietary Cancer-Chemopreventive Compounds: From Signaling and Gene Expression to Pharmacological Effects," *Trends Pharm. Sci.,* 26(6):318-26, 2005.

Chung, F.L., et al. "Chemoprevention of Colonic Aberrant Crypt Foci in Fischer Rats by Major Isothiocyanates in Watercress and Broccoli," *Proc. Am. Assoc. Cancer Res.,* 41:660, 2000.

Devling, T.W.P., et al. "Utility of siRNA against KEAP1 as a Strategy to Stimulate a Cancer Chemopreventive Phenotype," *Proc. Nat. Acad. Sci.,* 102:7280-85, 2005.

Duncan, S.J., et al. "Isolation and Structure Elucidation of Chlorofusin, a Novel p53-MDM2 Antagonist from a *Fusarium sp.*," *J. Am. Chem. Soc.,* 123:554-60, 2001.

El-Deiry, W.S. "The p53 Pathway and Cancer Therapy," *Cancer J.,* 11:229-36, 1998.

Gao, Z., et al. "Controlled and Targeted Tumor Chemotherapy by Micellar-Encapsulated Drug and Ultrasound," *J. Contr. Release,* 102:201-23, 2005.

Gerster, H. "The Potential Role of Lycopene for Human Health," *J. Am. Coll. Nutr.,* 16:109-26, 1997.

Hendrick, A.M., et al. "Verapamil with Mitoxantrone for Advanced Ovarian Cancer: A Negative Phase II Trial," *Ann. Oncol.,* 2(1): 71-72, 1991.

Jain, K.K. "Drug Delivery in Cancer," *Jain Pharmabiotech Publications, Basel,* 1-433, 2005.

Johnstone, R.W. "Histone-Deacetylase Inhibitors: Novel Drugs for the Treatment of Cancer," *Nat. Rev. Drug Disc.,* 1:287-99, 2002.

Kent, O.A., and MacMillan, A.M. "RNAi: Running Interference for the Cell," *Org. Biomol. Chem.,* 2(14):1957-61, 2004.

Kornek, G., et al. "Phase I/II Trial of Dex-verapamil, Epirubicin, and Granulocyte-Macrophage-Colony Stimulating Factor in Patients with Advanced Pancreatic Adenocarcinoma," *Cancer,* 76:1356-62, 1995.

Lane, D.P. and Hupp, T.R. "Drug Discovery and p53," *Drug Disc. Tod.,* 8:347-55, 2003.

Lemoine, N.R., and Cooper, D.N. (eds.). *Gene Therapy.* Oxford: BIOS Scientific Publishers, 1996.

Lesniak, M.S. "Novel Advances in Drug Delivery to Brain Cancer," *Technol. Cancer Res. Treat.,* 4:417-28, 2005.

Levy, D.E., et al. "Matrix Metalloproteinase Inhibitors: A Structure-Activity Study," *J. Med. Chem.,* 41:199-223, 1998.

Motzer, R.J., et al. "Phase I/II Trial of Dex-verapamil Plus Vinblastine for Patients with Advanced Renal-Cell Carcinoma," *J. Clin. Oncol.,* 13:1958-65, 1995.

Pardal, R., et al. "Applying the Principles of Stem-Cell Biology to Cancer," *Nat. Rev. Cancer,* 3:895-902, 2003.

Reya, T., et al. "Stem Cells, Cancer, and Cancer Stem Cells," *Nature,* 414:105-11, 2001.

Stahl, W., and Sies, H. "Lycopene: A Biologically Important Carotenoid for Humans?" *Arch. Biochem. Biophys.,* 336:1-9, 1996.

Vassilev, L.T., et al. "*In Vivo* Activation of the p53 Pathway by Small-Molecule Antagonists of MDM2," *Science,* 303:844-48, 2004.

Workman, P. "Altered States: Selectively Drugging the Hsp90 Cancer Chaperone," *Trends Mol. Med.,* 10:47-51, 2004.

10 Personalized Treatments

Interest is growing in the possibility of predicting both the efficacy and likely extent of adverse reactions in all types of drug classes by screening patients for the presence or absence of certain genes or their expression products (often referred to as *biomarkers*). This so-called "personalized medicine" approach to drug therapy, sometimes known as *oncogenomics* in the area of oncology, is predicted to grow significantly in the next decade and is best illustrated by the following example. In 2005, Genaissance Pharmaceuticals Inc. (now part of Clinical Data) was granted a patent for a process of screening individuals for a genetic predisposition for a reduced ability to metabolize certain drugs. The process involves testing for the presence of a common genetic variant of the CYP3A4 gene (and a mutation in the GSTM1 gene) that may predict for this reduced ability. The test is based on the assumption that more than 50% of all drugs, including anticancer agents and a wide range of over-the-counter products, are metabolized by CYP3A4. The company claims that this so-called *pharmacogenetic test* provides a selection process for chemotherapeutic agents for individual patients based on the presence or absence of the variant gene CYP3A4*1B.

Many more screens and assays of this type can be expected to be introduced during the next decade. Some view this as scientific progress and an opportunity to tailor drugs to individual patients (e.g., identification of *responders*), to enhance activity and reduce side effects. However, more cynical observers believe that the advantages to patients will be minimal and that pharmaceutical companies will benefit by marketing costly screening kits alongside their drugs. The drug companies respond to this allegation by pointing out that this personalized approach to treatment will lead to reduced sales of particular agents as the medical community shifts away from the "one-size-fits-all" approach that has operated since the dawn of modern therapeutics.

It is worth noting that, in the U.S., prescribing guidance for a number of drugs including imatinib (Glivec™), trastuzumab (Herceptin™), tretinoin (for leukemia), and cetuximab (Erbitux™) now includes information on the use of *predictive pre-prescription genetic testing* to help identify potential responders (or nonresponders), or those who are likely to suffer adverse drug reactions (ADRs), before treatment is initiated.

10.1 INTRODUCTION

It is widely agreed that there is a desperate need for new ways to control and treat cancer. As our understanding of the genetics of cancer has progressed, so advances have been made in predicting the risk of disease development and recurrence, in prognosis and staging, in treatment selection and prediction of side effects, and in the facilitation of clinical trials of new agents.

A tumor from any given individual will contain a number of different genetic alterations. In particular, these genetic changes can predict whether patients may be more or less at risk for developing advanced disease and whether they might respond to certain therapies. At present, there is usually very limited insight into these genetic differences at initial diagnosis, so proposing treatments tailored to the individual patient is problematic. Therefore, *genomics* (the study of complex sets of genes) and *proteomics* (the study of their expression and function) are becoming increasingly important. In particular, new genetic services are being developed and offered that provide clinically-validated individual tumor profiles at the nucleic acid or protein level. In the future, such information may significantly improve the quality of treatment decisions for cancer patients (see Figure 10.1).

The application of the science of oncogenomics to cancer diagnosis and treatment may appear relatively straightforward; however, the characteristics of a tumor are highly dependent on many different genes, the way they interact, and the environment they create for disease progression. Although at present it is possible to

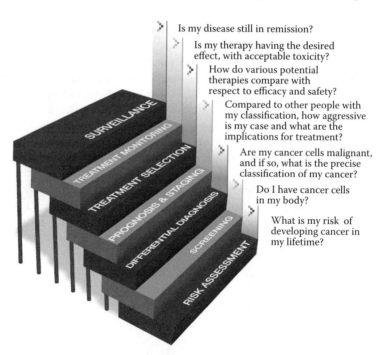

FIGURE 10.1 Potential roles of genomics in cancer (From Genomic Health, Inc.).

TABLE 10.1
Examples of Genes Associated with Specific Tumor Types

Cancer Type	Associated Gene(s)
Breast	BRCA1, BRCA2, ATM, Her2/neu
Burkitt's lymphoma	c-MYC
Colon (Bowel)	MLH1, MSH2, MSH6
Lung	EGFR (ErbB-1), HER2/neu (ErbB-2), HER3 (ErbB-3), HER4 (ErbB-4)
Chronic myelogenous leukemia	BCR-ABL
Malignant melanoma	CDKN2, BCL-2
Endothelial	VEGFR, VEGFR2
Various	P53 tumor suppressor protein, PKA, VEGFR, VEGFR2, PDGF, PGGFR

identify the up-regulation of single genes that may be associated with more-aggressive cancers (see Table 10.1), in the future it should be possible to study sets of key relevant genes and the way they interact which should more reliably inform prognosis and treatment decisions. The way in which genes interact, and the consequences of this for both normal and cancer cells, has given rise to new areas of research such as *systems biology* and the study of *gene networks*.

In order to use genomics in cancer diagnostics, prognostics, and treatment, it is necessary to determine which sets of genes and gene interactions are important in different types and subsets of cancers. Analyses can be carried out that link the pattern of gene expression in tumor cells to the individual's response to therapy or the likelihood of recurrence of the cancer. Results from studies of this type can be used to create a genomic profile (or signature) of an individual's tumor that, in the future, should allow physicians to better predict the most beneficial treatments and gain insight into how the cancer is likely to develop clinically. In summary, genomics is ultimately expected to play a role in each step of the cancer management process.

A number of genes are now known to be overexpressed in specific cancers and can be used for oncogenomic purposes. Some examples are given in Table 10.1.

Finally, it should be noted that caution is required when screening for the presence of either single or multiple gene biomarkers (at either the nucleic acid or protein level) in solid tumor biopsy material, due to the heterogeneity of many tumors.

10.2 SCREENING FOR RISK OF DISEASE DEVELOPMENT

It is possible to screen for the presence of *single nucleotide polymorphisms* (SNPs), or changes involving more than one base pair, in certain genes (see Table 10.1) that may predict the risk of a cancer occurring. An example of this would be the BRCA1 and BRCA2 genes whose presence can predict the occurrence of breast cancer in women. There are many other genetic biomarkers increasingly gaining recognition, such as the N-acetyltransferase 2 (NAT2) gene. Individuals with the slow metabolism variant of NAT2 are thought to be at higher risk of developing bladder cancer.

There is also interest in identifying patterns of nucleic acid or protein markers (often known as *biomarkers*) in blood, urine, or saliva that may predict the risk of cancer development or recurrence. Saliva is ideal for this purpose because it is not a segregated bodily fluid but, in effect, part of the circulation. Saliva can also be obtained noninvasively, making it a preferable source of biomarkers. Physiologically, it is a filtrate of the blood, moving through the salivary glands and into the oral cavity. Analytes present in blood are also present in saliva, albeit at lower magnitudes and often at concentrations below the limits of detection of an enzyme-linked immunosorbent assay such as ELISA. However, advances in mass spectrometry mean that very low levels of proteins can now be detected and identified. Once the salivary proteome is decoded, it can be used as a diagnostic tool to identify individuals with diseases such as diabetes, Alzheimer's disease, and major cancers.

With regard to nucleic acids, the saliva of normal individuals contains approximately 3,000 species of ribonucleic acid (RNA). Recent studies using Affymetrix™ DNA microarrays demonstrated that approximately 185 of these are common in all healthy individuals. However, the main challenge with microarray technologies is that they require complex statistical analysis methodologies and data validation techniques (e.g., self-organizing maps, Fisher discriminant analysis, and logistic regression models) to validate the "signatures." Hence, choosing the appropriate algorithm is critical, a task that is becoming increasingly more challenging due to the accelerating introduction of new tools and techniques. Furthermore, DNA array chips continue to improve, such as the introduction of the Human U133 Plus 2.0 Set (analyzing over 47,000 transcripts) and the new all-exon array.

First, a completely independent data set from the microarray experiments is required to demonstrate that the findings are reproducible based on a number of statistical criteria, such as p-value fold-change and percent present in a number of individuals. Next, these candidate markers must be validated in an independent population. Then a biochemical technique, such as quantitative reverse transcription polymerase chain reaction (RT-PCR) or immunohistochemistry, must be used to confirm that the findings are real. Furthermore, an additional level of validation at the proteomics level is ideally required. Current methodologies can scan for transcriptional activity in nearly all known coding genes. However, by looking at the genome in greater detail — for example, by analyzing splice variation — it is possible to develop even more specific signatures and more robust classifiers.

One good example of this approach is provided by the work of researchers at the University of California at Los Angeles School of Dentistry who are studying the use of high-density microarrays and proteomic technologies to identify salivary biomarkers that can distinguish patients with oral squamous cell carcinoma (OSCC) from healthy controls. OSCC is a good model for work of this kind because it does not have the complicating factor of metastatic disease. With other cancer types, such as breast and lung, if a signature for metastasis is found, it is difficult to confirm whether it is associated with a lymphatic or distant metastasis. As more researchers produce data for different cancer types, it is likely that, eventually, specific signatures will be found for all the different metastatic types.

These researchers started by stratifying saliva into the five sources from the three major salivary glands (parotid, submandibular, and sublingual), from the space between the tooth and gum (the gingival crevicular fluid), and from desquamated epithelial cells. Each source was profiled and RNA found in every sample. In a small pilot study using a population of just 32 cancer patients with the same number of matched-controls, a small number of salivary RNA biomarkers were found that could distinguish whether a saliva sample originated from an oral cancer patient or a control with 91% accuracy. Such studies such as these are encouraging in suggesting that these RNA signatures are disease-associated, and researchers are now validating this approach by looking for human RNA transcripts in xenograft models transplanted with human tumors.

It is worth noting that, at present, this approach captures only a fraction of the information available from saliva. Because about 50% of the transcripts are devoid of a poly-A tail, using a T7-oligo dT primer in linear amplifications causes half of the information to be missed. Further loss of information may occur due to other degradative mechanisms associated with RNA, such as decapping or clipping of a length of RNA in the middle. Alternative amplification methodologies may improve this situation in the future. In particular, studies of this type will be helped by advances in technology such as the new all-exon array, in which every coding sequence has multiple 25-mers tiled to it. Thus, theoretically, every fragment of RNA in a biological fluid or tissue can be registered or captured. Coupled with improved amplification procedures not reliant on RNA poly-A tails, this should allow a significant improvement in analytical ability.

10.3 PROGNOSIS AND STAGING

A number of commercial laboratory tests are becoming available that claim to help with the prognosis and staging of a tumor for an individual patient. For example, a test called MammaPrint™ (www.agendia.com) for breast cancer prognosis was recently launched in the U.S. The Molecular Profiling Institute (MPI), a specialty reference laboratory formed to apply discoveries from the Human Genome Project to the analysis of individual patients' cancers, licensed the test from Agendia and claims that MammaPrint™ is the first commercially available microarray diagnostic to provide an individual DNA expression profile. The test is claimed to predict a patient's likelihood of surviving 10 years based on a gene-expression signature formed from 70 genes. A study published in 2002 in the *New England Journal of Medicine* reported an accuracy level of 96.7% for this test.

Besides MammaPrint™, Agendia also offers CupPrint™ to patients and health care professionals in Europe, which is claimed to assist in identifying the primary tumor in patients with primaries of unknown origin. Finally, Genomic Health (www.genomichealth.com) released its physician-requested Oncotype DX™ gene expression-based breast cancer prognosis test in 2004, which tracks the expression levels of 21 genes using RT-PCR.

Both MammaPrint™ and Oncotype DX™ are complex assays that are not widely available and can only be carried out by MPI and Genomic Health, respectively. In 2005, MPI justified the cost of MammaPrint™ (>$3000) by pointing to its complexity and accuracy, and claiming that the cost is low compared to the economic and personal cost of unnecessary treatment or of not treating a patient at high risk of recurrence.

10.4 SCREENING FOR RISK OF RECURRENCE OF DISEASE

A number of genomics-based commercial assays are now available that claim to be able to predict the risk of cancer recurrence. For example, Oncotype DX from Genomic Health (as discussed in Section 10.3) is a physician-requested laboratory test for women with newly diagnosed, early-stage invasive breast cancer that is designed to predict the risk of disease recurrence, as well as facilitating choice of the best chemotherapy strategy for the patient. Clearly, information regarding the risk of a cancer returning is potentially distressing for patients and so must be handled sensitively by physicians. It is the view of most clinicians that predictive tests of this type should never be freely available to the public.

10.5 SELECTING BEST TREATMENTS FOR PATIENTS

Much research is ongoing to develop methods to predict the best treatments for individual cancer patients. For example, there is widespread agreement that traditional methods for determining which women need the most aggressive therapy for breast cancer are unsatisfactory. For example, it is particularly difficult to decide which patients whose lymph nodes are clear of cancer should still have follow-up therapy. As a result of this uncertainty, many more patients receive chemotherapy than actually need it, and some who might benefit from follow-up treatment do not get it. The hope is that genomic and proteomics tests can be developed that will help in this decision-making process. An alternative approach is *chemosensitivity testing* in which the most appropriate chemotherapeutic agent (or combination of agents) is selected based on the sensitivity of biopsied tumor cells from individual patients grown *in vitro* and exposed to a wide range of single agents and combinations (see Section 10.5.2).

10.5.1 Genomic and Proteomic Tests

One example of progress in this area is the discovery of epidermal growth factor receptor (EGFR) mutations in some tumors, a discovery that has ushered in a new era linking diagnostics with targeted therapeutics. Recently, researchers at the Dana-Farber Cancer Institute (Boston, MA) and the Massachusetts General Hospital (Boston, MA) reported that mutations in the EGFR gene help to predict which patients respond to drugs that target the EGFR receptor, including such high-profile products as Iressa™ (AstraZeneca) and Tarceva™ (Genentech and Roche). Based on these observations, a test has now been developed to identify patients most likely to respond to these EGFR-targeting cancer drugs, and a commercial organization (Genzyme Genetics — www.genzymegenetics.com) signed an agreement in 2005 with these two cancer centers to put this approach into practice.

Similarly, KuDOS Pharmaceuticals Ltd. has identified stand-alone activity of their poly-ADP-ribose polymerase (PARP) inhibitors against certain common tumor genetic backgrounds that lack the functional tumor suppressor genes BRCA1 and BRCA2. Therefore, PARP inhibition has the potential to help the specific subset of patients who have mutations in these genes that predispose them to early-onset cancer (e.g., typically breast, ovarian, prostate, and pancreatic cancers). However, in the shorter term it is likely that PARP inhibitors will be combined in the clinic with agents such as temozolomide or irinotecan to enhance the treatment of a number of tumor types such as melanoma and glioma, and colorectal and gastric cancers, respectively.

10.5.2 CHEMOSENSITIVITY TESTING

An alternative approach to personalized treatment known as *chemosensitivity testing* that does not require sophisticated genomic or proteomic methodologies is being evaluated by a number of researchers. This simple approach has a parallel in antibiotic treatment, whereby samples from patients (e.g., blood, urine, sputum, throat or wound swabs) are incubated in growth medium in the laboratory to establish which microorganisms are present and the best antibiotic to use. Chemosensitivity testing involves taking a biopsy from the patient (including blood or bone marrow in the case of leukemia patients) and growing the cells in the laboratory, usually in the form of a primary culture. A number of different anticancer agents can then be incubated with the cells to discover which is the most effective. Although there is anecdotal evidence that this approach is beneficial, several clinical trials are underway to try to validate it.

A major advantage of this approach is that it can be readily automated in the laboratory and results can be obtained quickly, even for combinations of drugs. Also, an assay of this type may be less expensive than sophisticated genetic-based screens. One disadvantage is that biopsies may be difficult to obtain for tumors deep in the body, although this is also the case for genomic and proteomic screens. There are also political, organizational, and financial barriers to the introduction of this type of screen in countries such as the U.K., where national protocols of best current practice exist to treat patients with specific tumor types with particular drugs or drug combinations.

10.6 PREDICTING SIDE EFFECTS OF CHEMOTHERAPEUTIC AGENTS

In addition to predicting which patients might be suitable for a particular drug, it is also possible to use a genomics-based approach to predict which patients may develop serious side effects. One example of this is a commercially available laboratory test that can predict which patients are likely to have severe adverse reactions to 5-fluorouracil (5-FU) and capecitabine (Xeloda™). The assay is based on the observation that a small percentage of individuals (3% to 5%) are deficient in dihydropyrimidine dehydrogenase (DPD), an enzyme that is important in the metabolism of fluoropyrimidines. Due to a build-up of the drug during treatment at what

FIGURE 10.2 The AmpliChip™ device marketed by Roche and used to evaluate CYP450 activity.

would be a normal dose for most patients, these individuals experience severe vomiting, diarrhea, mouth sores, and other dangerous side effects; in some cases, these problems can lead to death. Occasionally, even the first dose can be too late, with severe reactions leading directly to a fatal outcome. Although testing for low levels of DPD is not always routine before chemotherapy with fluoropyrimidine agents, patients who have severe side effects after starting 5-FU or Xeloda™ should be tested and chemotherapy should be discontinued if patients lack the enzyme or have low levels of it.

The company Genzyme is marketing an assay that identifies patients who may be at increased risk for severe ADRs after treatment with the topoisomerase I inhibitor irinotecan (Camptosar™), by detecting variations in the UGT1A1 gene that have been associated with that risk. A prospective clinical study has demonstrated that patients with one of these variations have a greater than 9-fold risk of decreased white blood cell counts compared to patients without it. Camptosar's labeling in the U.S. has been recently updated to include dosing recommendations based on a patient's UGT1A1 status.

A further example is a pharmacogenetic test developed for use with the purine antimetabolites 6-mercaptopurine (Puri-Nethol™; 6-MP) and 6-thioguanine (Lanvis™) which have a number of potentially serious side effects including a sometimes fatal myelosuppression. Metabolism of these agents is carried out primarily by the enzyme thiopurine S-methyltransferase (TPMT), and a number of common polymorphisms in the corresponding gene determine its level of activity. Individuals with low or intermediate activity are at risk of ADRs unless the drug dose is reduced to approximately 10% of the standard dose. Pre-treatment genetic testing for TPMT has been carried out in the U.S. for about 10 years, and there is evidence to suggest that it is cost effective in some health care settings.

Finally, a number of drugs used in the direct (tamoxifen) or indirect (ondansetron, codeine, dextromethorphan, tramadol) treatment of cancer are metabolized by the enzyme CYP2D6, and the activity of this enzyme in individuals can affect the dose of drug required. One company, Roche, has developed a microarray-based device known as the AmpliChip™ which has been approved for in vitro diagnostic use in the U.S. and EU (Figure 10.2). It allows the activity of both CYP2D6 and the related CYP2C19 to be evaluated (including deletions and duplications) which play a major role in the metabolism of an estimated 25% of all prescription drugs. The AmpliChip™ allows more accurate determination of patient's genotype and predicted phenotype (i.e., poor, intermediate, extensive, or ultra rapid metabolizers), and can help with the choice of both starting and maintenance doses for drugs affected.

10.7 PHARMACOGENOMICS IN CLINICAL TRIALS

The rapidly growing introduction of so-called "personalized medicine" and "pharmacogenomics" will almost certainly lead to changes in the way clinical trials of anticancer agents are designed and carried out. This claim is based on the principle that it should be possible, even at the early Phase I stage, to design trials to identify subgroups who should respond rather than simply establishing toxic side effects and maximum-tolerated doses (MTDs). A number of prominent U.S. and European cancer organizations are now calling for the traditional clinical trials format to be redesigned, and are advocating additional testing before and during clinical trials to establish differences in drug responses between individual patients more rapidly. This has coincided with the U.S. Food and Drug Administration's (FDA's) March 2005 release of guidelines for personalized medicine.

Traditional clinical trials test initially for safety (i.e., Phase I) and later for efficacy (i.e., Phase II), but some argue that this format fails to take advantage of continuing advances in pharmacogenomics which, with a redesign of clinical trial procedures, could allow investigators to find out whether a new drug is working within the first small group of patients. Also, according to this new model proposed by the American Association for Cancer Research (AACR), the American Society of Clinical Oncology, and the Association of American Cancer Institutes, early clinical trials should include ongoing analysis of patients' tissue and blood samples so that, if a drug fails, scientists can then determine whether the target is appropriate or whether genetic differences prevent the drug from hitting the target in some individuals. Such a design may have prevented the overall failure of Iressa™ in large numbers of lung cancer patients, as the trials were at an advanced stage by the time researchers identified a mutation in the EGFR gene underlying the positive response in a small subset of individuals. It is also suggested that, in the future, before entering the clinic, scientists should understand the biology of a new drug and its target so that, in principle, it should be possible to predict how a mutation could alter that interaction. Notwithstanding this, some drug designers are trying to create resistance-free inhibitors that will bind to a target regardless of mutations.

The guidelines also propose so-called "unbiased testing," such as blood-based proteomics assays, to establish whether those who respond to the drug have a particular genetic profile. Once likely "responders" are identified, a second clinical trial can then be designed using only this population. It is also suggested that funds should be designated in all clinical trials for blood and tissue analyses to allow for follow-up studies if necessary. There are plans to test the new clinical trial design ideas with small-scale trials of EGFR inhibitors in non-small-cell lung cancer, and one such trial is currently underway with Iressa™ at Massachusetts General Hospital in Boston.

To facilitate this pharmacogenomic approach, it is hoped that drug companies will develop diagnostic tools for biomarkers along with new drugs. However, the business model and regulatory path for this development are not yet clear. At present, the FDA's new guidelines encourage, but do not require, companies to submit data on the impact of genetic variations on drug responses. A further complication is illustrated by one example from the noncancer area. In March 2004, the U.S. FDA issued dosing recommendations for Asians taking the cholesterol-loweringdrug Crestor™, after research showed that this group metabolizes the drug differently from the general population. Although such differences are becoming increasingly apparent, the FDA currently has no plans to ask for specific studies in ethnic subgroups in the oncology areas unless there is a reason to suspect significant clinical differences.

FURTHER READING

AmpliChip CYP450 Test. *Medical Letter on Drugs and Therapeutics* 2005, 47, (1215-6), 71-72.

Bullinger, L., Dohner, K., Bair, E., Frohling, S., Schlenk, R. F., Tibshirani, R., Dohner, H., Pollack, J. R., Use of gene-expression profiling to identify prognostic subclasses in adult acute myeloid leukemia. *New England Journal of Medicine* 2004, 350, (16), 1605-1616.

Cheng, C., and Evans, W.E., "Cancer Pharmacogenomics May Require Both Qualitative and Quantitative Approaches," *Cell Cycle,* 4(11):1506-09, 2005.

Dalton, W. S., Friend, S. H., Cancer Biomarkers - An Invitation to the Table. *Science* 2006, 312, (5777), 1165-1168.

Ginsburg, G. S., Haga, S. B., Translating genomic biomarkers into clinically useful diagnostics. *Expert Review of Molecular Diagnostics* 2006, 6, (2), 179-191.

Golshan, M., Miron, A., Nixon, A. J., Garber, J. E., Cash, E. P., Iglehart, J. D., Harris, J. R., Wong, J. S., The prevalence of germline BRCA1 and BRCA2 mutations in young women with breast cancer undergoing breast-conservation therapy. *American Journal of Surgery* 2006, 192, (1), 58-62.

Lee W., et al., "Cancer Pharmacogenomics: Powerful Tools in Cancer Chemotherapy and Drug Development," *Oncologist,* 10(2):104-11, 2005.

Maitland, M. L., Vasisht, K., Ratain, M. J., TPMT, UGT1A1 and DPYD: genotyping to ensure safer cancer therapy? *Trends in Pharmacological Sciences* 2006, 27, (8), 432-437.

Murphy, N., Millar, E., Lee, C. S., Gene expression profiling in breast cancer: towards individualising patient management. *Pathology* 2005, 37, (4), 271-277.

Nagler, R., Bahar, G., Shpitzer, T., Feinmesser, R., Concomitant Analysis of Salivary Tumor Markers - A New Diagnostic Tool for Oral Cancer. *Clinical Cancer Research* 2006, 12, (13), 3979-3984.

Nakada, S., Aoki, D., Ohie, S., Horiuchi, M., Suzuki, N., Kanasugi, M., Susumu, N., Udagawa, Y., Nozawa, S., Chemosensitivity testing of ovarian cancer using the histoculture drug response assay: sensitivity to cisplatin and clinical response. *International Journal of Gynecological Cancer* 2005, 15, (3), 445-452.

Need, A. C., Motulsky, A. G., Goldstein, D. B., Priorities and standards in pharmacogenetic research. *Nature Genetics* 2005, 37, (7), 671-681.

Oh, K. T., Anis, A. H., Bae, S. C., Pharmacoeconomic analysis of thiopurine methyltransferase polymorphism screening by polymerase chain reaction for treatment with azathioprine in Korea. *Rheumatology* 2004, 43, (2), 156-163.

Paik, S., Shak, S., Tang, G., Kim, C., Baker, J., Cronin, M., Baehner, F. L., Walker, M. G., Watson, D., Park, T., Hiller, W., Fisher, E. R., Wickerham, D. L., Bryant, J., Wolmark, N., A multigene assay to predict recurrence of tamoxifen-treated, node-negative breast cancer. *New England Journal of Medicine* 2004, 351, (27), 2817-2826.

Sequist, L. V., Haber, D. A., Lynch, T. J., Epidermal growth factor receptor mutations in non-small cell lung cancer: Predicting clinical response to kinase inhibitors. *Clinical Cancer Research* 2005, 11, (16), 5668-5670.

Simon, R., Wang, S. J., Use of Genomic Signatures in Therapeutics Development in Oncology and Other Diseases. *Pharmacogenomics Journal* 2006, 6, (3), 166-173.

11 Adjunct Therapies

11.1 INTRODUCTION

Adjunct therapies are given in combination with cancer chemotherapeutic agents either to counteract side effects, such as nausea and vomiting, or to enhance their therapeutic effect. Types of adjunct therapies considered in this chapter include anti-emetics, steroidal agents, and adjuvant enzymes. Although analgesics are also considered to be adjunct therapies, they are not considered in this chapter.

The role of anti-emetics in cancer chemotherapy is of great importance because many of the cytotoxic agents in clinical use cause profound nausea and vomiting. Uncontrolled vomiting may outweigh the benefits of treatment and can lead to poor patient compliance. Traditionally, drugs such as metoclopramide were used but, with the addition of the 5-hydroxytryptamine type 3 (5-HT$_3$) serotonin antagonists such as ondansetron and granisetron, the incidence and severity of emesis has been substantially reduced. Steroids are also important in the palliative care of terminal cancer patients for whom prednisolone, for example, can produce a feeling of well-being and also enhance appetite. In addition, prophylactic dexamethasone is given to reduce both fluid retention and the possibility of hypersensitivity reactions associated with docetaxel (Taxotere™).

11.2 ANTI-EMETIC AGENTS

11.2.1 Oral and Parenteral Dosing

Nausea and vomiting cause great discomfort and distress to many patients receiving chemotherapy and, to a lesser extent, certain types of radiotherapy, particularly treatment to the abdomen. In some cases, the symptoms may be so severe that patients refuse further treatment. Symptoms are categorized as *acute* (i.e., occurring within 24 hours of treatment) or *delayed* (i.e., first occurring more than 24 hours after treatment). However, *anticipatory* symptoms may also manifest prior to further doses of treatment. Delayed and anticipatory symptoms are more difficult to manage than acute ones and require different strategies.

The susceptibility of individuals to chemotherapy-induced nausea and vomiting can vary greatly and depends on the type of drug (or drug combinations) used. Thus, it is possible to classify different types of drugs according to their *emetogenic potential*. Through many years of experience with cancer therapies, it has been established that female patients and those of both genders who are younger than age 50 tend to be more susceptible. Interestingly, patients who develop a greater degree of anxiety than most, and those who are prone to motion sickness, tend to develop symptoms more easily. Also, susceptibility to nausea and vomiting may increase with each repeated dose of a drug. Therefore, in order to attempt to predict the likelihood of nausea and vomiting, clinicians consider a number of factors, including the type of drug, the dose given, and the number of times it has previously been administered along with individual susceptibility. In addition, the type and dose of other drugs coadministered are taken into account.

Drugs with weak emetogenic potential include 5-fluorouracil, etoposide, methotrexate, and the vinca alkaloids. Abdominal radiation is also classed as mildly emetogenic. Agents classed as moderately emetogenic include mitoxantrone and doxorubicin, high doses of methotrexate, and low and intermediate doses of cyclophosphamide. Dacarbazine, cisplatin, and high doses of cyclophosphamide are considered to be highly emetogenic.

Metoclopramide   Domperidone (Motilium™)

STRUCTURE 11.1 Structures of metoclopramide and domperidone (Motilium™).

To try to prevent acute symptoms, patients at low risk can be pretreated with either metoclopramide or domperidone (Motilium™) (see Structure 11.1), and this treatment is continued for up to 24 hours after the initial dose of chemotherapy. Dexamethasone (see Structure 11.4) may also be added to enhance effectiveness. For patients at high risk of nausea and vomiting, or when other antinauseant treatments have proved inadequate, a specific 5HT$_3$ serotonin antagonist such as ondansetron (Zofran™), granisetron (Kytril™), or tropisetron (Navoban™) (usually given orally) often works well to alleviate symptoms, particularly when given with dexamethasone (see Structure 11.2 and Structure 11.4).

Ondansetron (Zofran™)   Granisetron (Kytril™)   Tropisetron (Navoban™)

STRUCTURE 11.2 Structures of ondansetron (Zofran™), granisetron (Kytril™), and tropisetron (Navoban™).

STRUCTURE 11.3 Structures of prochlorperazine, lorazepam, and nabilone (Cesamet™).

Delayed symptoms of nausea may be treated with oral dexamethasone, some-times in combination with prochlorperazine (Structure 11.3) or metoclopramide. The $5HT_3$ antagonists are not as useful for this purpose.

The best strategy for preventing anticipatory nausea is to provide good symptom control. Therefore, the addition of the benzodiazepine lorazepam (Structure 11.3) to the anti-emetic therapy can be very useful due to its anxiolytic, sedative, and amnesic properties. The cannabinoid derivative nabilone (Cesamet™), which has both anti-emetic and CNS activity, is also occasionally used to control chemotherapy-associated emesis (see Structure 11.3).

11.2.2 SUBLINGUAL DOSING

Sublingual dosing is highly desirable for the acute treatment of symptoms of any disease. Apart from the obvious benefits of noninvasive, patient self-administration, once the symptoms have subsided, delivery of the drug can be immediately halted. Progress has been made in this area in the treatment of nausea and vomiting, and encouraging results from a preliminary clinical trial designed to evaluate the pharmacokinetics of a sublingual dosage form of ondansetron hydrochloride (Zof-ran™ [GlaxoSmithKline]) were announced in 2005. The trial evaluated 8 mg of ondansetron spray compared to an 8-mg oral dose of Zofran™ (in tablet form). The study was successful in demonstrating that the spray was well tolerated and produced a similar pharmacokinetic profile to the oral route but with faster delivery, a clear advantage for patients. Measurable drug concentrations in the blood were obtained

STRUCTURE 11.4 Structures of prednisolone and dexamethasone.

20 minutes sooner with the sublingual dose compared to the tablets and, during the first 20 minutes after dosing, the spray achieved both significant increases in the total amount of drug delivered and in the highest mean concentration of drug. However, the maximum plasma concentration and bioavailability (as evaluated from the area under the curve [AUC]) achieved during the first 12-hours did not exceed that of the oral route, an important finding for safety considerations. These results suggest that ondansetron lingual spray may become an established treatment in chemotherapy clinics in the future.

11.3   STEROIDAL AGENTS

11.3.1   PREDNISOLONE AND DEXAMETHASONE

Prednisolone is an orally administered synthetic corticosteroid widely used in oncology for its significant antitumor effect in Hodgkin's disease, non-Hodgkin's lymphomas, and acute lymphoblastic leukemia, where it is often used in combination with other anticancer agents (see Structure 11.4). However, it is also used in the palliative treatment of the final stages of cancer to stimulate appetite and induce a feeling of well-being.

Prophylactic dexamethasone is given to reduce both fluid retention and the possibility of hypersensitivity reactions associated with docetaxel (Taxotere™) and other experimental agents that are prone to cause edema (see Structure 11.4).

11.4   ADJUVENT ENZYMES

11.4.1   CHEMOPHASE™

Hyaluronidases are enzymes that have in common the ability to cleave glycosidic bonds of hyaluronic acid and, to some extent, other acid mucopolysaccharides of connective tissue. Although hyaluronidase is found in sperm and has a physiological role in fertilization, the skin probably contains the largest store in the body. Hyaluronidase is also produced by many bacteria and is thought to help them and their toxins invade healthy tissue. For similar reasons, it is found in high concentrations in the heads of leaches and in snake venoms.

The enzyme, although generally present in an inactive form, is thought to regulate the rate of water and metabolite exchange in the body by decreasing the viscosity of the intercellular matrix. For this reason, it is used as a pharmaceutical aid, for example, as a diffusing agent in subcutaneous injections, as a spreading agent, and to promote diffusion, absorption, and resorption. Therefore, it is known under other names such as "spreading factor," "diffusing factor," and "Invasin," and is produced by a large number of companies under a variety of trade names.

One possibility is to use hyaluronidase to help anticancer agents penetrate tumor tissue. Recently, one company has developed and commercialized a product consisting of recombinant human enzymes (known as Chemophase™) for use as a chemo-adjuvant in certain tumor types. An investigational new drug application (NDA) was recently submitted to the U.S. FDA to seek permission to carry out the first evaluation of Chemophase™ in humans by intravesical administration in combination with mitomycin in patients with superficial bladder cancer.

11.5 OTHER THERAPIES

11.5.1 RAZOXANE

Razoxane is a cyclized analog of ethylenediaminetetraacetic acid (EDTA) that was first synthesized in the late 1960s (see Structure 11.5). It has limited activity in the leukemias and is now little used in the clinic. Not surprisingly, line EDTA, razoxane possesses intracellular iron-chelating activity.

STRUCTURE 11.5 Structure of razoxane.

Razoxane has also been coadministered with doxorubicin on the grounds that it plays a role in preventing the formation of doxorubicin-iron complexes that generate radical oxygen species associated with cardiotoxicity.

FURTHER READING

Harris, D.; Noble, S., A practical approach to symptom management in palliative care. *British Journal of Hospital Medicine* 2006, 67, (8), 404-408.

Kobrinsky, N.L. "Regulation of Nausea and Vomiting in Cancer Chemotherapy: A Review with Emphasis on Opiate Mediators," *Am. J. Ped. Hematol. & Oncol.*, 10(3):209-13, 1988.

Rizk, A.N., and Hesketh, P.J. "Anti-emetics for Cancer Chemotherapy-Induced Nausea and Vomiting: A Review of Agents in Development," *Drugs Res. Develop.*, 2(4):229-35, 1999.

Stephenson, J.; Davies, A., An assessment of aetiology-based guidelines for the management of nausea and vomiting in patients with advanced cancer. *Supportive Care in Cancer* 2006, 14, (4), 348-353.

Zubairi, I. H., Management of chemotherapy-induced nausea and vomiting. *British Journal of Hospital Medicine* 2006, 67, (8), 410-413.

Index